HANDBOOK OF STRATA-BOUND AND STRATIFORM ORE DEPOSITS

Volume 2
GEOCHEMICAL STUDIES

HANDBOOK OF STRATA-BOUND AND STRATIFORM ORE DEPOSITS

Edited by
K.H. WOLF

I
PRINCIPLES AND GENERAL STUDIES

1. Classifications and Historical Studies

2. Geochemical Studies

3. Supergene and Surficial Ore Deposits ; Textures and Fabrics

4. Tectonics and Metamorphism

II
REGIONAL STUDIES AND SPECIFIC DEPOSITS

5. Regional Studies

6. Cu, Zn, Pb, and Ag Deposits

7. Au, U, Fe, Mn, Hg, Sb, W, and P Deposits

ELSEVIER SCIENTIFIC PUBLISHING COMPANY
Amsterdam — Oxford — New York   1976

HANDBOOK OF STRATA-BOUND AND STRATIFORM ORE DEPOSITS

I. PRINCIPLES AND GENERAL STUDIES

Edited by
K.H. WOLF

Volume 2
GEOCHEMICAL STUDIES

ELSEVIER SCIENTIFIC PUBLISHING COMPANY
Amsterdam — Oxford — New York   1976

ELSEVIER SCIENTIFIC PUBLISHING COMPANY
335 Jan van Galenstraat
P.O. Box 211, Amsterdam, The Netherlands

Distributors for the United States and Canada:

ELSEVIER/NORTH-HOLLAND INC.
52, Vanderbilt Avenue
New York, N.Y. 10017

ISBN: 0-444-41402-9

Printed in The Netherlands

LIST OF CONTRIBUTORS TO THIS VOLUME

E.K. DUURSMA
Technical University Twenthe, Enschede, The Netherlands

P. FRITZ
Department of Earth Sciences, University of Waterloo, Waterloo, Ontario, Canada

B. HITCHON
Physical Sciences Branch, Alberta Research Council, Edmonton, Alberta, Canada

C. HOEDE
Technical University Twenthe, Enschede, The Netherlands

V. KÖPPEL
Institut für Kristallographie and Petrographie, Eidgenössische Technische Hochschule, Zürich, Switzerland

W. MERCER
Department of Geology, University of Alberta, Edmonton, Alberta, Canada

E. ROEDDER
Geological Survey, United States Department of the Interior, Reston, Virginia, U.S.A.

R. SAAGER
Institut für Kristallographie and Petrographie, Eidgenössische Technische Hochschule, Zürich, Switzerland

D.F. SANGSTER
Geological Survey of Canada, Ottawa, Ontario, Canada

J.D. SAXBY
Division of Mineralogy, CSIRO Minerals Research Laboratories, North Ryde, New South Wales, Australia

P.A. TRUDINGER
Baas Becking Geological Laboratories, Canberra City, Australian Capital Territory, Australia

D.J. VAUGHAN
Department of Geological Sciences, University of Aston in Birmingham, Birmingham, Great Britain

CONTENTS

Chapter 1. MINOR ELEMENTS IN METAL DEPOSITS IN SEDIMENTARY ROCKS – A
REVIEW OF THE RECENT LITERATURE
by W. Mercer

Chapter 2. PRINCIPLES OF DIFFUSION IN SEDIMENTARY SYSTEMS
by E.K. Duursma and C. Hoede

Chapter 6. MICROBIOLOGICAL PROCESSES IN RELATION TO ORE GENESIS
by P.A. Trudinger

Chapter 7. OXYGEN AND CARBON ISOTOPES IN ORE DEPOSITS IN SEDIMENTARY ROCKS
by P. Fritz

Chapter 8. SULPHUR AND LEAD ISOTOPES IN STRATA-BOUND DEPOSITS
by D.F. Sangster

Chapter 9. URANIUM-, THORIUM- AND LEAD-ISOTOPE STUDIES OF STRATA-BOUND
            ORES
by V. Köppel and R. Saager

Chapter 10. SEDIMENTARY GEOCHEMISTRY AND MINERALOGY OF THE SULFIDES
             OF LEAD, ZINC, COPPER AND IRON AND THEIR OCCURRENCE IN SEDI-
             MENTARY ORE DEPOSITS
by D.J. Vaughan

Chapter 1

MINOR ELEMENTS IN METAL DEPOSITS IN SEDIMENTARY ROCKS – A REVIEW OF THE RECENT LITERATURE

WILLIAM MERCER

INTRODUCTION

Metal deposits in sedimentary rocks are frequently of controversial origin. This being so, any new approach that has bearing on the genesis of ores in sediments is important. Both epigenetic and syngenetic theories have been invoked for most stratabound deposits. Also, the relative importance of normal sedimentary processes and unusual processes, such as volcanogenic exhalations, has been debated at length. Trace- or minor-element studies of most types of metal deposit have been undertaken by a number of geologists with varying success. This paper reviews some of the more recent geochemical work on certain classes of ores in sediments and relates the geochemical data to the prevalent theories of origin of the deposits.

CLASSIFICATION OF DEPOSITS

This compilation is limited to certain classes of metal sulphide deposits found in sediments. Not discussed are Precambrian iron formations and uranium–gold occurrences, recent placers and the numerous minor examples of metal concentrations of suspected sedimentary origin, such as those scheelite bands described from the European Alps.

The major deposits in sedimentary rocks, apart from hydrothermal veins, may be classified into four major groups for the purpose of minor-element data presentation. These are:

(1) Stratabound galena–sphalerite in carbonate rocks. The deposit is often restricted to one stratigraphic horizon as discrete massive sulphide bodies, but is not concordant (Mississippi Valley deposits, U.S.A.; Pine Point, Canada).

(2) Stratiform massive Fe–Cu, Fe–Cu–Zn or Fe–Cu–Zn–Pb sulphide deposits, which are distinguished from (1) above in being distinctly lensoid and not occurring within the host rocks as cross-cutting veins, except in cases of extreme deformation. Chemically, they are characterised by a predominance of iron (Besshi-type and Kuroko-type, Japan; Cyprus-type).

(3) Metalliferous argillaceous rocks which may grade laterally into normal sedimentary rocks. An individual shale unit, often organic-rich, may have high enough metal-content to become locally economic to mine (Kupferschiefer, Germany; White Pine, U.S.A.; Zambian Copperbelt).

(4) U, V and Cu within arenites which may vary from U–V rich (Colorado Plateau, U.S.A.) to Cu–U rich (Copperbelt, Zambia). The major-element similarities of many of these deposits, however, may not infer similar geneses.

MINOR-ELEMENT DATA COMPILATION

Two types of study are discussed below; they are minor elements in bulk ores and minor elements in minerals. It is not proposed to define strictly what is covered by the term "minor element"; for when applied to bulk ores it may have a completely different meaning for the mine geologist and the geochemist, respectively. A minor element cannot be defined as that occurring below a certain concentration, for mineable grades of metals vary significantly according to a large number of factors, many of which are non-geological. However, the implications are simply that minor elements do not constitute the principal economic metals of a deposit.

The minor element may occur as sparse, discrete grains of a specific mineral, or as a trace constituent in other minerals. For example in Mississippi Valley deposits, Cd which may occur as greenockite or in solution in sphalerite, respectively.

Major rock-forming elements are not often considered as minor elements; for example, Na is not usually considered in minor-element analyses, even though it may be present in low concentrations. For a Kuroko deposit, Ag and Au might be termed minor elements even though they contribute to the net worth of the deposit. On the other hand, Fe and Si, which are not of value, and Cu, Zn and Pb, which are economic, constitute major elements.

Most ore deposits are zoned with respect to their elemental composition. An element that is "major" in one part of an orebody may be "minor" in another part. Thus, severe restrictions are placed on the techniques and significance of sampling in such a case.

There are also difficulties inherent in comparing the work of numerous authors in the field of geochemistry. Firstly, analyses for minor elements vary considerably according to the methods used and materials analysed. One author may have performed electron-microprobe analyses of zoned mineral grains in contact with each other, whilst another might discuss the composition of bulk ore of a mine.

Secondly, there is by no means a consensus of opinion as to what elements it is advisable to analyse for in particular ores. As a result, one study may cite the As, Ag, Co and Se contents of pyrite and another the Co, Ni, V and Ti contents of the mineral. Certain elements are, however, commonly analysed, such as Co and Ni in pyrite, Cd, Mn and Fe in sphalerite and Ag and Bi in galena.

Apart from the difficulties mentioned above, the investigation of trace elements in minerals has other drawbacks. Any method of analysis that uses mineral separates is open to the possibility that traces of impurities are included. This would cause serious errors in analyses of elements at very low concentrations, though an exact level of accuracy cannot, of course, be set. It must be borne in mind that most mineral analyses discussed in this paper may suffer from this error.

The electron microprobe seems to offer an opportunity for avoiding the above error, but here the limit of accuracy is set by the tool itself, which has a detectability of 50–1000 ppm for most elements (Beaman and Isasi, 1972). In most cases of partition between sulphides better sensitivity than the microprobe can offer is required to produce meaningful results (D. Smith, personal communication, 1974).

Other reviews of interest concerning minor elements in sulphide minerals are by Hegemann (1943) and Fleischer (1955). In addition, Cambel and Jarkovsky (1967) have published an extensive review of minor elements in most types of pyrite found in Czecho-slovakia.

Stratabound galena–sphalerite in carbonates[1]

Deposits of this type have been reviewed in the monograph edited by Brown (1967). In these, the major elemental constituents are Pb, Zn, Ba and F. The deposits range from Pb-rich to Zn-rich and carry subordinate amounts of pyrite and lesser amounts of chalco-pyrite.

Hall and Heyl (1968) and McKnight and Fischer (1970) have analysed Mississippi Valley deposits for minor elements (Table I). In comparison to other types of lead–zinc ores, these are generally deficient in precious metals; the Ag-content is usually less than 20 ppm and Au has never been detected. High Cd- and Ge-contents are recorded.

In detail, Hall and Heyl (1968) found a marked dissimilarity between deposits of the Illinois–Kentucky area, where the main constituents are F, Zn and Ba and a high minor element content exists, and those of the upper Mississippi Valley, a main source of Pb and Zn, where ores exhibited lower minor elements. In addition, the Illinois–Kentucky area shows a distinct zonation of the minor elements.

The sphalerite of these deposits is distinctive in having a low Fe-content, but is rich in other elements such as Cd, Cu, Ge and Ga (McKnight and Fischer, 1970). Interestingly, in contrast with stratiform massive sulphides, the galena is often depleted in Ag and the sphalerite enriched in this element.

Hall and Heyl (1968) presented analyses of pyrite of various generations with Co-content between 10,000 and 20 ppm and Ni-content between 10,000 and 300 ppm. The high levels are almost certainly due to Co- and Ni-minerals intergrown with the pyrite. The Co/Ni ratio varies from 1.5 to 0.1.

---

[1] For details, see several chapters in this multi-volume publication devoted to these ore deposits.

TABLE I

Stratabound lead–zinc deposits in carbonate rocks (analyses of minerals in ppm *)

Reference	Area	Mineral	No. of samples	Cd	Ag	Ge	Co	Ni	Fe	Mn	Others
1	Upper Mississippi Valley	pyrite	5		3–70		20–10,000	300–10,000			Cu = 10–15,000 / As = 0–13,000
		sphalerite	6	14,000	14	102	15	42		49	
		galena	22		18						
1	Illinois–Kentucky	sphalerite	5	6900	5	290	11	4	20,900	8	
2	Oklahoma–Kansas	sphalerite		7300	1	470	6				
3	West Canada	sphalerite	20	1800					23,000	350	Se = 160
		galena	12								Se = 1085
4	Gorno	sphalerite	10	1400	25	110			3500	123	Sb = 517

* Blank means element not analysed for.

References: 1 = Hall and Heyl (1968); 2 = McKnight and Fischer (1970); 3 = Evans et al. (1968); 4 = Fruth and Maucher (1966).

Sphalerite in Zn—Pb deposits of the Alps (Landinian Plateau Reef) also shows high Cd, Ga and Ge and low Fe-contents (Fruth and Maucher, 1966).

Numerous small occurrences of Pb—Zn enrichments are present in carbonate lithologies of western Canada. Here sphalerite shows lower Cd but richer Fe levels relative to Mississippi Valley ores (Evans et al., 1968). The content of Se, up to 1000 ppm in galena, is of some significance in discussion of Se/S ratios in ore genesis.

Stratiform massive sulphide deposits [1]

These deposits are distinguished from the other sediment-associated types in being almost wholly composed of sulphide minerals. The usual shape is lensoid with the larger dimensions in the plane of stratification of the host rocks. The general features have been reviewed by numerous authors (Anderson, 1969; Sangster, 1972; Stanton, 1972; Chapter 5 by Sangster and Scott in Vol. 6).

Geochemically, there are a number of combinations of major elements recognised in stratiform massive sulphides (Stanton, 1972): Fe; Fe—Cu; Fe—Cu—Zn; Fe—Cu—Zn—Pb. There is a tendency for the Fe—Cu deposits to be associated with mafic volcanics (Besshi-type and Cyprus-type deposits) and the Fe—Cu—Zn—Pb to be with felsic volcanics (Kuroko-type).

The deposits of Mount Isa, Australia, and Sullivan, Canada, where volcanic rocks are minor or absent, may be examples in which the role of volcanism was less important.

Zonation across the stratification is frequently present in the Fe—Cu—Zn—Pb deposits. Undeformed orebodies show a concentration of Cu at the stratigraphic bottom and Zn and Pb at the stratigraphic top. In deformed areas, where way-up criteria are absent, this zonation is frequently taken as proof of way-up of the beds enclosing the ore deposit.

Minor elements in bulk ores of massive sulphides. Minor-element studies on bulk ore samples of cupriferous massive sulphides have been published for Besshi deposits in Japan (Doi, 1962; Yamaoka, 1962) Cyprus deposits (Constantinou and Govett, 1973; Govett and Pantazis, 1971) and Ergani in Turkey (Griffiths et al., 1972). The results are summarised in Table II.

The Turkish deposits are high in Co and Ni according to Birgi (quoted by Griffiths et al., 1972) compared with other deposits of this type, but no method of recovery of the Co has so far been found. The Co/Ni ratio is about 7.4.

Low Co- and Ni-contents and a ratio of Co/Ni of 2.1 to 1.2 are found in the Cyprus deposits (Constantinou and Govett, 1973). Hutchinson (1965) comments that the Mavrovouni ore (Cyprus) contains 0.3—0.8 ppm Au and 8—35 ppm Ag. The highest Au accompanies the Cu-rich ore. High precious metal contents are often found in these deposits.

---

[1] For more details, see Chapter 12 by Lambert in Vol. 6, Chapter 5 by Sangster and Scott in Vol. 6 and Chapter 4 by Vokes in Vol. 6.

TABLE II

Stratiform massive cupriferous sulphide deposits (analyses of minor elements in bulk ores in ppm, except * in %[1])

Reference	Mine	Ore type	No. of analyses	As	Co	Cu*	Ag	Au	Fe*	S*	Ni	Zn*	Se
1	Cyprus (various)	massive	45		94	2.7				41	45	0.7	
		stockwork	70		69	0.3				18	37	0.07	
2	Ergani, Turkey			1200	5200	10	22	1.4	39	48	700		
3	Besshi, Japan		5	90	630	10	30	0.6	38	37	42	0.92	45

References: 1 = Constantinou and Govett (1973); 2 = Griffiths et al. (1972); 3 = Doi (1962).
[1] Blank means element not analysed for.

Pb-content is usually very low in the cupriferous deposits, although Zn is on occasion important. In the Precambrian of Canada, there are a number of massive pyritic bodies with economic Cu and Zn, but little Pb. The subsidiary elements commonly recorded are Cd, Ag and Au with lesser amounts of Se, Te, Bi and Sn (Sangster, 1972; Pollock et al., 1972).

Ore of stratiform deposits with significant Pb-content shows a wide range of minor elements (Table III). Analyses are published of Kuroko-type, Japan (Takahashi, 1966; Tatsumi and Oshima, 1966; Shiikawa, 1970; Matsukuma and Horikoshi, 1970; Lambert and Sato, 1974), the McArthur River deposit, Australia (Croxford and Jephcott, 1972; Lambert and Scott, 1973) and Mount Isa, Australia (Smith and Walker, 1971).

The distinct zonation present in the main metals of most stratiform massive sulphides is also manifest in many minor elements. In the Kuroko deposits five types of ore are distinguished in terms of metal zonation (Matsukuma and Horikoshi, 1970). It is apparent that as Cu decreases and Zn and Pb increase stratigraphically upwards in the deposits, As, Sb, V, Ag and Au increase and Ni and Co decrease (Tatsumi and Oshima, 1966; Lambert and Sato, 1974). A number of other elements including Mo, W, Tl, Hg and Bi show contradictory or unclear trends in various deposits.

It has been shown that Au and Ag frequently correlate with the Pb–Zn ore in the Bathurst, New Brunswick, stratiform sulphides (Mercer and Crocket, 1972), whilst Pd has no preferential enrichment in the ore.

The Mount Isa and McArthur River deposits of Australia have complex patterns of minor-element concentrations within a succession of lithologically heterogeneous ore horizons. Metals associated with the sulphides at Mount Isa are primarily Fe, Cu, Pb and Zn and at a lower concentration Cd, Co and Mo, whilst Ni-, Ti- and V-contents are higher in the surrounding sediments. McArthur River shows mainly Fe, Zn and Pb with some Cu and concentrations of As, Sb, Co, Cd, Hg, Ni, Ag, Tl and Ge in ore horizons. In both mines, there is a strong correlation of Cd with Zn and Ag with Pb. Croxford and Jephcott (1972) also demonstrate correlations of Sb–Ag, Tl–Fe and Ag–Cu. Note that Se, where detected, occurs in very low concentrations in these ores.

Minor elements in minerals. The analyses from studies of minerals of stratiform massive sulphides are condensed in Table IV. The commonest approach is a comparison of the Co/Ni ratios of pyrite. The ratio is greater than unity in all cupriferous pyrite deposits (Itoh and Kanehira, 1967; Itoh, 1971a,b).

Values quoted by Loftus–Hills and Solomon (1967) from Tasmania show that pyrites associated with Cu-mineralization and the accompanying Pb–Zn rich ore of Fe–Cu–Zn–Pb deposits have Co/Ni ratios greater than and less than unity, respectively. There is little data in the literature to support or reject this as a generalization. Lambert and Sato (1974) do show a change from a Co/Ni ratio of 2.7–0.1 in the bulk ore from the bottom to the top of the Kosaka mine in Japan (Kuroko-type).

Generally As, Co, Cu, Ni and Zn reach more than 100 ppm in these pyrites. In the

TABLE III

Stratiform massive sulphide deposits (Fe–Cu–Zn–Pb-type) (analyses of minor elements in bulk ores in ppm, except * in %[1])

Reference	Mine	Ore type	No. of analyses	Sb	As	Ba*	Bi	Cd	Cu*
1	Kosaka and Hanaoka,	black-ore	7	300	1100	23	20	1300	2.3
	Japan	semi-black	5	60	280	36	170	200	4.5
		yellow-ore	18	230	280	0.4	180	80	7.4
		pyrite-ore	3	320	320	Tr.	80	Tr.	0.4
		silica-ore	7	80	170	0.2	140	Tr.	1.9
2	Kano, Japan	black-ore	9	325	555		25		3.0
		pyrite-ore			250				7.4
		yellow-ore	4	500	1300	1.9			15.5
3,4	McArthur River, Australia			0–160	0–2300			0–600	0–0.45
5	Mount Isa, Australia	Urquhart Shale	10			0.04		12	0.03
		dolomite	3			0.004		5	0.4
6	Meggen, Germany		3	157	444	0.05			0.003

[1] Blank means element not analysed for.

Reference: 1 = Lambert and Sato (1974); 2 = Takahashi (1966); 3 = Croxford and Jephcott (1972); 4 = Lambert and Scott (1973); 5 = Smith and Walker (1971); 6 = Gasser (1974).

Canadian Precambrian deposits, the Sn-content of pyrite is notably high. Se data is sparse; it seems to vary from less than 3 ppm in New Brunswick ores (Sutherland, 1967) to 1000 ppm in the Horne Mine, Ontario (Hawley and Nicol, 1961). Hawley and Nicol conclude that there is no regular concentration of Se, and no correlation between the Se-content of ores and the presence or absence of Pb or Zn. The Se-content simply reflects the Se-content of the whole ore. A limited amount of data on Se is summarised by Stanton (1972, p. 171) suggesting the possibility of some fractionation of Se and S between minerals, but the variation of the data is very great.

From numerous analyses of pyrites occurring in the same orebody (Hawley and Nicol, 1961; Farkas, 1973) a wide range in Co-, Ni- and Se-contents is apparent and this must reflect pyrites of different origins within the deposits.

Distribution of the minor elements between the various sulphides of stratiform sulphides can be exhibited in two ways. Firstly, some elements may show no consistent partitioning between minerals and are enriched or impoverished according to the general content of the ore-forming solution, for example Se. Secondly, some elements are always

Au	Fe*	Pb*	Mn	Mo	Co	Ni	Ag	S*	V	Zn*
3	5.6	17		200	Tr.	10	300	24	80	24
0.9	19.9	0.6		400	10	20	92	27	20	7.0
0.8	38.9	0.2		1200	30	40	36	45	Tr.	2.0
0.6	42.4	0.1					15	48		0.2
0.3	17.9	0.1		300	40	15	11	20	Tr.	0.2
	12.6	5.7		78			169			22
	32.7			5			7			1.9
	31.6	0.06					32	39		1.3
	0–22	0–11			0–140	0–110	0–140	0–29		0–23
	4.3	0.2	2600		19	19	10		33	0.04
	4.7	0.01	1500		70	20	10		19	0.004
	33.4	0.70	1340		31	183		41		6.1

high in certain minerals, if present at all, for example Ag in galena, Cd in sphalerite and As in pyrite.

Many authors have concentrated on the level of the above elements in the respective mineral phases. The work of Both (1973) on the Broken Hill sphalerite and galena and Farkas (1973) on the Kidd Creek deposit, Ontario, Canada, can be cited, though the objectives of these two researchers were quite different.

Metal deposits in argillaceous rocks

Transition metals are usually enriched in shales compared to sandstones and limestones. Exceptions to this are, for example, high Sr and Mn in limestones and Zr and rare-earth elements in sandstones. In addition to transition metal elements (Sc, Ti, V, Cr, Mn, Co, Ni, Cu, Zn, etc.), many other elements characteristic of sulphide deposits (As, Sb, Se and S) are rich in shales (Turekian and Wedepohl, 1961). The mechanisms for enrichment of trace metals in shales include ionic substitution in clay minerals, adsorption on clay minerals, adsorption on metallic hydroxides, precipitation of insoluble com-

TABLE IV

Stratiform massive sulphide deposits (analyses of minor elements in sulphide minerals in ppm[1])

Reference	Mine	Mineral	No. of analyses	As	Cd	Co	Cu	Ni	Ag	Au	Sn	Others
1	Various Pre-cambrian mines, Canada	pyrite	4	137		600		93	33	6.1	197	
1		pyrrhotite	4	58		540		75	39	1.4	263	
2		pyrite	39			900		34	40		690	
2		pyrrhotite	50			600		14	38		400	
2		chalcopyrite	44			190		29	220		270	
3	Texas Gulf, Precambrian, Canada	pyrite	30			940	430	140	210		180	
3		chalcopyrite	13			400		150				
3		sphalerite	26		2050	828		61	135		73	Mn = 145
3		galena	15		71							Mn = 62
4	Besshi-type, Japan	pyrite	4	ND	2	39	188	ND	44		ND	Mn = 43
4		chalcopyrite	4	177	23	20		ND	19		ND	Mn = 108
4		sphalerite	5	43	2800	36	810	1	1		23	Mn = 213
5		pyrite	19			196		9.3				Se = 195
6	Shirataki, Japan	pyrite	77		13	1440	39,000	69	140			Mn = 84 Zn = 2000
7 + 8	Various Cyprus-type,	pyrite (massive ore)	11	1590		416	220	3	4			
7 + 8	Cyprus	pyrite (stockwork ore)				450	200	160	1			

9	York Harbour, Newfoundland, Canada	pyrite (massive ore)	8			183		289			
9		pyrite (disseminated ore)	7			282		142			
10	Bathurst, New Brunswick, Canada	pyrite	7	2200	50	700	4500	40	33	22	Se = 3
11		galena	8	3400			4100		850	600	Se = 370
12		zinc-concentrate (sphalerite)			1000						

[1] ND: means not detected, blank means not analysed for.

References: 1 = Roscoe (1965); 2 = Hawley and Nicol (1961); 3 = Farkas (1973); 4 = Yamaoka (1962); 5 = Yamamoto et al. (1968); 6 = Itoh (1971a); 7 = Hutchinson and Searle (1971); 8 = Johnson (1972); 9 = Duke and Hutchinson (1974); 10 = Sutherland (1967); 11 = Boorman (1968); 12 = The Canadian Mineral Industry in 1971 (Dept. Energy, Mines and Resources, Ottawa, Canada), Mineral Bull. MR 122 (1972), 114 pp.

pounds (such as sulphides) and adsorption on or reaction with organic matter. In the majority of cases of enrichment, correlation with organic matter is indicated (Vine and Tourtelot, 1970 — see Chapter 5 by Saxby, this Vol.). This correlation could be due to: (1) incorporation in a living organism, and subsequent deposition with the organic remains of the organism; (2) incorporation of the metal by reaction with organic matter after death of the organism; (3) adsorption on organic matter; and (4) reduction of the metal by the organic matter causing precipitation of an insoluble compound.

Economic grades of Cu, Pb and Zn occur in shales in, among others Germany (the Kupferschiefer)[1] and U.S.A. (the Nonesuch Shale — White Pine deposit). The variation of major and minor elements in these deposits have been discussed by Wedepohl (1964, 1971), White and Wright (1966), Haranczyk (1970) and Brown (1971). The data is summarised in Table V.

The features apparent are that the ores are enriched in certain characteristic elements, such as V, Cr, Ni and U and that there is a dominance of Ni over Co. This latter ratio is similar to that found in pyrites in recent sediments, for example in the Black Sea (Butuzova, 1969). The metal association is reminiscent of that correlated with bituminous carbon in shales (Vine and Tourtelot, 1970).

There is a lateral zoning of Cu, Pb and Zn (Wedepohl, 1971) in the Kupferschiefer, geographically related to the synchronous palaeo-shoreline.[1] The higher contents of Cu are found in the near-shore facies.

The Zechstein shales discussed by Haranczyk (1970) were formed in part of the same depositional basin as the Kupferschiefer. Here also, there is a distinct separation of the Cu- and Pb-rich parts. Peaks in Cu and Pb values coincide with peaks for Mo, V, Cr, Ni and Co. There seems to be a vertical zonation also, with Zn occurring higher in the succession than Cu.

The White Pine copper deposit, in the Nonesuch Shale, although grossly stratiform, is clearly discordant to the host rocks (Brown, 1971). Cu is concentrated in the lower part of a shale horizon and there is a distinct anomalous zone from Cu-mineralization to overlying barren pyrite. At this anomalous zone there is a concentration of Cd and Pb. Zn- and Mo-contents, on the other hand, show almost no change from mineralized to barren rock.

Principal component analysis of minor elements in a large number of shales from the U.S.A. (Vine and Tourtelot, 1969) has shown that in the Nonesuch Shale Cu, Ag, Pb, Ti, Mn and Y are abnormally abundant, and that Cu and Ag do not correlate with any other group of major or minor elements.

The lack of analyses of sulphide minerals from this type of ore is presumably due to practical difficulties of obtaining separates from such fine-grained material.

---

[1] See Chapter 7 by Jung and Knitzschke in Vol. 6.

TABLE V

Metalliferous shales (analyses of bulk ores for minor elements in ppm, except * in %[2])

Reference	Area	No. of analyses[1]	Cd	C	Co	Cu*	Fe*	Zn*	Pb*	Mo	Ni	Ag	S*	V
1	Mansfeld, Kupfer-schiefer	18	500	64,000	87	1.4	3.6	1.5	0.59	280	110	73	1.9	740
2	Zechstein, Poland	est.			900	8			12	200	100			1000
3	White Pine, U.S.A., cupriferous shale	est.	20			1	5.5	0.012	0.002	22		50	0.25	
3	White Pine, U.S.A., pyritic zone	est.	75			0	5.5	0.012	0.004	27		40	0.3	

[1] Total number of analyses, or estimated ("est.") from reference cited.
[2] Blank means element not analysed for.
References: 1 = Wedepohl (1964); 2 = Haranczyk (1970); 3 = Brown (1971).

Stratabound Cu–U–V deposits in arenaceous rocks

The western U.S.A. has a large variety of deposits of Cu, U or V in fluviatile sedimentary rocks (Fischer, 1970). Similar deposits may occur in Australia (Stanton, 1972) and the Precambrian of Canada (Morton et al., 1973). Occurrences, such as the Zambian Copperbelt ores (Mendelsohn, 1961a), which are in sandstones in certain cases, and Dzhezkazgan, U.S.S.R. (Magak'yan, 1968), may not be the same type of deposit, but have some geochemical and geological similarities.[1]

The deposits in the western U.S.A., despite overall similarities, exhibit a range from U-rich deposits with considerable Se, some Mo, and little V and Cu in Wyoming, through those U-rich, V-rich or Cu-rich with less Se and Mo in the Colorado Plateau, to the Cu-rich types often with notable Ag in Utah and New Mexico (Fischer, 1970). Other trace elements recorded are Be, As and Co (Harshmann, 1973).

The problem of obtaining representative analyses of such ores is illustrated by the variation of Se through orebodies in Wyoming (Harshmann, 1966). Within an individual U orebody, the Se ranges between 2–15 ppm as compared with less than 0.5 ppm in the host rock. There are areas of alteration within the host sandstone with 10–60 ppm Se and on the contact of ore and alteration Se can reach 1200 ppm.

Deposits in the Zambian Copperbelt occur in three types of host rocks. Minor orebodies are found in quartzites stratigraphically below the main ore zone. The deposits of the main ore zone are in argillaceous rocks in the western part and arenaceous rocks in the eastern part of the belt (Mendelsohn, 1961b).

Mendelsohn stresses the importance of the accompanying Co (Mendelsohn, 1961c) and that the ore also has significant U, Mo, W, Be, Ag, As and Ni (Table VI). It is stated that U and W occur immediately beneath the Cu at Chibuluma on the west side of the Copperbelt. Another U-rich area is found at Mindola where U and Cu are inversely correlated.

Co in the ores is frequently present as zones within the pyrite (Bartholome et al., 1971; Brown and Bartholome, 1972). Mendelsohn (1961c) mentions cases (e.g., Chibuluma) where thin bands of ore can carry linnaeite, and Co is present up to 19%. Pyrite can contain up to 4% Co in zones in these ores. Bartholome et al. (1971) found that the Ni-content, although low in the ores, parallels the distribution of Co, up to 0.2%.

The sandstone deposits of Dzhezkazgan (Volfson and Arkhangelskaya, 1972) appear to carry an unusual suite of minor elements including Re, Mo, Os, Bi, Au, As, Ag, Cd, Se and Te in addition to the Cu, Zn and Pb. Magak'yan (1968) gives values of 1.5% Cu, 0.5–1% Pb and records localised Zn and Ag.

The conclusion can be made that there are a number of different types of deposits with little in common except that they have varying amounts of uranium and copper and

---

[1] Cf. Chapter 3 by Rackley in Vol. 7 for uranium deposits of the western U.S.A. and Chapter 6 in Vol. 6 by Fleischer et al. for details on the Zambian Copperbelt.

TABLE VI

Zambian Copperbelt orebodies * (analyses of bulk ores from minor elements in ppm)

Ore type	No. of samples	Pb	Zn	Bi	Co + Ni	As	Ag	Au
Central Lode	5	Tr.– 60	Tr.–1570	Tr.	nd	Tr.	Tr.–9	Tr.–2
Hanging-Wall Lode	1	200	nd	Tr.	140	Tr.	7	Tr.
Footwall Lode	1	300	nd	Tr.	250	Tr.	5	Tr.
Combined Foot-wall and Hanging-Wall		Tr.–540	Tr.	Tr.	nd	Tr.	9	Tr.–2.5

* Mendelsohn, 1961a, Table 26.
nd = not determined; Tr. = trace amount present.

occur in sandstone. The minor-element contents of the deposits of sections (3), argillaceous rocks, and (4), arenaceous rocks, certainly suggest divergent types and this is reflected in different genetic interpretations suggested. These are, for example, syngenetic origin for the Kupferschiefer and the Zambian Copperbelt, epigenetic origin with metals from connate fluids or hydrothermal fluids for White Pine deposit, and ground water supplying the metals for the Western U.S.A. U–V deposits.

SIGNIFICANCE OF MINOR ELEMENTS

In any ore forming process involving an "ore solution", the chemical behaviour of the major transition-metal elements may be expressed in terms of the thermodynamics of the solution and with respect to the formation of metal sulphides from the solution. If chemical equilibrium is assumed, then the presence of a metal sulphide, put simply, implied saturation of the solution with respect to that compound. Using the thermodynamic data of Helgeson (1969) in conjunction with information on, for example, the temperature of formation of an orebody from fluid inclusion data (Chapter 4 by Roedder, this Vol.) or sulphur isotope information (Chapter 8 by Sangster), and the overall metallic composition of orebodies, physicochemical constraints may be applied to the conditions of ore deposition (Sato, 1973).

Most minor elements can occur in a variety of locales within a sulphide deposit, including among others: (1) minute grains of minerals; (2) exsolution blebs or lamellae within sulphide minerals; (3) solid solution components in the host sulphide; (4) adsorbed on surfaces or absorbed in defects of sulphide minerals; and (5) in the non-sulphide fraction, e.g., adsorbed on clay minerals, etc.

A major element in an ore, for example Fe, may be present in a number of compounds, in this case pyrite, pyrrhotite or chalcopyrite, but the amount bound as the main sulphide mineral, for Fe pyrite, usually accounts for a large proportion. However, a minor element can be distributed such that no one position accounts for a large proportion. As a result, we have two levels of complication, i.e., firstly, to find where a trace element is and, secondly, to reach a conclusion as to its genetic significance in that position.

Conventional methods of analysis for minor elements suffer from the inability to distinguish different sites except by indirect methods. Plotting a minor element's distribution in one phase against a coexisting phase (partition diagram) can result in three types of pattern: (1) a linear relationship suggesting dilute solution and equilibrium conditions between the two phases with respect to the minor element; (2) a curve, which does not equivocally establish equilibrium; or (3) random scatter.

The amount of a metal that may be held in solid solution in a sulphide of another metal is dependent upon, among other factors, the amount of the first element in the ore-solution, and temperature and pressure.

Bethke and Barton (1971) have attempted to experimentally define the partition of Cd, Mn and Se between sphalerite and galena in the temperature range 600–900 °C. The thermodynamics involved are briefly described in the above-mentioned paper. As Bethke and Barton (1971) pointed out, there are numerous problems involved in applying their data to palaeo-temperature estimation in ore deposits. For example, the purity of the analysed mineral must be ensured because of the large fractionation effects between sphalerite and galena. The electron microprobe is one possible answer to the above problem and the results of Farkas (1973) using wave-length dispersive analysis will be discussed below.

The analysis of bulk ore samples for minor elements has severe limitations in interpretation except in a purely empirical sense. It is possible to obtain some indication of the partitioning of the element between the major mineral-phases by use of multivariate statistics. The analyst is thus hoping that large-scale processes of element enrichment, such as volcanic exhalations, groundwater circulation or biogenic sulphide precipitation, may show convergence in major element concentration but divergence in minor elements. Thus, certain elements can empirically infer a certain process.

Loftus-Hills and Solomon (1967), following Goldschmidt's (1954) example, have suggested the use of Se/S ratios in sulphide minerals to distinguish volcanic from sedimentary processes. A large number of authors have discussed Co/Ni ratios as being distinctive according to the enrichment process (Hegemann, 1943). The suggestion is that Co and Se are concentrated most highly in ores derived from volcanic processes.

DISCUSSION OF ORE GENESIS

For each group of ore deposits mentioned above one of two theories have at one time or another held sway, i.e., epigenetic or syngenetic. In certain cases, the syngenetic theory at present is generally accepted (the volcanogenic massive sulphides) and in others the epigenetic (the uranium deposits of the western U.S.A.). Nevertheless, for the stratabound deposits in carbonates and many of the copper enrichments in shales and sandstones, for example, the Mississippi Valley deposits, the White Pine deposit and the Dzhezkazgan deposits, the relative importance of epigenesis and syngenesis are heatedly disputed.

Minor elements may have the potential of providing information concerning temperatures of ore formation, initial composition of the ore solution and changes in physicochemical conditions with time. The problem faced is whether the minor-elements studies tabulated above have contributed much to the knowledge of the deposits for which much is already understood, or helped resolve anything for those of controversial origin.

Stratabound lead–zinc in carbonates

The recent consensus of opinion concerning the Mississippi Valley-type was discussed adequately by Brown (1970). The opinions may be summarised as:

(1) The ore fluids were connate-marine water or compaction fluids with some epigenetic component.[1]

(2) The deposits were derived by diagenetic concentration of metals precipitated syngenetically.

(3) The deposits are of magmatic replacement origin.

(4) The deposits were precipitated from circulated ground water.

The metal association can be described as Pb and Zn with subsidiary Cu, Cd, Ni, Ge and Ga, but low Ag and Au-contents. The reason for the high Cd-content of these ores has not been clarified. Conventionally high Se/S ratios and Co/Ni ratios greater than unity have been considered indicative of volcanic origin. Most deposits of this class have low Se/S ratios, but some in western Canada are high (Evans et al., 1968). The Co/Ni ratio of the pyrite in these deposits varies from 1.5 to 0.1 and, hence, is also inconclusive.

Sphalerite is characteristically low in Fe-content, but due to the absence of coexisting pyrrhotite no detailed conclusions can be reached from this, except that temperatures were probably low.

Hall and Heyl (1968) concluded from the zonation in minor elements in Illinois–Kentucky district that ore deposition was connected with the emplacement of an igneous explosion breccia at "Hick's Dome".

Stratiform massive sulphides

These deposits are termed "polymetallic" in the Russian literature and this is supported by the great variety of minor elements found. The most economically important are undoubtedly Ag and Au.

---

[1] For a review article on the influence of compaction on ore genesis, see Wolf (1976).

An understanding of the processes of formation of those massive sulphides intimately associated with volcanism has reached a very advanced stage in Japan. The zoning and amount of metals has been reconciled with the other data from the deposits, e.g., fluid inclusion composition and homogenisation temperatures, sulphur isotope compositions and hydrothermal alteration, in order to construct sophisticated models of metal transport within chloride systems (Sato, 1973), using thermodynamic data of Helgeson (1969).

The Kuroko deposits are envisaged as forming when a hot brine, dense relative to sea water, was exhaled on the ocean floor and there precipitated the metals which were initially saturated in the solution (Sato, 1973) (see Chapter 5 by Sangster and Scott in Vol. 6). From the metal-content of the deposit, the metal-content of the ore solution may be calculated, making certain assumptions. The zoning of the deposit is then explicable in terms of brine upon the ocean floor undergoing gradual changes in fugacity of oxygen and hydrogen sulphide, temperature and pH.

The model of Sato (1973) is dependent upon considering equilibria in the ore solution in terms of the mineral phases of the major metals (Fe, Zn, Pb and Cu). The predicted distribution of the Ag and Au was found to be incorrect. This may be due to argentite being used for silver solubility calculations and an incorrect assumption that gold is transported as the chloride complex described by Helgeson (1969).

An alternative model for Kuroko-deposit formation, employing different assumptions from Sato (1973), has been developed by Kajiwara (1973) but suffers from a number of drawbacks. In fact, it cannot predict the zoning of the deposits except to suggest that the ore solution changed in composition from Cu-rich to Pb–Zn-rich in every case.

The origin of the metals in the Kuroko deposits has two possible explanations, namely: they are products of fractional crystallization of the parent dacitic magma or the result of leaching of metals from diverse rocks by heated water, probably of oceanic origin. The role of ocean water is supported by the work of Ohmoto and Rye (1974) on stable isotopes. Major or minor element abundances do not seem to give much indication as to the ultimate source of the metals. Se/S ratio of the Kuroko deposits are not available and Co/Ni ratios of the Cu-rich ore are similar to those of hydrothermal vein deposits. It is apparent that Co is fractionated relative to Ni during the formation of these deposits.

Unlike most massive sulphides in the world, the Kuroko deposits are unmetamorphosed. In metamorphosed ores,[1] due to the lack of indicators, such as fluid inclusion compositions, palaeo-temperatures from fluid inclusion homogenisation or sulphur isotope partition, and wall rock alteration mineralogy, conditions of ore formation are more difficult to model. It is possible, however, that the difficulties may soon be overcome. Minor-element geochemistry is not, at present, sufficiently predictable thermodynamically to be used for clarification of physicochemical conditions.

---

[1] See Chapter 5 by Mookherjee in Vol. 4, among several others, where metamorphism of ores is discussed.

Farkas (1973) attempted to use the partition curves of Bethke and Barton (1971), in conjunction with the previously mentioned electron microprobe determinations on the Texas Gulf deposit of Ontario, to calculate temperatures of equilibration of the sulphide minerals. The temperatures obtained were about 550°C for the partitioning of Cd between sphalerite and galena. It is considered highly unlikely that either during initial ore deposition, or subsequent metamorphism to greenschist facies, the ore ever approached such a temperature.

The possible reasons for the very high temperature obtained for the Texas Gulf deposit are twofold. Firstly, the Cd-content of galena is usually very low leading to analytical difficulties with the electron microprobe. Secondly, the expected temperature of a deposit like the Texas Gulf, by analogy with the Kuroko deposits, would be expected to be below 300°C. This is out of range of Bethke and Barton's (1971) experimental data. In addition, the significance of the experimental results may be questioned when the difficulties in reconciling various experimental sulphur isotope fractionations are considered (Rye, 1974). The study of Farkas (1973) illustrates the problems involved in utilizing trace element data to obtain quantitative results.

Farkas does, however, reach some conclusions concerning the Co/Ni ratio of the ore fluid. Chalcopyrite tends to include Ni rather than Co in its structure. The fact that the Co/Ni ratio is greater than unity for chalcopyrite suggests that the ratio was very large in the ore fluid.

Miyazaki et al. (1974) have noted the change in Co-content of pyrite and pyrrhotite and the Fe-content in sphalerite in the Besshi deposit, Japan. The ratio of pyrrhotite to pyrite, the FeS-content of the sphalerite and the Fe-content of the pyrrhotite increase and the Co-content of the pyrite decreases as cobaltite and mackinawite become present, with increasing depth in the mine. This is taken by Miyazaki et al. (1974) to indicate a decrease in the Co/Ni ratio of pyrite with increasing grade of metamorphism. Itoh and Kanehira (1967) and Itoh (1971a,b) have suggested that the Co/Ni ratio of pyrite in cupriferous massive sulphide deposits increases with increasing metamorphism. The high Co-content may be due to inclusion of cobaltite and mackinawite in the pyrite separates of Itoh (1971a,b). This indicates the problems involved in interpreting the minor-element partition between minerals of metamorphosed deposits.

It is possible that, as thermodynamic knowledge of the minor elements progresses, they will become extremely useful in metamorphosed deposits for elucidating the recrystallization history of the deposit. Eventually, analytical techniques and thermodynamic knowledge permitting, minor elements may be utilised as one more set of parameters to limit the physicochemical conditions of ore formation.

Metal deposits in argillaceous rocks

The geochemical features of the Kupferschiefer are described elsewhere in these volumes (by Jung and Knitzschke, Chapter 7, Vol. 6) so discussion will be limited for this particular group of deposits.

The dispute centering around this class of deposits is partly a matter of whether the normal processes that enrich shales in metals might cause sufficient concentration to form ore deposits. The alternative to normal sedimentary processes is either to suggest submarine hydrothermal springs, leaching from rocks below the sediments and diagenetic enrichment, or derivation of the metals from erosion of metal rich deposits on the nearby continent.

The fact that the Cu and Ag in the White Pine deposit seem to be statistically independent of the other metals suggests a dual origin. Brown (1971) on this basis has postulated that the Cu-zone originated by upward circulation of groundwaters from Cu-rich sandstones below. Thus, the normal minor element suite for shales of Zn, Ni, Co, Pb, Mo, As, etc., along with pyrite would be from normal sedimentary processes, but the Cu and Ag would be the result of oxidizing groundwaters reaching a still permeable shale, rich in sulphide in the form of pyrite and H_2S. This would explain the change in minor elements at the anomalous zone described above. Burnie (personal communication, 1974) has suggested similar processes connected with diagenesis of the shale itself.

Davidson (1965) has pointed out the frequent association of stratabound Cu-deposits and evaporites. This concept has been used by Renfro (1974) to construct a genetic model for stratiform metalliferous deposits underlain by red-beds and overlain by evaporites as part of the cyclic sedimentation involved in the sabkha process. The ore is visualised as forming from circulated groundwaters passing through the reduced shales as the evaporites are forming above. This is compatible to a certain degree with Brown's model (1971). Thus, the minor-element distribution, the metal zoning and the proximity to the palaeo-shoreline are reconciled.

On the other hand, Brongersma-Sanders (1968) has taken the closeness to the shoreline, the association of evaporites, and the suite of major and minor elements, to indicate that the deposits formed syngenetically. The author points out that many areas of upwelling at present are adjacent to arid coastlines with red-bed and evaporitic sediments. Upwelling causes high organic productivity, which can result in anomalous contents of minor elements (nutrients) in the ocean water.

The Zambian Copperbelt deposits (see Chapter 6 by Fleischer et al. in Vol. 6) which are found in *carbonaceous* rocks have some palaeo-geographical similarities to the Kupferschiefer. Renfro (1974) has applied the sabkha process to these ores also. Geochemical information is sparse. The association of the Cu with U in arenaceous rocks in certain instances suggests similarities with the Colorado Plateau-type of ore. The very high Co in these ores is also puzzling, especially as in general high Co/Ni ratios seem indicative of a connection with hydrothermal activity.

Stratabound Cu−U−V deposits in arenaceous rocks

The deposits of the western U.S.A. are perhaps the best documented, but may be unique. The mineralisation is localised in certain stratigraphic horizons within fluviatile sediments and shows a strong correlation with organic matter. Whole tree trunks are sometimes replaced by U-minerals. The accepted theory is that circulating groundwaters largely of meteoric origin, have deposited the metals in permeable beds, where decaying plant material locally effected reducing environments (Fischer, 1970; Chapter 3 by Rackley in Vol. 7).

The abundance of Se in certain deposits creates problems for the significance of Se/S ratios considering the apparent lack of association with igneous rocks. Harshmann (1966, 1973) has described the minor elements in the Wyoming deposits pointing out the zoning of the elements around the ore masses. Particularly noticeable is the zoning of Mo, Se and Be. Se is not as easily transported by surface or subsurface waters as sulphur and thus is usually low in the sedimentary environment. However, in the alkaline oxidising environment Se is soluble as the selenate (SeO_4^{2-}) ion and Be may behave in a similar fashion (Harshmann, 1966).

MODERN METALLIFEROUS SEDIMENTS

As knowledge of the ocean floor progresses, more and more localities of metal-enrichment are being discovered. These may be divided into five cases:

(1) Fe- and Mn-enrichment associated with hot springs around andesitic volcanoes; these have been described from the Mediterranean (Butuzova, 1966; De Bretizel and Foglierini, 1971; Honnorez et al., 1971), Indonesia (Zelenov, 1964) and New Britain (Ferguson and Lambert, 1972). A characteristic of all these deposits is that, despite the great quantities of iron and manganese, very little other metal is present. This may be due to high detrital sedimentation rates, low concentration in the hydrothermal fluids and no mechanism to concentrate particular metals.

(2) Fe- and Mn-enrichment occurs in many areas of the open ocean. In addition, anomalous V, As, U, Cd, Zn, B, Hg, Ba and P occur (Bostrom et al., 1971). The mechanisms that have been credited with providing these elements are magmatic hydrothermal solutions, sea water heated and circulated through ocean floor igneous rocks or adsorption on precipitating iron and manganese hydroxides (Corliss, 1971). There is, to a certain degree, a spatial association between the mid-ocean rises and the metal-rich sediments.

(3) In the western Atlantic, multicoloured clays enriched in Zn, Cu, Sr, Ni and V are found (Lancelot et al., 1972). Zn reaches 3.6% overall, 50% locally. There is evidence of local volcanic activity and extremely low sedimentation rates. The metal-enrichment is believed to be due to hot springs in an area of low sedimentation rate.

(4) The Red Sea floor in most cases has normal pelagic sediments, but metalliferous clays of two types have been found to be widespread (Ross et al., 1973; Chapter 4 by Degens and Ross, Vol. 4). In one instance, dark shales carried more than 5% pyrite with 1000 ppm V, 500 ppm Mo and 200 ppm Cu (compare with the Mansfeld district) and in the other black shales had 0.5—5% Zn. No explanation has been given for these sediments by Ross et al. (1973).

TABLE VII

Red Sea hot brine sediments * (analyses for minor elements in ppm[1])

Core No.	Description of samples	No. of samples	As	Se	Co	Cd	Ni	Cu	Cr	Fe	Mn	Pb	Zn	Mo	Au	Ag	S
84K	selected metal-rich	4	210	1	130	150	50	9000	20	80,000	2000	300	44,000	500	3.0	46	140,000
95K	outside brine area	1			33	14	65	80	46	60,000	24,000	40	2800	39	0.04	–	17,000
118K	outside brine area	4	62	9.5													

* Data from Hendricks et al. (1969) and Kaplan et al. (1969)
[1] Blank means no analysis for that element.

(5) The most spectacular metal-enrichment so far found on the ocean floor is also in the Red Sea namely, the Red Sea, hot brine, metalliferous sediments. Many distinct parallels with certain Fe—Cu—Zn deposits exist both in element abundances and Pb- and S-isotopes and several chapters in this multi-volume publication have included comparisons of ancient ore deposits with the Red Sea mineralization. See Table VII for summary of geochemical abundances.

The minor-element content of the Red Sea deposits has been discussed by Kaplan et al. (1969) and Hendricks et al. (1969). Kaplan et al. point out that Cd, Ag, Cu, As and Pb tend to follow Zn and to be enriched in the sulphide phase. However, it is noteworthy that Se is more concentrated in the sediments outside the deeps, as is Ni. The Co/Ni ratio of the hot brine sediments is around 2.5 and that of the normal sediments in the area about 0.5. It can be seen that the two criteria of hydrothermal—magmatic origin are thus contradictory.

CONCLUSIONS

(1) Ore deposits in sedimentary rocks may be characterised by their minor element suites. These, in conjuction with data on the distribution of major elements, can frequently be utilized to define conditions of ore genesis.

(2) Co/Ni ratios of pyrite greater than unity are apparently indicative of an hydrothermal environment (Hegemann, 1943) with the possibility that a change in the Co/Ni ratio upwards stratigraphically in Fe—Cu—Zn—Pb deposits is common.

(3) Co/Ni ratios of pyrite increase during metamorphism according to Cambel and Jarkovsky (1967) and Miyazaki et al. (1974), but decrease according to Itoh (1971a,b). Thus, caution should be exercised during the investigation of metamorphosed ores.

(4) Cambel and Jarkovsky (1967) confirm the finding of Hegemann (1943), which quantitatively agrees with this compilation, that sedimentary pyrite has about 0.02% Ni and 0.002% Co.

(5) From the sparse data that are available, high Se/S ratios do not give proof of magmatic hydrothermal origin of ore deposits; this is confirmed by the work of Cambel and Jarkovsky (1967). The Se-content of sulphides may be partly a regional feature. Se and S, though separated in the weathering cycle, travel together in biological processes (Brobst and Pratt, 1973). Se-concentration is possible by plants, hence its high concentration in pyrite of black shales.

(6) Considerable problems still seem to exist for the utilization of minor-element partition between sulphide minerals as a geothermometer. This is confirmed by studies of Rye et al. (1974) on vein-type deposits.

(7) Minor-element studies in the past have often been limited in their usefulness due to poor selection of material, poor choice of elements to analyse, the presentation of large amounts of data without reduction techniques, such as factor analysis, and to poor

analytical procedures. The introduction of new rapid and sensitive techniques of micro-probe analysis (D. Smith, Univ. of Alberta, personal communication, 1974) may herald a new era in trace element studies. However, analysis of minerals without separation, for elements at less than 100—50 ppm must wait for new technologies such as the ion-probe (Beaman and Isasi, 1972).

ACKNOWLEDGEMENTS

I am grateful to Dr. Takeo Sato for providing papers that were otherwise unavailable describing trace elements of Japanese massive sulphide deposits. Thanks are also due to Dr. R.D. Morton and Say-Lee Kuo for critical reading of the manuscript and many helpful suggestions.

REFERENCES

Anderson, C.A., 1969. Massive sulphide deposits and volcanism. *Econ. Geol.*, 64: 129—146.
Bartholome, P., Katekesha, F. and Ruiz, J.L., 1971. Cobalt zoning in microscopic pyrite from Kamoto, Republic of Congo (Kinshasa). *Miner. Deposita*, 6: 167—176.
Beaman, D.R. and Isasi, J.A., 1972. Electron-beam microanalysis. *Am. Soc. Test. Mater.*, STP 506: 80 pp.
Bethke, P.M. and Barton, P.B., 1971. Distribution of minor elements between coexisting sulphide minerals. *Econ. Geol.*, 66: 140—163.
Boorman, R.S., 1968. Silver in some New Brunswick galenas. *N. B. Res. Prod. Coun., Res. Note*, 11: 11 pp.
Bostrom, K., Eyl, W. and Farquharson, B., 1971. Submarine hot springs as a source of active ridge sediments. *Chem. Geol.*, 10: 189—203.
Both, R.A., 1973. Minor element geochemistry of sulphide minerals in the Broken Hill Lode (NSW) in relation to the origin of the ore. *Miner. Deposita*, 8: 349—369.
Brobst, D.A. and Pratt, W.P., 1973. United States mineral resources. *U.S. Geol. Surv., Prof. Pap.*, 820: 722 pp.
Brongersma-Sanders, M., 1968. On the geographical association of stratabound ore deposits with evaporites. *Miner. Deposita*, 3: 286—291.
Brown, A.C., 1971. Zoning in the White Pine copper deposit, Ontonagon County, Michigan. *Econ. Geol.*, 66: 543—573.
Brown, A.C. and Bartholome, P., 1972. Inhomogeneities in cobaltiferous pyrite from the Chibuluma Cu—Co deposit, Zambia. *Miner. Deposita*, 7: 100—105.
Brown, J.S., 1967. Genesis of stratiform lead—zinc—barite—fluorite deposits. *Econ. Geol. Monogr.*, 3.
Brown, J.S., 1970. Mississippi Valley-type lead—zinc ores. *Miner. Deposita*, 5: 103—119.
Butuzova, G.Y., 1966. Iron-ore sediments of the fumarole field of Santorin volcano, their composition and origin. *Dokl. Akad. Nauk. SSSR, Earth Sci.*, 168: 215—217.
Butuzova, G.Y., 1969. The mineralogy and chemistry of iron sulphides in Black Sea sediments. *Lithol. Miner. Resour.*, 4: 401—411.
Cambel, B. and Jarkovsky, J., 1967. *Geochemie der Pyrite einiger Lagerstätten der Tschechoslowakei.* VSAV Bratislava, 493 pp. (with English summary).
Constantinou, G. and Govett, G.J.S., 1973. Geology, geochemistry and genesis of the Cyprus sulphide deposits. *Econ. Geol.*, 68: 843—858.

Corliss, J.B., 1971. Origin of metal bearing hydrothermal solutions. *J. Geophys. Res.*, 76: 8128–8138.

Croxford, N.J.W. and Jephcott, S., 1972. The McArthur lead–zinc–silver deposit. *N.T. Aust. Inst. Min. Metal., Proc.*, 243: 1–26.

Davidson, C.F., 1965. A possible mode of origin of stratabound copper ores. *Econ. Geol.*, 60: 942–954.

De Bretizel, P. and Foglierini, F., 1971. Les gites sulfures concordants dans l'environnement volcanique et volcano-sédimentaire. *Miner. Deposita*, 6: 65–76.

Doi, M., 1962. Geology and cupriferous pyrite deposits (Besshi-type) of the Sanbagawa metamorphic zone including the Besshi and the Sazare mines in central Shikoku (III). *Min Geol. (Japan)*, 12: 63–83.

Duke, N.A. and Hutchinson, R.W., 1974. Geological relationships between massive sulphide bodies and ophiolite volcanic rocks near York Harbour, Newfoundland. *Can. J. Earth Sci.*, 11: 53–69.

Dymond, J., Corliss, J.B., Dasch, E.J., Field, C.W., Heath, G.R. and Veeh, H.H., 1973. Origin of metalliferous sediments from the Pacific Ocean. *Geol. Soc. Am. Bull.*, 84: 3355–3372.

Evans, T.L., Campbell, F.A. and Krouse, H.R., 1968. A reconnaissance study of some western Canadian lead–zinc deposits. *Econ. Geol.*, 63: 349–359.

Farkas, A., 1973. *A Trace-Element Study of the Texas Gulf Orebody, Timmins, Ontario*. Thesis, Univ. Alberta, 148 pp., unpublished.

Ferguson, J. and Lambert, I.B., 1972. Volcanic exhalations and metal enrichments at Matupi Harbour, New Britain, TPNG. *Econ. Geol.*, 67: 25–37.

Fischer, R.P., 1970. Similarities, differences and some genetic problems of the Wyoming and Colorado Plateau type of U-deposits in sandstones. *Econ. Geol.*, 65: 778–784.

Fleischer, M., 1955. Minor elements in some sulphide minerals. *Econ. Geol.*, 50th Anniv. Vol.: 970–1024.

Fruth, I. and Maucher, A., 1966. Spurenelemente und Schwefel-Isotope in Zinkblenden der Blei–Zink-Lagerstätte von Gorno. *Miner. Deposita*, 1: 238–250.

Gasser, U., 1974. Zur Struktur und Geochemie der stratiformen Sulfidlagerstätte Meggen. *Geol. Rundsch.*, 63: 52–73.

Goldschmidt, V.M., 1954. *Geochemistry*. Oxford Univ. Press, London, 730 pp.

Govett, G.J.S. and Pantazis, Th.M., 1971. Distribution of Cu, Zn, Ni and Co in the Troodos Pillow Lava Series, Cyprus. *Trans. Inst. Min. Metal.*, 80: B27–B46.

Griffiths, W.R., Albers, J.P. and Öner, Ö., 1972. Massive sulphide copper deposits of the Ergani–Maden area, S.E. Turkey. *Econ. Geol.*, 67: 701–716.

Hall, W.E. and Heyl, A.V., 1968. Distribution of minor elements in ore and host rock, Illinois–Kentucky fluorite district and upper Mississippi-Valley lead–zinc district. *Econ. Geol.*, 63: 655–670.

Haranczyk, C., 1970. Zechstein lead-bearing shales in the Fore Sudetian Monocline in Poland. *Econ. Geol.*, 65: 481–495.

Harshmann, E.N., 1966. Genetic implications of some trace elements associated with uranium deposits, Shirley Basin, Wyoming. *U.S. Geol. Surv., Prof. Pap.*, 550–C: C167–C173.

Harshmann, E.N., 1973. Distribution of some elements in roll type uranium deposits. *Min. Eng.*, 25: 57–58 (abstract).

Hawley, J.E. and Nicol, I., 1961. Trace elements in pyrite and pyrrhotite. *Econ. Geol.*, 56: 467–487.

Hegemann, F., 1943. Die geochemische Bedeutung von Kobalt und Nickel im Pyrite. *Z. Angew. Mineral.*, 4: 122–239.

Helgeson, H.C., 1969. Thermodynamics of hydrothermal systems at elevated temperatures and pressures. *Am. J. Sci.*, 267: 729–804.

Hendricks, R.L., Reisbick, F.B., Mahaffey, E.J., Roberts, D.B. and Peterson, M.N.A., 1969. Chemical composition of sediments and interstitial brines from the Atlantis II, Discovery and Chain Deeps. In: E.T. Degens and D.A. Ross (Editors), *Hot Brines and Recent Heavy Metal Deposits in the Red Sea*. Springer, New York, N.Y., pp. 407–440.

Honnorez, J., Honnorez-Gurstein, B., Valette, J. and Wauschkuhn, A., 1971. Present day formation of an exhalative sulphide deposit at Vulcano, Tyrrhenian Sea, Part 11. In: G.C. Amstutz and A.J.

Bernard (Editors), *Ores in Sediments (Int, Union Geol. Sci., Ser. A, 3)*. Springer, New York, N.Y., pp. 139–166.

Hutchinson, R.W., 1965. Genesis of Canadian massive sulphides reconsidered by comparison to Cyprus deposits. *Can. Inst. Min. Metal.*, 58: 972–986.

Hutchinson, R.W. and Searle, D.L., 1971. Stratabound deposits in Cyprus and relations to other sulphide ores. *Proc. IMA–IAGOD Meet. '70, IAGOD Vol. – Soc. Min. Geol., Japan*, Spec. Issue, 3: 198–205.

Itoh, S., 1971a. Minor elements in sulphide minerals from Shirataki mine, Kochi Prefecture, Japan. *Bull. Geol. Surv. Japan*, 22: 367–384.

Itoh,S., 1971b. Chemical compositions of country rocks and minor elements in sulphide minerals from bedded cupriferous pyrite deposits of Tenryu River basin. *Bull. Geol. Surv. Japan*, 22: 117–132.

Itoh, S. and Kanehira, K., 1967. Trace elements in sulphide minerals from the Tsuchikura mine, Shiga Prefecture, with special reference to the contents of cobalt and nickel. *Min. Geol., Japan*, 17: 251–260 (in Japanese with English abstract).

Johnson, A.E., 1972. Origin of the Cyprus pyrite deposits. *Int. Geol. Congr., 24th Sess., Montreal, Sect. 4, Miner. Deposits*, pp. 291–298.

Kajiwara, Y., 1973. Chemical composition of ore forming solution responsible for the Kuroko mineralization in Japan. *Geochem. J., Japan*, 6: 193–209.

Kaplan, I.R., Sweeney, R.E. and Nissenbaum, A., 1969. Sulphur isotope studies on Red Sea Geothermal brines and sediments. In: E.T. Degens and D.A. Ross (Editors), *Hot Brines and Recent Heavy Metal Deposits in the Red Sea*. Springer, New York, N.Y., pp. 474–498.

Lambert, I.B. and Scott, K.M., 1973. Implications of geochemical investigations of sedimentary rocks within and around the McArthur zinc–lead–silver deposit, Northern Territory. *J. Geochem. Explor.*, 2: 307–330.

Lambert, I.B. and Sato, T., 1974. The Kuroko and associated ore deposits of Japan: a review of their features and metallogenesis. *Econ. Geol.*, in press.

Lancelot, Y., Hathaway, J.C. and Hollister, C.D., 1972. Lithology of sediments from the Western North Atlantic; Leg XI, Deep Sea Drilling Project. In: A.G. Kaneps (Editor), *Initial Reports of the Deep Sea Drilling Project*. Nat. Science Foundation, Washington, D.C., XI: 901–949.

Loftus-Hills, G. and Solomon, M., 1967. Cobalt, nickel and selenium in sulphides as indicators of ore genesis. *Miner. Deposita*, 2: 228–242.

Magak'yan, I.G., 1968. Ore deposits. *Int. Geol. Rev. Suppl.*, 10 (8) and 10 (9): 202 pp.

Matsukuma, T. and Horikoshi, E., 1970. Kuroko deposits in Japan, a review. In: T. Tatsumi (Editor), *Volcanism and Ore Genesis*. Univ. Tokyo Press, Tokyo, pp. 153–180.

McKnight, E.T. and Fischer, R.P., 1970. Geology and ore deposits of the Picher field, Oklahoma and Kansas. *U.S. Geol. Surv., Prof. Pap.*, 588: 165 pp.

Mendelsohn, F., 1961a. *The Geology of the Northern Rhodesian Copperbelt*. MacDonald, London, 523 pp.

Mendelsohn, F., 1961b. Ore deposits. In: F. Mendelsohn, *The Geology of the Northern Rhodesian Copperbelt*. MacDonald, London, pp. 117–129.

Mendelsohn, F., 1961c. Ore genesis. In: F. Mendelsohn, *The Geology of the Northern Rhodesian Copperbelt*. MacDonald, London, pp. 130–165.

Mercer, W. and Crocket, J., 1972. Gold and palladium in host rocks and ores from the Heath Steele B-1 Deposit, New Brunswick. *Int. Geol. Congr., 24th Sess., Montreal*, Sect. 10: 180–185.

Miyazaki, K., Mukaiyama, H. and Izawa, E., 1974. Thermal metamorphism of the bedded cupriferous iron sulphide deposit at the Besshi Mine, Ehime Prefecture, Japan. *Min. Geol., Japan*, 24: 1–12 (Japanese, with English abstract).

Morton, R.D., Goble, E. and Goble, R.J., 1973. Sulphide deposits associated with Precambrian Belt-Purcell strata in Alberta and British Columbia. In: *Belt Geology Symposium, 1*. Idaho Bur. Mines Geol., pp. 159–179.

Ohmoto, H. and Rye, R.O., 1974. Hydrogen and oxygen isotopic compositions of fluid inclusions in the Kuroko deposits, Japan. *Econ. Geol.*, 69: 947–953.

Pollock, G.D., Sinclair, I.G.L., Warburton, A.F. and Wierzbicki, V., 1972. The Uchi orebody – a massive sulphide deposit in an Archaean siliceous volcanic environment. *Int. Geol. Congr., 24th Sess., Montreal, Sect.* 4: 299–308.

Renfro, A.R., 1974. Genesis of evaporite-associated stratiform metalliferous deposits – a sabkha process. *Econ. Geol.,* 69: 33–45.

Roscoe, S.M., 1965. Geochemical and isotopic studies, Noranda and Mattagami. *Can. Inst. Min. Metall. Trans.,* LXVIII: 279–285.

Ross, D.A., Whitmarsh, R.B., Ali, S.A., Bourdeaux, J.E., Coleman, R., Fleischer, R.L., Girdler, R., Manheim, F., Matter, A., Nigrini, C., Stoffers, P. and Supko, P., 1973. Red Sea drillings. *Science,* 179: 377–380.

Rye, R.O., 1974. A comparison of sphalerite–galena sulphur isotope temperatures with filling temperatures of fluid inclusions. *Econ. Geol.,* 69: 26–32.

Rye, R.O., Hall, W.E. and Ohmoto, H., 1974. Carbon, oxygen and sulphur isotope study of the Darwin lead–silver–zinc deposit, Southern California. *Econ. Geol.,* 69: 468–481.

Sangster, D.F., 1972. Precambrian volcanogenic massive sulphide deposits in Canada: a review. *Geol. Surv. Can., Pap.,* 72–22, 44 pp.

Sato, T., 1973. A chloride complex model for Kuroko mineralization. *Geochem. J., Japan,* 7: 245–270.

Shiikawa, M., 1970. Limonite deposits of volcanic origin in Japan. In: T. Tatsumi (Editor), *Volcanism and Ore Genesis.* Univ. Tokyo Press, Tokyo, pp. 295–300.

Smith, S.E. and Walker, K.R., 1971. Primary element dispersions associated with mineralization at Mount Isa, Queensland. *Aust. Bur. Miner. Res. Bull.,* 131: 80 pp.

Stanton, R.L., 1972. *Ore Petrology.* McGraw-Hill, New York, N.Y., 713 pp.

Sutherland, J.K., 1967. Chemistry of some New Brunswick pyrites. *Can. Mineral.,* 9: 71–84.

Takahashi, K., 1966. Geochemistry of some minor elements in ores and ore minerals in the Kuroko-type deposits. *J. Min. Metall. Inst. Japan,* 82: 1051–1064 (in Japanese with English abstract).

Tatsumi, T. and Oshima, K., 1966. Mineralogical composition of ores from "Black Ore" deposits of the Kosaka and Hanaoka mines. *J. Min. Metall. Inst. Japan,* 82: 1008–1014 (in Japanese with English abstract).

Turekian, K.K. and Wedepohl, K.H., 1961. Distribution of the elements in some major units of the earth's crust. *Geol. Soc. Am. Bull.,* 72: 175–192.

Vine, J.D. and Tourtelot, E.B., 1969. Geochemical investigations of some black shales and associated rocks. *U.S. Geol. Surv. Bull.,* 1314A: 43 pp.

Vine, J.D. and Tourtelot, E.B., 1970. Geochemistry of black shale deposits – a summary. *Econ. Geol.,* 65: 253–272.

Volfson, F.I. and Arkhangelskaya, V.V., 1972. Conditions of formation of cupriferous sandstone deposits. *Lithol. Miner. Resour.,* 7: 268–279.

Wedepohl, K.H., 1964. Untersuchungen am Kupferschiefer in Nordwestdeutschland: Ein Beitrag zur Deutung der Genese bituminoser Sedimente. *Geochim. Cosmochim. Acta,* 28: 305–364.

Wedepohl, K.H., 1971. "Kupferschiefer" as a prototype of syngenetic sedimentary ore deposits. *Proc. IMA–IAGOD Meetings '70, IAGOD Vol. – Soc. Min. Geol., Japan,* Spec. Issue, 3: 268–273.

White, W.S. and Wright, J.C., 1966. Sulphide mineral zoning in the basal Nonesuch shale, N. Michigan. *Econ. Geol.,* 61: 1171–1190.

Wolf, K.H., 1976. The influence of compaction on ore genesis. In: G.V. Chilingar and K.H. Wolf (Editors), *Compaction of Coarse-Grained Sediments, II,* Elsevier, Amsterdam, in press.

Yamamoto, M., Ogushi, N. and Sakai, H., 1968. Distribution of sulphur isotopes, selenium and cobalt in the Yanahara ore deposits, Okayama–Ken, Japan. *Geochem. J., Japan,* 2: 137–156.

Yamaoka, K., 1962. Studies on the bedded cupriferous iron sulphide deposits occurring in the Sanbagawa metamorphic zone. *Tohoku Univ., Sci. Rep., 3rd Ser.,* 8: 1–68.

Zelenov, K.K., 1964. Iron and manganese in exhalations from the submarine volcano Banu-Wuhu, Indonesia. *Dokl. Akad. Nauk., SSSR,* 155: 94–96 (in English).

Chapter 2

PRINCIPLES OF DIFFUSION IN SEDIMENTARY SYSTEMS

E.K. DUURSMA and C. HOEDE

INTRODUCTION

Ever since gaseous, liquid and solid media have existed, there has been migration of substances due to physical forces. Such migration occurs together with, or is superposed on the transport phenomena which the media have themselves. In the case of sedimentary rocks, the migration is a kind of diffusion process superposed on the movements of pore liquids of the rock systems (Duffell, 1937). The migration processes are solely diffusion processes when the fluid medium is completely at rest. The causes of diffusion processes are differences in concentration of substances in the pore liquids and also differences in concentration inside of the solid material. Diffusion, as defined in this way, leads to a levelling of concentrations, but might also be responsible for an accumulation of substances. In the latter case, ores might be formed, where the diffusion is called "negative" diffusion.

The diffusion takes place in the direction of increasing concentrations. Side reactions are responsible for the "negative" diffusion, since in the "ore" layer the substances are immobilized thus lowering the concentrations in the pore liquids. This causes decreasing concentrations in the pore liquid towards the ore layer, and therefore a diffusion occurs in that direction.

The term diffusion is used in a variety of ways in geochemical and physical studies, in each of them describing migration processes of substances. In order to avoid confusion, it is necessary in each case to define such migration processes and determine whether terms like molecular diffusion, self-diffusion, negative diffusion, apparent diffusion, thermo-diffusion, or other terms should be used. Since natural processes, and in our case migration processes in sedimentary systems, are rarely of a kind that can be simply described with a mathematical model, this terminology might lead sometimes to confusion.

It will be attempted here to define the different kinds of diffusion, while some mathematical models are presented that might be used for the description of migration processes in sediments and sedimentary rocks, where *no* migration of pore liquids occurs. Furthermore, some implications to geochemical processes will be discussed.

THEORY OF DIFFUSION

Diffusion is defined as the process by which material is transported from one part to another part of the system as a result of random molecular motions (Crank, 1964). Random molecular motions exist in very dilute solutions, where each dissolved molecule behaves independently of the other dissolved molecules, and constantly undergoes collisions with other dissolved and solvent molecules. As a result of these collisions the movement of individual dissolved molecules has no preferred direction. When two compartments of such a solution contain a different amount of dissolved molecules, and the boundary between the two compartments is only hypothetical or transmittant, there will be a transport of dissolved molecules from the compartment with higher concentration to that with lower concentration. The net transfer is the difference between the movement of molecules through the (hypothetical) boundary from one side to the other, and back. Since there are more molecules per unit volume on one side of the boundary than on the other side, the net transport is in the direction of the compartment containing less dissolved molecules.

Although it seems to be obvious that diffusion has to be characterized in this way, the description also shows immediately that the definition is only valid for very dilute solutions (molecular diffusion) and not applicable without adaptation to solutions or solids where the freedom of molecular movement is not guaranteed.

For molecular diffusion in solvents, the basic formula of diffusion has been developed by Fick (1855) by adoption of Fourier's (1822) mathematical equation of heat conduction. This equation is based on the hypothesis that the rate of transfer of diffusing substances in isotropic media through unit area of a section is proportional to the concentration gradient normal to the section:

$$F = -D(\partial C / \partial x) \tag{1}$$

where F is the rate of transport per unit area of the section, C is the concentration of the diffusing substance, x the space coordinate normal to the section, and D is called the molecular diffusion coefficient.

Regarding the possible migration processes in sedimentary rocks, it is clear from the beginning that the term molecular diffusion, as defined above, cannot be used in this sense. Any diffusion other than this one should be called apparent diffusion. Only with adaptations can the mathematical developments based on formula (1) be used. Since in natural processes diffusion is mostly "apparent" diffusion, for simplicity the term diffusion will be used in agreement with the usage of the term by Crank (1964). Such simplification has met criticism by Manheim (1970) but regarding the wide use of the term diffusion also for the eddy transport processes in sea water (Pritchard et al., 1971), the description of transport and migration in sedimentary systems by the term diffusion makes sense.

This diffusion will concern: the molecular diffusion of substances dissolved in pore water, the migration of substances in solid materials, either along or through the crystal structures, and finally, a combination of both being the migration through interstitial water combined with exchange of the dissolved substances with the solid grains (sorption + desorption).

Most mathematical relationships for such processes are basically developed from formula (1). In a system where the molecules have much less freedom than in a dilute solution, the flux through an area can be put proportional to the gradient of concentration, normal to this area, when the condition of isotropic media is applicable. This last condition is not really satisfactory in sedimentary systems containing liquids, pores and spaces in between crystal lattice layers, but still diffusion models can be used for calculations of the transfer processes.

A summary of different model calculations is presented in Table I, giving mathematical relationships for diffusion from a constant source, an instantaneous source and a constant inflow, for linear, cylindrical and spherical diffusion (linear only for constant source). The diffusion coefficient D in the formulae of Model 1 (Table I) might be composed of two components, being the diffusion coefficient of the transfer processes as they occur in reality (flux proportional to concentration gradient in interstitial water) and the factor that determines the reactions (either chemical or physical or both) between the diffusing substances and the solid part of the medium. The diffusion coefficient for the complete system, D', in a simple case can be taken to be equal to the diffusion coefficient in the interstitial water (D) divided by $(1 + K)$, when K is the distribution coefficient for the first-rate reactions occurring at the interstitial water/solid interfaces. This $D' = D/(1 + K)$ is valid for the most simple interface reactions where K is equal to the ratio of the concentrations on both sides of the interface, and the exchange reactions are rapid relative to the diffusion in the interstitial water. This model (Model 2 in Table I) is also applicable to diffusion of substances inside solid or semi-solid media, where the effective migration occurs separately also with additional exchange reactions in the system.

Another series of models can be developed for diffusion in solid systems, being the diffusion into layered systems (Model 4 of Table I) and diffusion into spherical solid particles (Model 5 of Table I). These models were calculated by Hoede (1964a,b) and their applicability was empirically verified by Duursma and Bosch (1970).

FACETS OF DIFFUSION IN SEDIMENTS AND SOLID SYSTEMS

The study of diffusion processes measurable in short-time periods (weeks, months or a few years), is much easier than the study of diffusion processes that have occurred during geological periods (1,000 to millions of years). It is, therefore, necessary to investigate the several facets that determine the speed of diffusion processes in order to be able to extrapolate results from short-time diffusion studies to the long-term migration processes in geologic periods.

TABLE I

Table of mathematical models of diffusion as given by Duursma and Hoede (1967), reproduced with permission of *Neth. J. Sea Res.*

Model	Boundary conditions	Basic formulae	Applied formulae
I Diffusion (no sorpt.)	**a** $C(o,t) = C_o$ $(C_o = \text{const.}, t > 0$ $C(x,o) = 0, x > 0$ const. source)	linear: → [1] $C(x,t) = C_o \, \text{erfc}\left\{ x(4Dt)^{-\frac{1}{2}} \right\}$ [2] $C(x,t) \simeq C_o \frac{2}{x}\sqrt{\frac{Dt}{\pi}} \exp\left[-\frac{x^2}{4Dt}\right]$ (for $t \ll \frac{x^2}{4D}$) (cyl.+spher. diff. imposs.)	$\frac{C(x,t)}{C_o} = \text{erfc} \frac{x}{2\sqrt{Dt}}$ * $\frac{C(x,t)\sqrt{\pi}}{C_o} \simeq \frac{\exp(-\xi)}{\sqrt{\xi}}$ $(\xi = \frac{x^2}{4Dt} \gg 1)$
	b $\int_{-\infty}^{+\infty} \tilde{C}(x,t)\,dx = s$ $(s = \text{const.}, t > 0$ $C(x,o) = 0, x > 0$ r instead of x inst. source)	lin.: → —(X for →) $C(x,t) = \frac{s}{\sqrt{4\pi Dt}} \exp\left[-\frac{x^2}{4Dt}\right]$ cyl.: $C(r,t) = \frac{s}{4\pi Dt} \exp\left[-\frac{r^2}{4Dt}\right]$ sper.: $C(r,t) = \frac{s}{\sqrt{(4\pi Dt)^3}} \exp\left[-\frac{r^2}{4Dt}\right]$	$^{10}\log C(x,t) = A - \frac{x^2}{4Dt} \, {}^{10}\log e$ ibid, but $x = r$ and $A = B$ ibid, but $x = r$ and $A = C$ (A, B and C = const. f. t = c.)
	c $\int_{-\infty}^{+\infty} \tilde{C}(x,t)\,dx = \int_0^t q\,d\tau$ $(q = \text{const. infl.}$ $C(x,o) = 0, x > 0$ r instead of x const. inflow)	lin.: → $C(x,t) \simeq \frac{2qt\sqrt{Dt}}{\sqrt{\pi}\, x^2} \exp\left[-\frac{x^2}{4Dt}\right]$ (for $t \ll \frac{x^2}{6D}$) cyl.: $C(r,t) \simeq \frac{qt}{\pi r^2} \exp\left[-\frac{r^2}{4Dt}\right]$ (for $t \ll \frac{r^2}{4D}$) spher.: $C(r,t) \simeq \frac{q\sqrt{t}}{2\pi\sqrt{D\pi}\, r^2} \exp\left[-\frac{r^2}{4Dt}\right]$ (for $t \ll \frac{r^2}{2D}$)	├———————┤ $D = \frac{r^2}{4t} \times \frac{{}^{10}\log e}{{}^{10}\log\left[\frac{qt}{\pi r^2 C(r,t)}\right]}$ ├———————┤
2 Diff.+ rapid sorpt.	**a** $\frac{C_{\text{sorbed}}}{C_{\text{solution}}} = K$ $(K = \text{const.})$	$D' = \frac{D}{1+K}$; see then 1	Valuable at low conc. for e.g. Langmuir ads. and ionexchange.
	b $\frac{C_{\text{sorb.}}}{C_{\text{sol.}}} = f(C_{\text{sol}})$	Not solved.	Occurs at high conc. for the same reactions.
3 Diff.+ slow sorpt.	$\frac{\partial C_{\text{sorb.}}}{\partial t} = a[KC_{\text{sol}} - C_{\text{sorb}}]$ $(a, K = \text{const.}$ const. source)	$\bar{C}(x,p) = \frac{B}{p} \exp(-qx)$ (Laplace transf., B=const. $q = \sqrt{\frac{p}{D}\left\{1 + \frac{aK}{a+p}\right\}}$)	For small t equal to 1 For very large t, see 2

* C_0 for the same medium conditions of C(x,t).
** Key to symbols (see also text): D = diffusion coefficient; K = distribution coefficient; C = concentration; x = distance; t = time; R, r = radius; erfc = complementary errorfunction; s = instantaneous source quantity; A, B, C in last columns = log values of terms of formulae in previous column.

(Table I continued)

Directions	Graphs
1 Plot $\left\{\frac{C(x,t)}{C_o}, x\right\}$ on a theor. erfc graph. Interpolate Dt, to give D. **2** Calc. $\frac{C(x,t)\sqrt{\pi}}{C_o}$ f. certain x and t; use as $f(\xi)$ on the graph to give ξ. Calc. D from $D = x^2 : 4\xi t$.	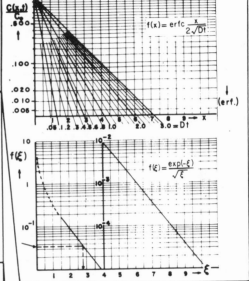
1 Plot $\{c(x,t), x^2\}$ on semi-log. paper (t=const.). Calculate D from the slope of the line ($tg\phi = \frac{^{10}log(\cdots)_2 - ^{10}log(\cdots)_1}{x_2^2 - x_1^2}$), or by overlapping with the theoretical graph. **2** Plot $\{c(x,t), t\}$ on graph paper (x = const.). D is simply found from x (or r) and t_{max}.	
The formulae for the linear and spherical diffusion are too difficult to allow simple calculations. For the cylindrical diffusion D can be derived directly from the formula for a certain r and t.	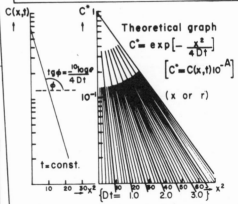
Det. f. sediments $K = \frac{C_{sorbed}}{C_{solution}}$ (f. $t=\infty$) in sorption exp.; then use the formulae of I to give D'. $D = (1+K)D'$.	
For solutions of higher concentrations the problem is not solved.	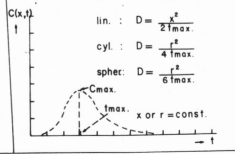
For short time phenomena the problem approaches to I; for very long term processes to 2.	

(Table I continued)

Model	Boundary conditions	Basic formulae	Applied formulae
4 Diffu- sion in two- layered medium	**a** $C_0 \mid D_1 \mid D_2$ $0 \mid a \to x$ (C_0 = const. $C(x,0)=0, x>0$)	$M(t) \simeq 2\sqrt{\dfrac{D_2 t}{\pi}} + a(1-\delta^2)$ (for term I >> term II; $M(t)$ p. unit surface $C_0 \equiv 1$, $\delta = \sqrt{\dfrac{D_2}{D_1}}$).	$\dfrac{M(t)}{C_0(1+K)} \simeq 2\sqrt{\dfrac{D_2 t}{\pi}} + a(1-\delta^2)$ (K = accumulation fact. $1+K = \dfrac{C_{medium}(t=\sim)}{C_0}$.)
	b $C_0 \mid D_1 \mid D_2$ $0 \mid a \to x \mid b$ (C_0 = const. $C(x,0)=0, x>0$)	$M(t) \simeq b - \dfrac{8}{\pi^2}\left[b-(1-\delta^2)a\right]\exp\left[\dfrac{-D_2\pi^2 t}{4(b-a)^2}\right]$ (for $t \gg \dfrac{4(b-a)^2}{D_2\pi^2}$; $C_0 \equiv 1$ $\dfrac{a\delta}{b-a} \ll 1$; $\delta=\sqrt{\dfrac{D_2}{D_1}}$)	$^{10}\log\left[1-\dfrac{M(t)}{M(\sim)}\right] \simeq {}^{10}\log\left[\dfrac{8}{\pi^2}\left\{1-(1-\delta^2)\dfrac{a}{b}\right\}\right]$ $- \dfrac{D_2\pi^2 t\,{}^{10}\log e}{4(b-a)^2}$ ($M(\sim)=bC_0(1+K)$)
	c (diagram: D_2, b, c, D_1) (N particles C_0 = const. $C(r<b,0)=0$)	⓵ $C_0 \equiv 1$ $M(t) \simeq \dfrac{4}{3}\pi Nb^3 - \dfrac{8Nc^3}{\pi}\left[1+\dfrac{\delta^2 a}{c}\right] \times$ $\exp\left[-\dfrac{D_2\pi^2 t}{c^2}\right]$ ($t \gg \dfrac{c^2}{D_2\pi^2}$, $\dfrac{a}{c} \ll 1$ $\delta \ll 1$) ⓶ $C_0 \equiv 1$ $M(t) \simeq 8\pi Nb^2\sqrt{\dfrac{D_1 t}{\pi}}$ (t small, $\dfrac{a}{c} \ll 1$ $\delta \ll 1$; $\delta=\sqrt{\dfrac{D_2}{D_1}}$)	$^{10}\log\left[1-\dfrac{M(t)}{M(\sim)}\right] \simeq {}^{10}\log\dfrac{N}{V_p}\left[\dfrac{8c^3}{\pi}\left\{1+\dfrac{\delta^2 a}{c}\right\}\right]$ $- \dfrac{D_2\pi^2 t\,{}^{10}\log e}{c^2}$ ($M(\sim)=V_p C_0(1+K)$) $\dfrac{M(t)}{M(\sim)} \simeq 8\dfrac{N}{V_p}b^2\sqrt{\dfrac{D_1 t}{\pi}}$ (V_p = vol. N particles)
5 Diffu- sion in homog. spheres	(diagram: D, R, V_p, V_s) ($C_0 \ne$ const. V, V_s, V_p = vol. tot., sol., part.)	$M(t) \simeq \dfrac{4}{3}\pi R^3 N\left[\dfrac{1}{1+\frac{\rho R}{3}} - \dfrac{6}{\pi^2}\sum_{n=1}^{\sim}\dfrac{1-\frac{5\rho R}{n\pi}}{n^2} \times\right.$ $\left. \exp\left\{\dfrac{-Dt\left(n\pi+\frac{\rho R}{n\pi}\right)^2}{R^2}\right\}\right]$ (N = numb. of part., C_0=1, $\rho R = \dfrac{\frac{4}{3}\pi R^3 N}{V-\frac{4}{3}\pi R^3 N} = \dfrac{V_p}{V_s}$)	$^{10}\log\left[1-\dfrac{M(t)}{M(\sim)}\right] \simeq A - \left\{\dfrac{D\pi^2}{R^2}{}^{10}\log e\right\} t$ (for $n=1$, $t \gg \dfrac{R^2}{4\pi^2 D}$, $V_p \ll V_s$; A = const. $M(\sim) = V_p C_0 \dfrac{1+K}{1+V_p/V_s\,K}$)
6 Diff. fr. laminar current	(diagram: C_0, $C(x,t)$, h, D) ($C(0,t)=C_0$, $C(x,0)=0, x>0$ C_0, v = const.)	$C(x,t)=C_0\left[1-\dfrac{\sqrt{\frac{D}{\pi}}}{hv} \times \dfrac{x}{\sqrt{t}}\right]$	$\dfrac{C(x,t)}{C_0} = 1 - \dfrac{\sqrt{\frac{D}{\pi}}}{hv} \times \dfrac{x}{\sqrt{t}}$

* See below A. ** The validity is indicated in the graph by the straightness of the line.

(Table I continued)

Directions	Graphs
Det. f. sediments K in sorpt. exp. Plot $\left\{\frac{M(t)}{C_o(1+K)}, \sqrt{t}\right\}$ on graph paper. Extrap. the line f. $\sqrt{t} \to o$ to the ordinate $\to a(1-\delta)$. The slope $\to D_2$. Now D_1 can be found fr. these two values.	
Det. f. sedim. K or M($\sim$) in sorpt. exp.. Plot $\left\{\left[1-\frac{M(t)}{M(\sim)}\right], t\right\}$ on semi-log. paper, the slope $\to D_2$ $\left[tg\,\phi = \frac{{}^{10}log(\cdots)_2 - {}^{10}log(\cdots)_1}{t_2 - t_1}\right]$. Fr. th. extrap. graph for $t \to o$, the value on the ordinate $\to D_1$.	
$\overline{1}$ Det. f. susp. M($\sim$) on graph paper fr. a $\{M(t), t\}$ diagr.. Plot $\left\{\left[1-\frac{M(t)}{M(\sim)}\right], t\right\}$ on semi--log. paper. The slope $\to D_2$ $\left[tg\,\phi = \frac{{}^{10}log(\cdots)_2 - {}^{10}log(\cdots)_1}{t_2 - t_1}\right]$. Extr. $t \to o$: D_1. $\boxed{2}$ Det. again M($\sim$). Plot $\left\{\frac{M(t)}{M(\sim)}, \sqrt{t}\right\}$ o. gr. paper. The slope $\to D_1$. D_2 can be found fr. (1).	
Det. for suspensions M($\sim$), in a $\{M(t), t\}$ diagram. Plot $\left\{\left[1-\frac{M(t)}{M(\sim)}\right], t\right\}$ on semi-logarithmic paper. The slope of the line gives D, by calculation from $\left[tg\,\phi = \frac{{}^{10}log(\cdots)_2 - {}^{10}log(\cdots)_1}{t_2 - t_1}\right]$. (Only for a limited source.)	
Plot $\left\{\frac{C(x,t)}{C_o}, \frac{x}{\sqrt{t}}\right\}$ on graph paper ; the slope of the line gives D.	

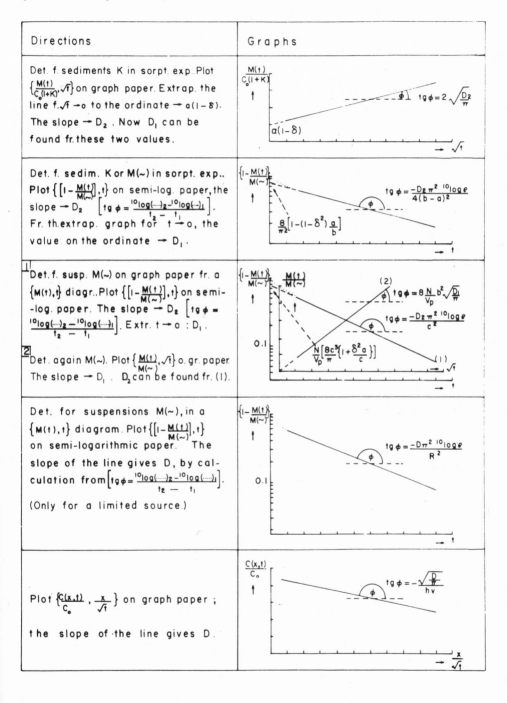

Diffusion of substances in a medium is a migration of molecules, clusters of molecules or parts of molecules (ions). Temperature, viscosity of the medium, interfacial tensions, free path, electrical potentials and chemical reactions play a role in the determination of the residual speed of the diffusion. All these facets will be treated along with a simplified example of a sediment containing immobile interstitial water in all its pores (saturated sediment) (Duursma and Bosch, 1970).

For uncharged molecules or clusters of molecules the diffusion is not dependent on the co-diffusion of other substances, unless some reactions occur. For ions, there should be a co-diffusion of oppositely charged ions, since otherwise a charge barrier is built up that prohibits the diffusion. This implies that for ionic diffusion, the diffusion coefficient will be different when the ion is accompanied by different oppositely charged ions. In Table II this is illustrated for chloride ions in the electrolytes, KCl, $CaCl_2$ and HCl, diffusing into interstitial water of a marine sand and fine sediment, to which these ions have almost no affinity of being adsorbed. Since diffusion of ions in dilute solutions is characterized by the equivalent conductance at infinite dilution, this physical factor can be taken as comparison for the relationships that diffusion should have. Hydrogen ions have the most elevated equivalent conductance (a H^+ ion acts as a single proton), and the diffusion coefficient of Cl^- as co-ion in HCl, is much higher than of Cl^- as co-ion in KCl or $CaCl_2$. The most realistic diffusion coefficient of Cl^- would be the one measured for KCl, since potassium and chloride ions have practically the same equivalent conductance. This diffusion coefficient would count for Cl^- ions as if they were uncharged, and diffusion is related to their mobility only.

For sedimentary rock systems, this ionic-charge effect might only play a role when one of the ions is more or less fixed in the solid system. This means that the existence of a gradient in concentration does not necessarily cause a transport by diffusion.

TABLE II

Diffusion coefficients of Cl^- ions with K^+, Ca^{2+} and H^+ as co-ions in the interstitial water of a sandy and clay sediment (Duursma and Bosch, 1970); table reproduced with permission of *Neth. J. Sea Res.*

Source	Interstitial water	Sediment	Equivalent conductance at infinite dilution (ohm^{-1} cm^2; 25°C)	Diffusion coefficients (10^{-5} cm^2/sec)	
				cation	chloride
KCl	distilled water	North Sea	K^+: 74.5 Cl^-: 75.5	0.79	0.75
KCl	sea water	North Sea		0.77	0.78
$CaCl_2$	distilled water	North Sea	$\frac{1}{2}Ca^{++}$: 60	0.28	0.45
$CaCl_2$	sea water	North Sea		0.32–0.50	0.38–0.72
HCl	distilled water	Mediterr.	H^+: 350	1.20	1.17
HCl	sea water	Mediterr.		1.30	1.21

Free path

When migrating through pores of a sedimentary system, the diffusing material will take a sinuous path in contrast to its path in a pure solution. The measured diffusion coefficient will be lower than for diffusion of substances without obstacles. The decrease of the diffusion coefficient due to the obstacles is equal for sediments consisting of uniform spheric particles in a close-packed system which also have an equal pore volume. Both diffusion coefficient, conductivity and pore volume are theoretically independent of the diameter of the spheres (Duursma and Bosch, 1970).

An experimental verification with marine sands sieved for different grain sizes, confirmed the independence of these factors from the grain size. The factor of decrease for the diffusion coefficient relative to a free solution was equal to about 2, the theoretical value. When small and big grains of different forms are present together, small particles fill up the holes between bigger grains, and the factor of decrease will be somewhat lower than 2. However, for packed unconsolidated well-sorted marine sediments, the diffusion coefficients of chloride (KCl) rarely decrease more than a factor of 4 as compared to molecular diffusion in free water.

For porous rocks the free-path factor is given by a lithologic factor (Klinkenberg, 1961), L, and the porosity, f. The factor of decrease of the diffusion coefficient is given by the ratio f/L, which is called the "effective directional porosity". $D_{effective} = D_{liquid} \times f/L$. Since f/L can be determined by conductivity measurements, the free-path effect on diffusion in porous rocks can be determined.

Interface surface action, viscosity and zeta potential

Other physical factors might also influence the diffusion of dissolved substances in the interstitial water of sediment. These are the interfacial tension, responsible for the capillary forces, the viscosity and the forces in the electrical double layers around the solid surfaces characterized by the zeta potential. With quartz sands, sieved for grain sizes from 100 to 2,000 microns, the interfacial tension as well as the viscosity is not changing due to the size of the grains (and the size of the pores) (Duursma and Bosch, 1970), but because the specific surface is larger for the smaller grain sizes, the effect of interfacial tension is measurable by a slight decrease in the diffusion coefficient of KCl. For the electrical double layer, the impact of changes in zeta potential (also called electrokinetical potential) on a variety of electrolytes present in sea water, brackish and fresh water, is small. As far as is detectable, the diffusion coefficients in sediments with these electrolytes seem to be constant for ionic diffusion as long as there exists either a positive or negative zeta potential, independent of the value of this potential above + 50 mV or below −50 mV. Probably the influence of the zeta potential is limited to the film around solid particles. It does not matter whether positive or negative charged ions are adsorbed — the effects are identical. This confirms what has been mentioned above about the co-diffusion of ions. Almost zero zeta potential means that there is little adsorption and, therefore, also little retardation of ionic diffusion. About 30% higher values for the diffusion coefficient may then be found (Duursma and Bosch, 1970).

Temperature

The mobility of molecules, the equivalent conductance of ions and the viscosity of liquid media are affected by temperature fluctuations. This will result in a variation of diffusion being related to each of these three physical factors (Jackson, 1963). In a free solution of water, the equivalent conductance doubles for each 25°C increase (Fig. 1), resulting in the doubling of the molecular diffusion coefficient for the same temperature increase. Since fluidity (reciprocal of viscosity) is also doubling (or even something more) for each 25°C, this will be the main cause of the increase in the equivalent conductance. Hence, the increase of diffusion coefficients with temperature (Fig. 2) is also mainly caused by the decrease of the viscosity of the liquid medium.

Pressure

The physical properties such as viscosity, specific volume, electrical conductance and dielectric constance of interstitial water undergo slight changes up to 8% for a pressure of 1,000 atm. (Horne et al., 1969). For diffusion this was smaller, and the diffusion coefficient decreases about 2% in value for a rise of 1,000 atm.

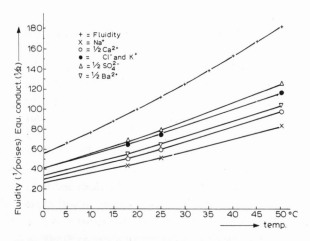

Fig. 1. Equivalent conductance values of some cations and anions at infinite dilution in relation to temperature. Relationship of fluidity (reciprocal of viscosity) of distilled water with temperature (Hodgman, 1962; cf., Duursma and Bosch, 1970). (Figure reproduced with permission of *Neth. J. Sea Res.*)

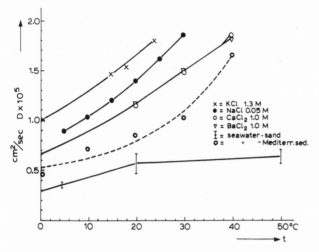

Fig. 2. Temperature dependence of the diffusion coefficients of some chlorides in free solution (Washburn, 1929), and of chloride in Mediterranean sediment and Dutch Wadden Sea sand (Duursma and Bosch, 1970). (Figure reproduced with permission of *Neth. J. Sea Res.*)

Compaction

Compacting sediments lose water and have, therefore, a displacement of the interstitial water caused by compaction. Mostly, there are the combined effects of both sedimentation and compaction, where the sediment/water interface is moving upwards. This problem related to early diagenesis is discussed by Anikouchine (1967) for recent marine sediments. His model calculations are based on the space coordinate, which is moving relative to the sediment/water interface. For diffusion, relative to this model it is possible to arrive at diffusion equations where the advection term is omitted, since the advection factor disappears because movement of water toward the interface is countered by interface movement during sediment accumulation.

For sedimentary rocks, this problem is probably of another relevance. Compaction, due to better packing of the sedimentary grains, relative to displacement of pore waters, occurs only in early to late diagenesis. In sedimentary rocks under great pressure in later stages another process might occur, namely that related to metamorphism. Sedimentary rock, such as quartzite, is deformed by pressure solution, which also obeys a kind of diffusion-flow law, although it is difficult to propose a unique flow law (Weyl, 1959). The migration of substances concerns a kind of material creep due to pressure. The microstructure of grain boundaries, at which the diffusion occurs, resembles a channel network one or two atomic spaces thick, having a high diffusivity with micron-wide zones (Elliot, 1973). In reality, it is possible to call this kind of migration diffusion creep, a process enhanced when recrystallization might occur at the interfaces. At elevated temperatures, mobility starts far below the melting point of the minerals (Gill, 1960), thus migration of material could be partly or wholly effected by diffusion.

Although it is questionable whether metamorphisms would lead to ore formation, the replacement processes, where minerals are transformed into other minerals in the host rocks, might do so (Garrels and Dreyer, 1952). These authors verified experimentally the replacement processes of copper, iron and other metals in sulphides and limestone.

Diffusion and side reactions

Although the previously mentioned interfacial factors can be classified as side reactions occurring with diffusion, the impact of non-physical reactions, like physico-chemical and chemical adsorption is easily several orders of magnitude larger. Such adsorption reactions occur with almost any elements from the periodic system and with almost all natural sedimentary materials, with the exception of the elements of some halogens and alkali metals.

These side reactions can be classified both as rapid adsorption/desorption reactions between pore waters and surfaces of the sedimentary particles and as slow adsorption/desorption reactions taking into account the migration into or out of the crystal lattices, including the formation and dissolution of insoluble compounds. Surface adsorption processes appear to obtain equilibrium in days or sometimes weeks (Duursma and Eisma, 1973), the inter-lattice processes are much slower and are only measurable after weeks, months or even years. The quantities absorbed, however, might exceed the amounts adsorbed on the surfaces (Ros Vicent et al., 1974).

The question remains as to which part of the sorbed material can actually take part in the apparent migration when gradients of concentrations exist. Another point is whether, for geological periods, the results from short-term measurements can be extrapolated to long-term processes. These problems still have to be solved.

As measured for a number of radionuclides and marine sediments, the diffusion coefficients are indeed several orders of magnitude lower than the diffusion coefficients of molecular diffusion of these radionuclides in free solutions of water (see Fig. 3). Measurements of such diffusion coefficients in the laboratory are very difficult to make since the diffusion proceeds slowly even over a period of years. During such periods it is very difficult to keep the conditions constant and, for example, avoid convection and bacterial decomposition. For this reason an indirect method might be applied where the impact of the side reaction is separately measured from the diffusion inside the interstitial water (Duursma and Bosch, 1970).

This indirect method applies to Model 2 (Table I), where $D' = D/(1 + K)$ as mentioned before. D is the diffusion coefficient in the interstitial water of the element in question; K is the distribution coefficient of the side reaction and D' is the diffusion coefficient for the complete system. D can be determined from the diffusion coefficient of a non-reacting element, for example chloride-ion using KCl as salt. Then D_{Cl} has to be multiplied by a factor giving the difference in mobility between the "reacting" element and chloride. This factor is just the ratio of the two equivalent conductances (at infinite dilution), which can be found in handbooks. The determination of the distribution coefficient K needs an experiment, where the sediment is exposed to the "reacting" element in solution. The ratio of concentrations in sediment and liquid, given in a dimensionless way, is called the distribution coefficient:

$$K = \frac{\text{amount "reacting" element/ml of sediment (dry)}}{\text{amount "reacting" element/ml of liquid}}$$

For a number of radionuclides, the results of the indirect method of determination of the diffusion coefficient have been compared with the results of a direct method, applying Model 1b of Table I where linear diffusion took place from an instantaneous source. The results are given in Table III showing some relatively good agreement between the determined and calculated diffusion coefficients.

From these results it is obvious that the diffusion coefficients are very much lower than the molecular diffusion coefficients of substances in a free solution. In free solution, the values are about $0.5 \cdot 10^{-5}$ cm^2/sec, while for most of the investigated elements the diffusion coefficients are as low as 10^{-9} and 10^{-10} cm^2/sec. What such figures mean for an effective transport can be seen from Fig. 3. For a constant source, the migration of a 10% concentration front is for $D = 10^{-10}$ cm^2/sec only about 4.5 cm for a time of 10 centuries.

Such a diffusion is slow for processes taking place in recent sediments of aquatic

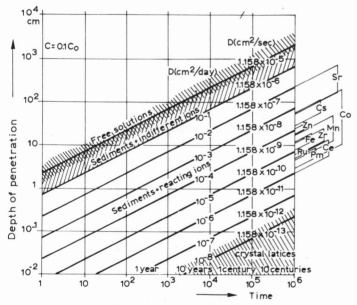

Fig. 3. Depth penetration of a 10% concentration front into a marine sediment from an overlying layer with constant concentration. The range of diffusion coefficients found for various radionuclides and sediments is shown on the right. (After Duursma and Eisma, 1973, and reproduced with permission of *Neth. J. Sea Res.*)

TABLE III

Diffusion coefficients of 11 radioisotopes in Mediterranean sediment, determined by "indirect" and by direct method (after Duursma and Bosch, 1970, reproduced with permission of *Neth. J. Sea Res.*)[1]

Isotope	Carrier quantity (μg/l)			Distribution coefficients · 10^4			Correction factor (25°C)	Diffusion coefficients (10^{-10} cm²/sec)	
	susp. techn.	thin-l. techn.	sed. techn.	susp. techn (4°C)	thin-l. techn. (4°C)	sed. techn. (20°C)		calculated	determined
^{90}Sr	25	33	20	±0.001	±0.0001	±0.001	0.88	5000–30000	7000–11000
^{90}Y	12.5	17	10	±0.1	±0.1	±0.1	0.97	60	?
^{137}Cs	0.8	1.1	0.64	?	0.5	0.25	1.15	14–30	80–110
^{106}Ru	–	–	–	1.3	1.1	1	0.8*	4–5	15–150
^{59}Fe	45	60	36	30	>1	5	1.01	1.2–6	20–130
^{65}Zn	130	173	104	0.2	2	0.4	0.78	2.4–24	?–10
^{60}Co	0.8	1.1	0.64	<0.1	5	2	0.82	1->50	10–20
^{147}Pm	–	–	–	2	3	3	0.8*	0.6–2.4	4–10
^{54}Mn	2.0	2.7	1.6	0.8	<10	3	0.79	2–6	3–5
^{95}Zr/Nb	0.75	1.0	0.6	3	1–10	3	0.8*	0.5–5	6–60
^{144}Ce	5.8	7.7	4.6	2	6	3	1.04	1–3	5–20

[1] The indirect method used the distribution coefficients as determined by three techniques and $D_{Cl^-} = 6 \cdot 10^{-6}$ cm²/sec; the direct method is in accordance with Model 1a of Table I.
* Correction factor estimated.

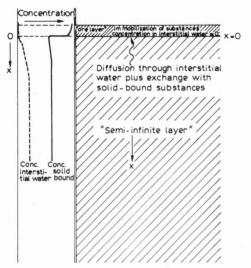

Fig. 4. Scheme of "negative" diffusion as possible mechanism for the accumulation of substances in selected horizons.

systems. The substances seem to be fixed to the sediments. For geological periods, however, the diffusion of substances through interstitial waters (being sorbed and desorbed by the sedimentary particles with distribution coefficients between 10^4 and 10^5) is certainly not a negligible process even not when the extrapolation is hypothetically made. Extrapolating from Fig. 3 for a time of, for example, 10^9 years, and $D = 10^{-10}$ cm^2/sec, the 10% concentration front has moved for about 45 m. Such distances might be sufficient for the formation of ores in layered systems.

As mentioned already in the introduction, this diffusion is a kind of negative diffusion, since the migration occurs in the direction of increasing concentrations (Fig. 4). This negative diffusion is only an apparent negative diffusion, which is the logical consequence of having included side reactions in the diffusion model. In reality, there exists a gradient of decreasing concentrations inside the interstitial water, which is the determining factor for the migration, while the very high distribution coefficient in the ore layer is the determining factor of the amounts that can be bound by the sedimentary particles.

A mathematical model that describes this kind of diffusion into an ore layer, is given by Crank (1964) for diffusion out of a semi-infinite medium with an initial concentration C_0 throughout, and the concentration at the interface with the absorbing layer (ore layer) maintained at the constant concentration C_1 (Fig. 4):

$$\frac{C(x,t) - C_1}{C_0 - C_1} = \text{erf} \frac{x}{2\sqrt{Dt}} \tag{2}$$

where C_0 is the initial concentration in the medium (and later far from the interface), $C(x,t)$ is the concentration in the semi-infinite medium at time t and depth x, measured with respect to the interface ($x = 0$) of this medium and the ore layer. C_1 is the concentration at the interface which is maintained constant, erf is the abbreviation of the error function and D is the diffusion coefficient in the semi-infinite medium.

Supposing C_1 is small relative to C_0 and $C(x,t)$, the form (2) will read (cf. also Garrels et al., 1949):

$$\frac{C(x,t)}{C_0} = \text{erf} \frac{x}{2\sqrt{Dt}} \tag{3}$$

This form can also be used when $C(x,t)$ and C_0 are the overall concentrations in the sediment (concentration in interstitial water and on the sediment particles, supposing sorption and desorption occur), and having D as the apparent diffusion coefficient for the semi-infinite layer, supposing the substances are completely immobilized in the ore layer, which makes C_1 of formula (2) negligible. According to Crank (1964), the total amount M_t of diffusing substance which has left the semi-infinte medium at time t is found by integration with respect to x and is:

$$M_t = \int_0^\infty [C_0 - C(x,t)]\, dx = C_0 \int_0^\infty \left(1 - \text{erf} \frac{x}{2\sqrt{Dt}}\right) dx = C_0 \int_0^\infty \text{erfc} \frac{x}{2\sqrt{Dt}}\, dx = 2\,C_0 \sqrt{\frac{Dt}{\pi}}$$

$$\tag{4}$$

where erfc denotes the complimentary error function.

Supposing such a model calculation is valid, the amount of substance being taken up in the ore layer is given in Table IV expressed by means of the thickness of a layer of the semi-infinite medium in the initial state that contained this amount. A common value of $D = 10^{10}$ cm^2/sec would give 62 cm as equivalent for a period of 10^6 years and 20 m as equivalent for a period of 10^9 years.

The conditions as to whether the substances are built up into an ore layer, depend on the difference in the affinity of sorption given by the distribution coefficients. When the distribution coefficients in the ore layer are very much larger than those in the surrounding sediment layers, the calculation as given in (4) is the best approximation. Whether or not such processes have been found to occur, will be discussed later.

TABLE IV

Equivalent-distances in cm giving the amount of material diffused out of a semi-infinite layer according to formula (4); 200 cm equivalent-distance means the amount of material that was present in a column of 200 cm of the semi-infinite layer having a concentration of C_0

D (cm^2/sec)	Time in years								
	10	10^2	10^3	10^4	10^5	10^6	10^7	10^8	10^9
10^{-6}	20	62	200	620	2000	6200	20000	62000	200000
10^{-7}	6.2	20	62	200	620	2000	6200	20000	62000
10^{-8}	2.0	6.2	20	62	200	620	2000	6200	20000
10^{-9}	0.62	2.0	6.2	20	62	200	620	2000	6200
10^{-10}	0.20	0.62	2.0	6.2	20	62	200	620	2000
10^{-11}	0.06	0.20	0.62	2.0	6.2	20	62	200	620

DIFFUSION IN SOLID MEDIA

The replacement of carbonate fossil by SiO_2, FeS_2, etc., in sedimentary rocks during geological periods is proof that replacement of elements inside solid formations has been possible. Since the structures of the fossils have been maintained, there should have been a migration of the elements responsible for the fossil formation. Such a migration taking place in solid material might be called a kind of diffusion also.

The diffusion coefficients determined for diffusion in solid materials have been found to be extremely low: in the range from about 10^{-14} cm^2/sec (Duursma and Bosch, 1970) down to about 10^{-19} cm^2/sec (Lahav and Bolt, 1964). The theoretical displacement of a 10% concentration front as based on the graph given in Fig. 3 by extrapolation, leads to a distance of 45 cm for $D = 10^{-14}$ cm^2/sec and 0.045 cm for $D = 10^{-20}$ cm^2/sec for a period of 10^9 years. This calculation shows that for geological times diffusion can also have its influence in solid substances in order to produce changes in composition of the solid substances.

In some way or another, diffusion in sedimentary rocks has its parallel in the diffusion of water-saturated sediment. Also in sedimentary rocks there exists a kind of basic freedom of space through which atoms can move when they have no affinity for being fixed at certain positions. This phenomenon of freedom, also called the diffusivity (Newton and Round, 1961), can be characterized by the diffusion of helium. This diffusivity is also expressed as the diffusion coefficient and is for helium about 10^{-10} cm^2/sec (Barrer, 1952). For kinetics and mathematics, see also Lagerwall and Zimen (1963).

This implies that for any lower diffusion coefficient the diffusion can be regarded as diffusivity combined with side reaction, where the side reaction is determined by the affinity of the atoms to be fixed for certain positions in the crystal structure of the sedimentary rocks. A higher affinity or steric hindrance will result in a lower coefficient.

At elevated temperatures, it could be measured that the diffusion of O, Li, Na and K in quartz, for example, is probably related to the vacant oxygen lattice positions. The diffusion coefficient of oxygen in quartz at 500°C is about $3 \cdot 10^{-11}$ cm^2/sec, for the alkali metals even lower (Verhoogen, 1952). These values are of the order of the diffusion coefficients of water-saturated sediments as given in Fig. 3. In marble, the diffusion coefficient of sodium could be measured by ^{22}Na-radiotracer to be even lower (10^{-7} cm^2/sec) at 400°C and 300 bars pressure (Jensen, 1965). This value is about 10^6 times higher than at 20° and 1 bar pressure. The vapour pressure of this particular metal seems to be involved with this rigorous reduction of the diffusion coefficient.

FIELD OBSERVATIONS AS RELATED TO DIFFUSION

The literature on diffusion measured in the natural environment concerns mainly the processes occurring, or having occurred, in marine conditions. Processes occurring in terrestrial environments have mostly another basic principle, since water movements above and below the water table, even with very low velocities, do overlap easily with the diffusion. The migration processes of substances are then better described by percolation and leaching, for which the theories are similar to those developed for the ion-exchange kinetics (Helfferich, 1965).

During early diagenesis of marine sediments, the migration of iron and sulphur within anaerobic sediments was one of the most characteristic phenomenona and is linked to the formation of sediments containing organic matter. Especially in coastal waters rich in organic matter, as a result of upwelling due to favourable hydrographic and climatic conditions (Brongersma-Sanders, 1971), the accumulating sediments will also be rich in organic material, usually containing trace metals adsorbed on the fine particles. The metals may be remobilized during diagenesis. The migration of iron and sulphur connected to organically rich layers is illustrated in Fig. 5, showing the diffusion of SO_4^{2-} into the organically rich layer to supply the necessary energy for anoxic bacterial activities, and the diffusion of H_2S out of this layer consequently changes the conditions in the adjacent layers (Berner, 1969, 1971). Continuous diffusion of dissolved iron and sulphate towards the organically rich layer can result in the formation of iron sulphide concretions, surrounding and enclosing the organically rich layer. In this way, perhaps, many pyrite concretions and pyritized fossils are formed, containing also other metals which are accumulated by identical processes as related to the anoxic conditions (Brongersma-Sanders, 1970).

At least 0.1% dry-weight organic carbon is required to produce these so-called synsedimentary metal sulphide deposits containing more than 1% metal (Rickard, 1973), but an additional flux of metals is essential above that normally observed in sea water. Erosion of continental metal-rich rocks is a possible source, but is probably insufficient due to great dilution by non-metalliferous erosion products. The availability of the metal seems to occur in areas with hydrothermal activity for organic-rich, fine-grained sediments (e.g., shales).

Reducing conditions, alternated with oxidizing conditions, are also the cause of the formation of manganese concretions on the top of the sea bed in many sea areas. These

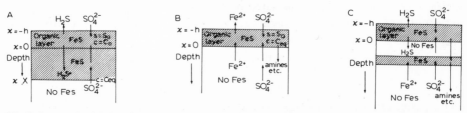

Fig. 5. Iron and sulphur diffusion models. Arrows indicate diffusion directions. Hachured areas represent darkened layers due to iron sulphide formation. A. Low-iron content model; B. high-iron content model; C. intermediate-iron content model. (After Berner, 1969, and reproduced with permission of *Am. J. Sci.*)

concretions are the best example of the fact that diffusion of materials in marine sediments does occur. The principles of this diffusion are the same as for the above-mentioned diffusion of H_2S and Fe. Manganese occurring as di-oxide is mobilized by reduction to Mn^{2+} inside the sediment and is diffusing in the pore water to the sea-water/sea-bed interface, where it is again immobilized. An example is given by Li et al. (1969) for a 3-m core of Arctic Basin sediment. In the sediment, the total manganese content is characterized by a general, if somewhat erratic, decrease of the concentrations with depth (from 2,000 p.p.m. at the surface to about 600 p.p.m. at 1 m depth, and then fluctuating between 400 and 800 p.p.m. deeper in the sediment), while the manganese in the pore water displays a remarkable continuous increase with depth from below 0.05 p.p.m. until 9 p.p.m. at 1.2 m depth, dropping at 4 p.p.m. deeper in the sediment (Fig. 6).

The diffusion which is related to this gradient, is a kind of diffusion with very slow desorption. This kind of diffusion is not represented by one of the models in Table I. Li et al. (1969) determined for manganese in the pore water a diffusion coefficient of $4 \cdot 10^{-7}$ cm^2/sec. This is about two orders of magnitude lower than the molecular diffusion coefficients. This means that additionally some sorption and desorption of the Mn^{2+} occurs as it migrates to the surface.

This phenomenon is reflected in the distributions found for various metals in the top 20 cm layer of a Pacific deep-sea core (Fig. 7) with an upper oxidized zone and a lower reduced zone. Manganese, Ni, Co, P and La are enriched in the upper zone while Cr, V, U and S are enriched in the lower zone (Bonatti et al., 1971).

The mechanism of the accumulation of manganese on the top of the sea bed is a reoxidation and, consequently, immobilization of the Mn^{2+} to MnO_2 on arrival at the sea-water/sea-bed interface. Other metals might be following, either scavenged by manganese or due to an identical diffusion process. The manganese concretions contain rather high quantities of iron, nickel, copper, cobalt, molybdenum and zinc (Mero, 1966).

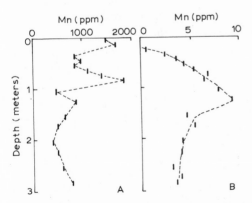

Fig. 6. Contents of manganese in an Arctic marine sediment core. A. Overall concentration; B. concentration in the pore water. (After Li et al., 1969, and reproduced with permission of North-Holland, Amsterdam.)

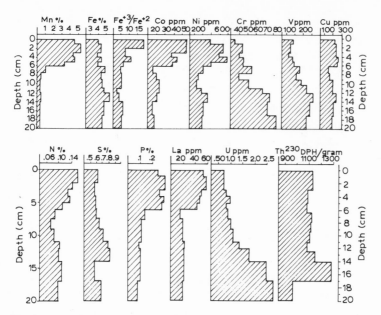

Fig. 7. Concentration distribution of various elements in a Pacific core off Panama, depth 3274 m, calculated on carbonate-free basis (except for nitrogen). (After Bonatti et al., 1971, and reproduced with permission of Pergamon Press, Oxford.)

The information which is available on the migration processes deeper in the sediments is obtained primarily by the investigations of the pore-water composition of deep-sea drill cores (Manheim and Sayles, 1974). The results confirm data from Shiskina et al. (1969) and Smirnov (1969) that, for most of the interstitial waters of ocean sediments, there are few changes in composition, as compared to normal sea water, which can lead to any ore formation. The changes observed are interchange reactions, such as the exchange of calcium for magnesium which leads to dolomitization, authigenic uptake of cations from the pore waters, in particular magnesium and potassium in the form of silicate formation; and replacement of iron in clays by magnesium, as a result of sulphate reduction. The gradient of Ca in the interstitial water, for example, can be as was found in some of the deep-sea drill cores, from 0.4‰ in the surface sediment to 1.4‰ at 300 m in the sediment, accompanied by an inverse gradual decrease of 1.2 to 0.7‰ for magnesium.

When deep drill cores contain brine or fresh-water sediments, the pattern of salinities in the layers above and below these horizons, agree with diffusion of the interstitial salt away from or to the horizons, respectively. For example, total interstitial salt increased from about 13‰ in surface sediments of Caspian Sea sediments to about 140‰ at about 650 m (cf., Manheim and Sayles, 1974), while off Florida even at 120 km from the coast the salinities in the interstitial water decreased from 20‰ Cl at the surface to almost zero at about 200 m in the sediment.

CONCLUSIONS

Although diffusion processes might have taken part in the formation of selected sedimentary horizons containing higher concentrations of metals, it is doubtful whether the large ore deposits have been formed by such a migration of substances, which is called "negative" diffusion. Most of the diffusion processes that can occur in water-saturated sedimentary media are very slow, and even submarine brine horizons underneath the sea bed have not completely lost the excess salt during the period of burial. For sea salt, the diffusion coefficient is about 10^{-6} to $5 \cdot 10^{-6}$ cm^2/sec, the highest values possible in sedimentary media. Any substances that have an affinity to sedimentary particles, diffuse at a much slower rate. A common value is 10^{-10} cm^2/sec for trace-metal diffusion coefficients and this would result in a 10% concentration front only 45 m from a constant source in 10^9 years. The quantities of material that can be accumulated in a sediment horizon with a great affinity for these materials, can be estimated also from a diffusion model. Supposing again a period of 10^9 years and the diffusion coefficient is $D = 10^{-10}$ cm^2/sec for the surrounding layers, the equivalent amount of material originally present in 20 m surrounding layer, will be accumulated in the sediment horizon. This is usually not sufficient to produce a large ore deposit.

Nevertheless, one of the largest potential ore layers of manganese, including some other metals, existing on the sea bed as manganese nodules, have their origin by diffusion processes. Post-depositional reduction of manganese oxide within the sediment to Mn^{2+}, which is relatively mobile and diffuses upwards to the sea water containing much less Mn^{2+}, and re-oxidation at the sea-water/sediment interface, are processes resulting in the formation of manganese nodules. These nodules grow on top as they lose underneath when buried by sedimentation, thus remaining at the interface.

Equally, a minor metallic concentration or a specific feature (e.g., zoning) within larger deposits might be caused by diffusion processes (e.g., Brown, 1974). However, it is essential that either by different redox potentials or other processes the metals are stronger bound in these features thus causing an effective reduction of the concentrations in the pore liquids, for example. This will cause a gradient of concentrations and diffusion.

Diffusion in solid media at room temperature occurs with diffusion coefficients below 10^{-10} cm^2/sec, being the diffusivity of helium, until 10^{-20} cm^2/sec for other elements. Over geological periods, diffusion can be the cause of crystal transformations which result in dolomitization and fossil replacement. At elevated temperatures, as present in deep structures, the mobilities of various substances are greatly increased, thus diffusion coefficients of even above 10^{-10} cm^2/sec might occur.

REFERENCES

Anikouchine, W.A., 1967. Dissolved chemical substances in compacting marine sediments. *J. Geophys. Res.*, 72: 505–509.

Barrer, R.M., 1952. *Diffusion in and through Solids*. Cambridge University Press, Cambridge, 464 pp.

Berner, R.A., 1969. Migration of iron and sulfur within anaerobic sediments during early diagenesis. *Am. J. Sci.*, 267: 19–42.

Berner, R.A., 1971. *Principles of Chemical Sedimentology*. McGraw-Hill, London, 240 pp.

Bonatti, E., Fisher, D.E., Joensuu, O. and Rydell, H.S., 1971. Postdepositional mobility of some transition elements, phosphorus, uranium and thorium in deep-sea sediments. *Geochim. Cosmochim. Acta*, 35: 189–201.

Brongersma-Sanders, M., 1970. Origin of trace-metal enrichment in bituminous shales. In: G.D. Hobson (Editor), *Adv. Organic Chem., 3rd Int. Conf.*, pp. 231–236.

Brongersma-Sanders, M., 1971. Origin of major cyclicity of evaporites and bituminous rocks: an actualistic model. *Mar. Geol.*, 11: 123–144.

Brooks, R.R., Presley, B.J. and Kaplan, I.R., 1968. Trace elements in the interstitial waters of marine sediments. *Geochim. Cosmochim. Acta*, 32: 397–414.

Brown, A.C., 1974. An epigenetic origin for stratiform Cd-Pb-Zn sulfides in the Lower Nonesuch Shale, White Pine, Michigan. *Econ. Geol.*, 69: 271–274.

Crank, J., 1964. *The Mathematics of Diffusion*. Oxford University Press, London 3rd ed., 347 pp.

Duffell, S., 1937. Diffusion and its relation to ore deposition. *Econ. Geol.*, 32: 494–511.

Duursma, E.K. and Bosch, C.J., 1970. Theoretical, experimental and field studies concerning diffusion of radio-isotopes in sediments and suspended particles of the sea, B. Methods and experiments. *Neth. J. Sea Res.*, 4: 395–469.

Duursma, E.K. and Eisma, D., 1973. Theoretical, experimental and field studies concerning reactions of radio-isotopes with sediments and suspended particles of the sea, C. Applications to field studies. *Neth. J. Sea Res.*, 6: 265–324.

Duursma, E.K. and Hoede, C., 1967. Theoretical, experimental and field studies concerning molecular diffusion of radio-isotopes in sediments and suspended solid particles of the sea, A. Theories and mathematical calculations. *Neth. J. Sea Res.*, 3: 423–457.

Elliott, D., 1973. Diffusion flow laws in metamorphic rocks. *Geol. Soc. Am. Bull.*, 84: 2645–2664.

Fick, A., 1855. Über Diffusion. *Ann. Phys. Leipzig*, 170: 59–86.

Fourier, J.B., 1822. Théorie analytique de la chaleur. In: *Oeuvres de Fourier Darboux, 1889–1890*, 2 vol.

Garrels, R.M. and Dreyer, R.M., 1952. Mechanism of limestone replacement at low temperatures and pressures. *Bull. Geol. Soc. Am.*, 63: 325–380.

Garrels, R.M., Dreyer, R.M. and Howland, A.L., 1949. Diffusion of ions through intergranular spaces in water-saturated rocks. *Bull. Geol. Soc. Am.*, 60: 1809–1828.

Gill, J.E., 1960. Solid diffusion of sulphides and ore formation. *Int. Geol. Congr., 21st*, 16: 209–217.

Helfferich, F., 1965. Ion-exchange in chromatography. *Adv. Chromatogr.*, 1: 3–60.

Hodgman, C.D. (Editor in Chief), 1962. *Handbook of Chemistry and Physics*. The Chemical Rubber Publ. Co., Cleveland, Ohio, 43rd ed., 3515 pp.

Hoede, C., 1964a. Diffusion in layered media. *Tech. Note Math. Centre, Amsterdam*, TN 37: 19 pp.

Hoede, C., 1964b. On the absorption by homogeneous spherical particles. *Tech. Note Math. Centre, Amsterdam*, TN 38: 9 pp.

Horne, R.A., Day, A.F. and Young, R.P., 1969. Ionic diffusion under high pressure in porous solid materials permeated with aqueous, electrolytic solution. *J. Phys. Chem.*, 73: 2782–2783.

Jackson, R.D., 1963. Temperature and soil-water diffusivity relations. *Soil. Sci. Soc. Am.*, 27: 363–366.

Jensen, M.L., 1965. The rational and geological aspects of solid diffusion. *Can. Mineralogist*, 8: 271–290.

Klinkenberg, L.J., 1961. Analogy between diffusion and electrical conductivity in porous rocks. *Bull. Geol. Soc. Am.*, 62: 559–564.

Lagerwall, T. and Zimen, K.E., 1963. The kinetics of rare-gas diffusion in solids. *Hahn-Meitner Inst. Kernforsch., Berlin, Rep.*, HMI-B25: 29 pp.

Lahav, H. and Bolt, G.H., 1964. Self-diffusion of Ca-45 into certain carbonates. *Soil Sci.*, 97: 293–299.

Li, Y.H., Bischoff, J. and Mathieu, G., 1969. The migration of manganese in the Arctic Basin sediment. *Earth Planet. Sci. Lett.*, 7: 265–270.

Manheim, F.T., 1970. The diffusion of ions in unconsolidated sediments. *Earth Planet. Sci. Lett.*, 9: 307–309.

Manheim, F.T. and Sayles, F.L., 1974. Composition and origin of interstitial waters of marine sediments, based on deep sea drill cores. In: E.D. Goldberg (Editor), *The Sea*, 5. Wiley, New York, N.Y., pp. 527–568.

Mero, J.L., 1966. Manganese nodules (Deep Sea). In: R.W. Fairbridge (Editor), *The Encyclopedia of Oceanography*. Reinhold, New York, N.Y., pp. 449–454.

Newton, R. and Round, G.F., 1961. The diffusion of helium through sedimentary rocks. *Geochim. Cosmochim. Acta*, 22: 106–132.

Pritchard, D.W., Reid, R.O., Okubo, A. and Charter, H.H., 1971. Physical processes of water movement and mixing. In: *Radioactivity in the Marine Environment*. Nat. Acad. Sci. NRC., Washington, D.C., pp. 90–136.

Rickard, D.T., 1973. Limiting conditions for synsedimentary sulphide ore formation. *Econ. Geol.*, 68: 605–617.

Ros Vicent, J., Costa Yague, F., Parsi, P., Statham, G. and Duursma, E.K., 1974. The ease of release of some trace metals and radionuclides being sorbed for prolonged periods by marine sediments. *Bul. Inst. Esp. Oceanogr.* (in press).

Shishkina, O.V., Pavlova, G.A. and Bikova, V.S., 1969. *The Geochemistry of Halogens in the Sea Sediments and Interstitial Waters*. Nauka, Moscow, 117 pp.

Smirnov, S.I., 1969. Geochemical history of interstitial waters of sediments of marine origin. *Int. Geol. Rev.*, 11: 993–1004.

Verhoogen, J., 1952. Ionic diffusion and electrical conductivity in quartz. *Am. Mineralogist*, 37: 637–655.

Washburn, E.W., 1929. *International Critical Tables of Numerical Data, Physics, Chemistry and Technology, 5.6*. McGraw-Hill, London, 5: 465 pp; 6: 471 pp.

Weyl, P.K., 1959. Pressure solution and the force of crystallization. A phenomenological theory. *J. Geophys. Res.*, 64: 2001–2025.

Chapter 3

HYDROGEOCHEMICAL ASPECTS OF MINERAL DEPOSITS IN SEDIMENTARY ROCKS[1]

BRIAN HITCHON

INTRODUCTION

Water, through its unique or extreme properties, is the fundamental fluid genetically relating all mineral deposits in sedimentary rocks. It is the vehicle for the transportation of materials in solution and suspension throughout the hydrologic cycle, and it takes part in reactions during the dissolution of minerals in chemical weathering, diagenesis and metamorphism of sediments and sedimentary rocks, and the remelting and crystallization of igneous and metamorphic rocks.

The intimate interplay between the hydrologic and the geochemical cycles is illustrated in Fig. 1. The geochemical cycle is not closed, either energetically or materially, and the cycle may be temporarily halted, short-circuited, or reversed. Nevertheless, it is a convenient concept and framework within which to evaluate and discuss the role of water with respect to mineral deposits in sedimentary rocks. The geochemical cycle comprises two interlinked environments. The primary environment embraces those parts of the earth's crust extending below the level of effective circulation of meteoric water into regions of deep-seated metamorphic and igneous processes. The secondary environment encompasses the relatively shallow processes of weathering, sedimentation, diagenesis and low-grade metamorphism, that occur in the environment of effective circulation of meteoric water; it includes the two most important short-circuits within the geochemical cycle between sedimentary and metamorphic rocks and soils and sediments, as illustrated by the broken arrows in Fig. 1. The term meteoric water is used in the same sense as Clayton et al. (1966); thus there is an isotopic component in all water in the secondary environment which originates in the atmosphere and this component is absent in water from the primary environment.

There is no clearcut demarcation between the primary environment and the secondary environment. The nuances of the chosen limit may be appreciated by considering that circuit of the hydrologic cycle which follows directly the path of the geochemical cycle.

---

[1] Contribution No. 689 from the Alberta Research Council.

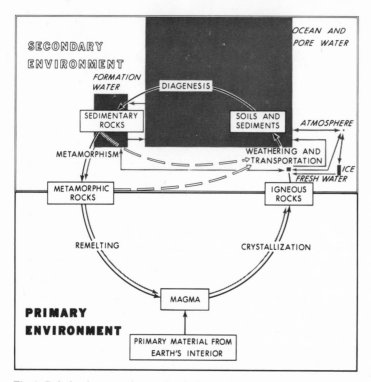

Fig. 1. Relation between the geochemical cycle (modified from Saxby, 1969) and the hydrologic cycle (screened arrows). The area of the rectangles in the hydrologic cycle are proportional to the volumes of water in the respective reservoirs, based on data from Horn and Adams (1966) and Penman (1970).

We know that sediments and sedimentary rocks lose part of their pore water by porosity reduction as a consequence of compaction and that water continues to escape during diagenesis and metamorphism. Dehydration reactions result in the transfer of "structural" water to the pore space as higher metamorphic grades are reached (White, 1957). Water (pore and structural) must remain in the metamorphic milieu because it ultimately becomes enriched in the magmatic residuum following remelting, magma generation and crystallization. Yet, the conditions of remelting and crystallization effectively remove that water from meteoric circulation. We are not yet able to examine water within the primary environment but the conceptual path outlined seems reasonable. It is doubtful, in my opinion, if water associated with recent volcanism represents anything more than relatively shallow circulating formation-waters that have contacted the ascending magma. Thus the tenuous limit of effective circulating meteoric water will be accepted as the demarcation between the primary environment and the secondary environment. Having now set this limit we will not be concerned further with the primary environment, except to note that geochemical processes that take place within the primary environment and that are specifically related to the search for mineral deposits have been described and evaluated in an excellent recent textbook by Levinson (1974).

Within the secondary environment the hydrologic cycle comprises two main circuits, a surface circuit and a subsurface circuit. The surface circuit is primarily driven by solar energy together with gravity flow of terrestrial surface water to the ocean. The main physical processes are evaporation and precipitation, and the important reservoirs are the atmosphere, terrestrial surface water (dominantly fresh water), the ocean, and glaciers and ice-sheets. The subsurface circuit is primarily driven by differences in fluid potential, and the major reservoirs are formation waters of sedimentary rocks and pore waters of sediments, with an important short-circuit by way of terrestrial surface waters. It is the result of chemical and physical processes occurring during movement of water through the subsurface circuit which is of direct concern to this review of hydrogeochemical aspects of mineral deposits in sedimentary rocks. (Editor's note: For a detailed summary of the influence of compaction on ore genesis, see Wolf, 1976, and on the importance of water during metamorphism consult Chapter 5 by Mookherjee in Vol. 4.)

HYDRODYNAMICS

If the movement of water in the subsurface circuit were to cease, chemical and physical equilibration between the fluids and the rocks would ultimately occur and there would be no further opportunities to generate new mineral deposits in sedimentary rocks. This hypothetical, and as we shall see, impossible; situation stresses the significance of hydrodynamics to the genesis of mineral deposits.

The flow of fluids through porous media is governed by a well recognized system of energy potential fields (fluid, thermal, electric and chemical). Of these, the fluid potential is the most important and the best known. The fluid potential is defined as the amount of work required to transport a unit mass of fluid from an arbitrary chosen datum (usually sea level) and state to the position and state of the point considered. Flow may be considered as a transient (unsteady-state) or steady-state phenomenon. Problems concerned with the flow of fluids released from storage, or from a borehole or small group of boreholes through confined layers, are probably best solved if they are analysed as transient phenomena. When consideration is given to studies of small drainage basins or to flow regimes across large sedimentary basins, it may be assumed that a relatively steady state has been reached. This does not imply static conditions, but rather that any recharge to the flow regime is negligible in relation to the vast amount of fluid in the system. In addition, the law of conservation of mass is operative, and the recharge is equal to the discharge. In short, we are considering a case of dynamic equilibrium.

The classic work of Hubbert (1940) on the theory of groundwater motion was the first published account of the basinwide flow of fluids that considered the problem in exact mathematical terms as a steady-state phenomenon. His concept of formation-fluid flow is shown in the top model in Fig. 2.

Starting with the standard equation for fluid potential derived by Hubbert (1940), Tóth (1962, 1963) developed an analytical method for the determination of the fluid potential at any point within a generalized flow region such as that illustrated in the centre diagram in Fig. 2, specifically for an isotropic, homogeneous case. Tóth represented the water table in two ways: as a simple straight line, and as a sine curve. Under general conditions the water table approximates to the topography, and certainly, on a regional scale, the relatively minor fluctuations of the water table can be neglected. It was

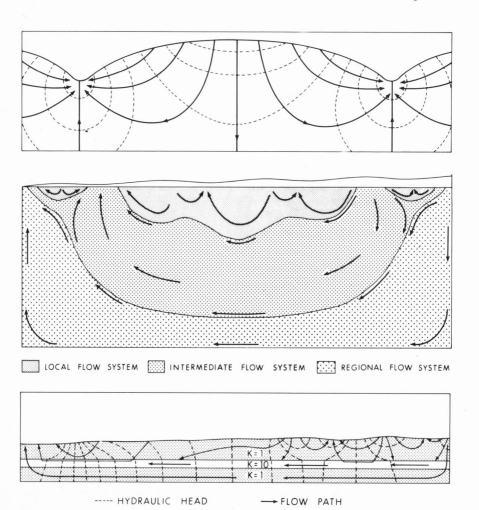

LOCAL FLOW SYSTEM   INTERMEDIATE FLOW SYSTEM   REGIONAL FLOW SYSTEM

----- HYDRAULIC HEAD   ⟶ FLOW PATH

Fig. 2. Development of fluid flow models. Top: Hubbert (1940); centre: Tóth (1962, 1963); bottom: Freeze and Witherspoon (1966, 1967, 1968). From Hitchon (1974), Chapter 13 in Levinson (1974). Courtesy Applied Publishing, Calgary, Alta.

necessary to assume theoretical vertical and horizontal impermeable boundaries if the model was to be meaningful. The horizontal boundary exists at some specified depth, above which all rock is permeable, even if only to the slightest degree. The two vertical boundaries are set to coincide with the highest (water divide) and lowest elevations in the entire flow system under consideration. Tóth (1962, 1963) showed conclusively that the fluid potential at any point in his model was related to several factors, including the topographic gradient and the elevation of the water table at the lowest point in the flow system, relative to the horizontal impermeable boundary. In every case, the local changes in topography controlled an upper zone of a local system of fluid flow, and the main regional trend in topography controlled a regional fluid flow regime extending around the extremities of the model, both systems being separated by an intermediate flow system.

Freeze and Witherspoon (1966, 1967, 1968) pointed out a number of restrictions to the Tóth model, and developed both analytical and numerical methods of analysis (Freeze, 1966, 1969) for a wide variety of different models, with up to three continuous or discontinuous layers of differing permeabilities, at various attitudes within the model, with a sloping basement, and with a variety of topographies including a general configuration. One of their models is illustrated in the bottom diagram in Fig. 2. Their work serves to confirm and strengthen the original flow systems demonstrated by Tóth. Because the parameters they used are dimensionless, the resulting flow nets apply to sedimentary basins of all sizes.

The Tóth—Freeze—Witherspoon model was applied to the western Canada sedimentary basin (Hitchon, 1969a,b). These studies show that the main variables affecting the fluid potential distribution are topography and geology (lithology and permeability). The fluid potential in any part of the basin corresponds closely to the fluid potential at the topographic surface in that part of the basin. Major recharge areas correspond to major upland areas, and major lowland regions are major discharge regions. Large river valleys commonly exert a drawdown effect on the fluid potential distribution, which may be observed to depths of up to 5000 ft. (Hitchon, 1971, fig. 4). Relatively highly permeable beds, if sufficiently thick, can significantly affect the regional fluid potential distribution by drawing down the fluid potentials, although the dominant features are topographically controlled. These observations apply to all sedimentary basins because of the dimensionless nature of the model.

Fluids in porous media move from regions of high energy to regions of low energy. The energy gradient to which the fluids react comprises an aggregate of potential differences resulting from elevation and pressure, thermal, electric and chemical forces coupled in a system as yet incompletely described mathematically (Van Everdingen, 1968). The elevation and pressure potential can be completely described by hydraulic head, using the same assumptions made by Hubbert (1940). Only when the thermal, electric, and chemical potentials influence the kinetic state of the fluid can their effect be seen in terms of hydraulic head, and then probably only a portion of that potential is observed. The electro-osmotic fluid movement is generated by electric potentials resulting from the existence of telluric currents and is assumed to have a negligible effect on fluid movement. The thermal potential is also probably negligible as a force to move fluids in most sedimentary basins, but it may be of significance in geothermal regions. Thus, the thermal and electric potentials probably contribute but a small fraction of the total energy of the system.

It is the chemical potential that most strongly influences the movement of the fluid, as described by the elevation and pressure potential, and it is most important in those sedimentary basins with both a wide range in content of dissolved material in formation waters and shales which act as semipermeable membranes. A large literature has built up on the subject of the semipermeable membrane capacity of shales, including significant recent reports by Coplen and Hanshaw (1973), Hanshaw and Coplen (1973), and Kharaka and Berry (1973, 1974), and it is now well documented that active chemical potentials manifest themselves by so-called anomalous pressures and salinities of formation waters and result in features such as apparently three-dimensionally closed hydraulic-head lows (Berry, 1959, 1960; Berry and Hanshaw, 1960; Hill et al., 1961; Back and Hanshaw, 1965; Coustau and Sourisse, 1967; Hanshaw and Hill, 1969; Hitchon, 1969b). The situation in these closed fluid potential lows is in part analogous to an osmotic pump acting within an elevation and pressure potential field.

Compaction of sediments produces a driving potential resulting from water squeezed by mechanical or provided by chemical (dehydration reactions) stresses, and is important in recent sediments and young sedimentary basins but should not be neglected as a potential source of fluid movement in ancient sedimentary basins. Under normal conditions most compaction which takes place soon after deposition results in the movement of fluids upwards out of the sediments towards the sea bottom, because porosity and permeability tend to decrease downwards. However, in thick sequences of argillaceous sediments, fluid expulsion may be inhibited because of low permeability, thus compaction will not be great, and a high porosity—high fluid pressure situation will result. This is termed under-compaction and is a temporary state. The higher-than-normal fluid pressures cause limited downward fluid flow in sections where thick, permeable sandstones or carbonates underlie the zone of under-compacted shale, with the boundary surface between the zones of upward and downward fluid flow existing in the lower middle part of the shales (Magara, 1969). This phenomenon has been observed in Japan (Magara, 1968) and in the western Canada sedimentary basin (Magara, 1973).

It should now be apparent to the reader that aqueous continuity within the subsurface circuit is a fundamental feature of the secondary environment. From laboratory and field evidence we now understand the essential attributes of the processes producing subsurface flow and can now appreciate why cessation of fluid flow in the subsurface circuit is impossible. Therefore, the subsurface situation is one of potential continuous fluid—rock interaction as the moving fluid is chemically and physically equilibrated and then re-equilibrated as it passes rocks and environments of differing chemical and physical properties. The potential for the generation of mineral deposits is persistently present.

HYDROCHEMISTRY

It is difficult, if not essentially impossible, to determine the source of a water sample based solely on its chemical and physical properties without prior information on that source. This statement is based on the fundamental premise that in the subsurface circuit we are considering a case of dynamic equilibrium, both with respect to fluid flow and fluid chemistry. The chemical and physical attributes of each water sample may be thought of as the result of static equilibrium with the rock from which it originated, and may be expressed in terms of the degree of saturation with respect to specific minerals, which need not be present in the rock. The difficulty expressed above may be illustrated by two relatively simple cases. First, it is easy to visualize a pure quartz arenaceous aggregate (sand, sandstone or quartzite) with a wide variety and range of aqueous interstitial fluid compositions, from fresh waters at recharge outcrops, through sea water as in a marine beach sand, to essentially saturated brines such as may be found deep in some sedimentary basins. The only useful common parameter might be, but not necessarily will be, the degree of saturation of each water sample with respect to quartz, and it is conceivable that each could be saturated with respect to quartz under the specific chemical and physical conditions at their source, thus providing possibly no useful parameter by which we may distinguish their source. Skinner and Barton (1973) provide the second example by noting that waters with high contents of sodium chloride may be found in a wide range of environments including oil fields, evaporite basins, fluid inclusions in pegmatite minerals, Colombian emeralds, and gangue minerals of porphyry copper deposits and base metal veins (Roedder, 1972). There may be no common feature to these waters except, fortuitously, the same degree of saturation with respect to halite. These two simple cases make it clear that if we are to fully understand the genesis of mineral deposits in sedimentary rocks we must approach the problem from the point of view of dynamic equilibrium of water—rock interaction.

The unique or extreme properties of water which allow tremendous variation in solution phenomena, together with the multitude of minerals that may be present in sedimentary rocks, result in constraints which must be considered in water—rock interaction studies. Some of the more important constraints are listed in Table I. Fundamentally, all are concerned with energy and mass transfers among solid, liquid and gaseous phases. For simplicity, usually only inorganic material is treated in an aqueous system, although as we become more sophisticated in our approach to fluid—rock interaction it will be possible to consider solid, liquid and gaseous organic phases reacting among themselves and with the inorganic phases. In both the inorganic and organic systems it should be possible, ultimately, to take into account stable and radioactive isotope transfers. In addition to the eight constraints in Table I, there exist the possibilities of equilibrium reactions incorporating irreversible changes, such as may occur in weathering, diagenesis and evaporative concentration, osmotic and ion exchange phenomena with clay minerals, the non-ideal behavior of the gas phase, incongruent reactions, chromatographic effects,

TABLE I

Constraints to consider in water–rock interaction studies

Constraint	Comment
Conservation of mass	In all ordinary chemical changes, the total mass of the reactants is always equal to the total mass of the products.
Conservation of energy	In a chemical change there is no loss or gain but merely a transformation of energy from one form to another.
Law of mass action	At a constant temperature and pressure, the product of the activities on one side of a chemical equation when divided by the product of the activities on the other side of the chemical equation is a constant. Thus if the reaction of b moles of B with c moles of C has come to equilibrium with the products d moles of D and e moles of E, then: $$bB + cC = dD + eE$$ and $$\frac{(a_D)^d (a_E)^e}{(a_B)^b (a_C)^c} = K$$ where a is the activity of the substance and K is the thermodynamic equilibrium constant.
Gibbs' phase rule	$$F = C + 2 - P$$ where F is the variance (number of "degrees of freedom") of the system, C is the number of components, and P is the number of phases existing together in equilibrium.
Diffusional transfer	If the concentration (mass of solid per unit volume of solution) at one surface of a layer of liquid is d_1 and at the other surface is d_2, the thickness of the layer h and the area under consideration A, then the mass of the substance which diffuses through the cross-section A in time t is: $$m = \Delta A \, \frac{(d_2 - d_1)t}{h}$$ where Δ is the coefficient of diffusion or diffusivity.
Oxidation–reduction reactions	These are conveniently considered as electron transfer reactions. For example, $$Fe^{2+}_{aq} + H^+_{aq} \rightarrow Fe^{3+}_{aq} + \tfrac{1}{2} H_{2g}$$ can be written as two half-cells or half-reactions $$Fe^{2+}_{aq} \rightarrow Fe^{3+}_{aq} + e$$

TABLE I (*continued*)

Constraint	Comment
	and $$H^+_{aq} + e \rightarrow \tfrac{1}{2} H_{2g}$$ The Eh, or oxidation potential is thus the potential of the half-cell or half-reaction referred to a standard hydrogen half-cell, which by convention is taken as zero.
Stoichiometric and nonstoichiometric reactions	The law of definite proportions states that for a given compound, the constituent elements are always combined in the same proportions by weight, regardless of the origin or mode of preparation of the compound. Reactions between such compounds are stoichiometric reactions. There are minor departures from stoichiometry with elements having more than one isotope, and major departures occur with some minerals, especially the sulphides and oxides of the transition metals, for example, pyrrhotite ($Fe_{1-x}S$), resulting in nonstoichiometric reactions.
Solid solution phenomena	Minerals are often solid solutions. Thus, for example, if x gram-atoms of Fe^{2+} replace the corresponding quantity of Mn^{2+} in rhodochrosite, $MnCO_3$, the formula of the resulting solid solution is $(Fe_xMn_{1-x})CO_3$. Because mutual solubility is greatly increased with increasing temperature, many minerals represent solid solutions that are thermodynamically unstable at lower temperatures. Solid solution minerals reacting irreversibly with an aqueous phase may be treated as minerals with fixed compositions, but when a solid solution mineral is a product of a reaction provisions must be included in the calculations for changes in its composition.

volume changes in the fluid phases or due to the precipitation or solution of reactants, and of course, the possibility of partial or local equilibrium. It is completely beyond the scope of this chapter to discuss any of these constraints in detail, and the interested reader is referred to standard textbooks on physical chemistry. Specific geochemically-oriented references include Garrels and Christ (1965), Helgeson (1964, 1968, 1970, 1971) and Helgeson et al. (1969, 1970).

Kharaka and Barnes (1973) have provided the first general purpose computer program for calculating solution—mineral equilibria, and specifically their program computes the equilibrium distribution of 162 inorganic aqueous species generally present in natural waters over the temperature range of 0–350°C from the reported chemical analysis, temperature and pH, and also calculates the states of reactions of the aqueous solutions with respect to 158 solid phases (minerals). Using this program, it is now possible to examine the departure from the static equilibrium between water and rock in terms of the degree of saturation with respect to specific minerals. The reader is cautioned to ensure the reliability of his analytical data before using the program, because it is the author's opinion that the overwhelming majority of available analyses are insufficiently complete to render using this program but a futile exercise; for example, few analyses include reliable contents for Al and SiO_2, which are essential parameters if equilibrium information on the silicates and aluminosilicates is desired. The fault lies not in the program but in the data.

Until the advent of general purpose computer programs for calculating mineral–solution equilibria it was not possible to comprehensively evaluate water–rock interaction with respect to formation waters in sedimentary basins, although some indication of possible water–rock interactions could be obtained by suitable use of conceptual models and advanced statistical techniques such as factor analysis. For example, in the western Canada sedimentary basin, the author and his colleagues (Billings et al., 1969; Hitchon and Friedman, 1969; Hitchon et al., 1971) have developed a mixing model for stable isotope composition and calculated the volume-weighted mean composition of formation waters in the basin. Using Q-mode, R-mode and second R-mode factor analysis they have shown that some of the water–rock and other processes active in the basin are dilution by freshwater recharge, concentration by membrane filtration, solution of halite, generation of chlorite, control of SO_4 by the solubility of $SrSO_4$ and $BaSO_4$, gain in Sr by the aragonite to calcite diagenetic recrystallization, redistribution of some alkali metals and alkaline earth metals through exchange on clays, desorption of Br and I from clays and a contribution from organic matter, and finally, $CaCO_3$ solubility equilibria have been demonstrated. Whatever the merits of these studies, they are no substitute for calculated mineral-solution equilibria data.

In the program of Kharaka and Barnes (1973), the computed equilibrium is independent of time and represents a static case; that is, given sufficient time, the equilibrium state they calculate will be reached eventually. However, as previously noted, the subsurface circuit is a system in dynamic equilibrium, both with respect to fluid flow and fluid chemistry. The kinetics of many of the reactions evaluated by Kharaka and Barnes (1973) are known, and a start has been made by Schwartz and Domenico (1973), to link hydrodynamics and hydrochemistry by developing a partial equilibrium simulation model of the chemical state of a regional groundwater flow system that incorporates mass transfer rates and reaction kinetics.

These valuable initial steps will undoubtedly lead, before the end of this decade, to comprehensive, powerful computer programs that can completely simulate the hydrodynamic and hydrochemical equilibria present in the subsurface circuit, both for inorganic and organic parameters.

APPLICATIONS

The perceptive reader will have noted that in the previous portions of this paper, the paragraphs on hydrodynamics contained no practical examples applicable to mineral deposits in sedimentary rocks, nor did the part on hydrochemistry include chemical analyses of waters pertinent to the study in question. These omissions were intentional, but in part unavoidable. Mineral deposits in sedimentary rocks usually represent the result of water—rock interaction, and at the time of deposition the aqueous phase was saturated with respect to the specific minerals of the deposit. Some of the factors which may cause migrating solutions to become supersaturated with respect to specific minerals include changes in temperature, pressure, pH, Eh, and concentration by membrane filtration at shale interfaces. If we seek the source of a deposit of galena (PbS), for instance, the problem can be considered in two ways. Either the lead and sulphur have always existed together, in which case the limiting factor is the saturation of the common carrier fluid with respect to PbS, or they have separate sources (two carrier fluids) and we must then determine the maximum amounts of lead and sulphur that can be carried in a form consistent with our general knowledge of the subsurface circuit. There are numerous excellent studies treating the theoretical aspects of this double dilemma, but no published practical report describing the potential fluid carrier(s) in terms of saturation with PbS, or lead and sulphur, separately. In part, this has been because of the difficulties of analysis of some trace components, but also because general purpose computer programs for mineral-solution equilibria have been unavailable until recently. Because it is the author's conviction that it is only through an understanding of water—rock interaction considered with respect to mineral saturations that we will comprehend the source and origin of mineral deposits in sedimentary rocks, then it is essentially superfluous to repeat the excellent reports of Lebedev and Nikitina (1968), White (1968), Degens and Ross (1969), Dunham (1970), Roedder (1972), Barnes and Hem (1973), and Skinner and Barton (1973). who provide possible examples of ore-bearing fluids. Rather, it is potentially more valuable to point out the extreme need for a new approach to the sampling and analysis of waters so that they are compatible to the sophistication of the present computer programs for mineral solution equilibria. It seems highly likely, in my opinion, that the most relevant approach to exploration for mineral deposits through examination of fluids might be statistical analysis of regional mineral solution equilibria data — for example, a trend surface analysis or map of factor scores using mineral saturation equilibria data as the input. This should be regarded as a minimal effort, which can be supplemented later, when the more advanced hydrodynamic—kinetic reaction models patterned after that of Schwartz and Domenico (1973) become available, as they surely will.

If one accepts a liberal definition of mineral deposit to include crude oil and natural gas, then hydrodynamics, hydrochemistry and mineral—solution equilibria are equally as important for these fluids as they are for the more conventional metalliferous mineral deposits. The essential difference is the mobility of the hydrocarbon mineral deposit in its own right, and it is the part played by mineral-solution equilibria in changing this mobility that is of interest here. In a sense, for metalliferous mineral deposits, the result of mineral-solution equilibria is the formation of the deposit itself, whereas for hydrocarbon deposits, the result of mineral-solution equilibria either enhances the probability of hydrocarbon accumulation through solution of minerals and consequent increase in porosity and permeability, or decreases the probability by plugging of pore space with secondary minerals. Common examples include the development of vuggy porosity in dolomites and the plugging of initial porosity by anhydrite or calcite cements. More difficult to interpret is plugging by secondary clay minerals, especially when it occurs at lithological boundaries such as sandstone—shale or carbonate—shale interfaces. Rumeau and Sourisse (1972) have studied the relation of abnormally high fluid pressures caused by under-compaction and suggested that they are important in diagenesis and in the movement of some components dissolved in formation waters resulting in deposition of cements. As with conventional metallic mineral deposits, the full potential of perceiving hydrocarbon deposits in terms of water—rock interaction has yet to be realized, and this is especially true for determining the location of hydrocarbon deposits through limitations set by mineral-solution equilibria.

The imaginative reader will doubtless conceive a considerable number of other ways in which the knowledge to be gained from a combined use of hydrodynamics, hydrochemistry and mineral-solution equilibria can be put to practical use, and the primary purpose of producing this paper will have been attained if that knowledge leads to economic benefits.

ACKNOWLEDGEMENTS

The author wishes to thank Dr. A.A. Levinson (Department of Geology, University of Calgary, Calgary, Alberta) and Dr. Y.K. Kharaka (U.S. Geological Survey, Menlo Park, California) who kindly reviewed the manuscript and who provided many useful comments.

REFERENCES

Back, W. and Hanshaw, B.B., 1965. Chemical geohydrology. *Adv. Hydrosci.,* 2: 49–109.
Barnes, I. and Hem, J.D., 1973. Chemistry of subsurface waters. *Ann. Rev. Earth Planet. Sci.,* 1: 157–181.
Berry, F.A.F., 1959. *Hydrodynamics and Geochemistry of the Jurassic and Cretaceous Systems in the San Juan Basin, Northwestern New Mexico and Southwestern Colorado.* Thesis, Stanford Univ., Calif.
Berry, F.A.F., 1960. Geologic field evidence suggesting membrane properties of shales. *Am. Assoc. Pet. Geol. Bull.,* 44: 953–954.

Berry, F.A.F. and Hanshaw, B.B., 1960. Geologic field evidence suggesting membrane properties of shales. In: *Volume of Abstracts – Int. Geol. Congr., 21st, Copenhagen*, p. 209.

Billings, G.K., Hitchon, B. and Shaw, D.R., 1969. Geochemistry and origin of formation waters in the western Canada sedimentary basin. 2. Alkali metals. *Chem. Geol.*, 4: 211–223.

Clayton, R.N., Friedman, I., Graf, D.L., Mayeda, T.K., Meents, W.F. and Shimp, N.F., 1966. The origin of saline formation waters – I. Isotopic composition. *J. Geophys. Res.*, 71: 3869–3882.

Coplen, T.B. and Hanshaw, B.B., 1973. Ultrafiltration by a compacted clay membrane – I. Oxygen and hydrogen isotopic fractionation. *Geochim. Cosmochim. Acta*, 37: 2295–2310.

Coustau, H. and Sourisse, C., 1967. Pressions anormales dans les réservoirs Infralias a Paléozoique du sud Aquitan (sud-ouest), *Bull. Centre Rech. Pau-SNPA*, 1: 143–152.

Degens, E.T. and Ross, D.A., 1969. *Hot Brines and Recent Heavy Metal Deposits in the Red Sea.* Springer, New York, N.Y., 600 pp.

Dunham, K.C., 1970. Mineralization by deep formation waters: a review. *Trans. Inst. Min. Metal.*, 79: B127–B136.

Freeze, R.A., 1966. *Theoretical Analysis of Regional Groundwater Flow.* Ph.D. Thesis, Univ. California, Berkeley, Calif.

Freeze, R.A., 1969. Theoretical analysis of regional groundwater flow. *Can. Inland Waters Branch, Sci. Ser.*, 3: 147 pp.

Freeze, R.A. and Witherspoon, P.A., 1966. Theoretical analysis of regional groundwater flow. 1. Analytical and numerical solutions to the mathematical model. *Water Resour. Res.*, 2: 641–656.

Freeze, R.A. and Witherspoon, P.A., 1967. Theoretical analysis of regional groundwater flow. 2. Effect of water-table configuration and sub-surface permeability variation. *Water Resour. Res.*, 3: 623–634.

Freeze, R.A. and Witherspoon, P.A., 1968. Theoretical analysis of regional groundwater flow. 3. Quantitative interpretations. *Water Resour. Res.*, 4: 581–590.

Garrels, R.M. and Christ, C.L., 1965. *Solutions, Minerals, and Equilibria.* Harper and Row, New York, N.Y., 450 pp.

Hanshaw, B.B. and Coplen, T.B., 1973. Ultrafiltration by a compacted clay membrane – II. Sodium ion exclusion at various ionic strengths. *Geochim. Cosmochim. Acta*, 37: 2311–2327.

Hanshaw, B.B. and Hill, G.A., 1969. Geochemistry and hydrodynamics of the Paradox basin region, Utah, Colorado and New Mexico. *Chem. Geol.*, 4: 263–294.

Helgeson, H.C., 1964. *Complexing and Hydrothermal Ore Deposition.* Macmillan, New York, N.Y., 128 pp.

Helgeson, H.C., 1968. Evaluation of irreversible reactions in geochemical processes involving minerals and aqueous solutions. I. Thermo-dynamic relations. *Geochim. Cosmochim. Acta*, 32: 853–877.

Helgeson, H.C., 1970. Description and interpretation of phase relations in geochemical processes involving aqueous solutions. *Am. J. Sci.*, 268: 415–438.

Helgeson, H.C., 1971. Kinetics of mass transfer among silicates and aqueous solutions. *Geochim. Cosmochim. Acta*, 35: 421–469.

Helgeson, H.C., Garrels, R.M. and Mackenzie, F.T., 1969. Evaluation of irreversible reactions in geochemical processes involving minerals and aqueous solutions. II. Applications. *Geochim. Cosmochim. Acta*, 33: 455–481.

Helgeson, H.C., Brown, T.H., Nigrini, A. and Jones, T.A., 1970. Calculation of mass transfer in geochemical processes involving aqueous solutions. *Geochim. Cosmochim. Acta*, 34: 569–592.

Hill, G.A., Colburn, W.A. and Knight, J.W., 1961. Reducing oil-finding costs by use of hydrodynamic evaluations. In: *Economics of Petroleum Exploration, Development, and Property Evaluation. Proc. 1961 Inst. Int. Oil Gas Educ. Center.* Prentice-Hall, Englewood Cliffs, N.J., pp. 38–69.

Hitchon, B., 1969a. Fluid flow in the western Canada sedimentary basin. 1. Effect of topography. *Water Resour. Res.*, 5: 186–195.

Hitchon, B., 1969b. Fluid flow in the western Canada sedimentary basin. 2. Effect of geology. *Water Resour. Res.*, 5: 460–469.

Hitchon, B. 1971. Origin of oil: geological and geochemical constraints. In: H.G. McGrath and M.E.

Charles (Editors), *Origin and Refining of Petroleum. Adv. Chem. Ser.*, 103. Am. Chem. Soc., Washington, D.C., pp. 30–66.

Hitchon, B., 1974. Application of geochemistry to the search for crude oil and natural gas. In: A.A. Levinson (Editor), *Introduction to Exploration Geochemistry*. Applied Publ. Ltd., Calgary, Alta, Chapter 13, pp. 509–545.

Hitchon, B. and Friedman, I., 1969. Geochemistry and origin of formation waters in the western Canada sedimentary basin. I. Stable isotopes of hydrogen and oxygen. *Geochim. Cosmochim. Acta,* 33: 1321–1349.

Hitchon, B., Billings, G.K. and Klovan, J.E., 1971. Geochemistry and origin of formation waters in the western Canada sedimentary basin. III. Factors controlling chemical composition. *Geochim. Cosmochim. Acta,* 35: 567–598.

Horn, M.K. and Adams, J.A.S., 1966. Computer-derived geochemical balances and element abundances, *Geochim. Cosmochim. Acta,* 30: 279–297.

Hubbert, M.K., 1940. The theory of groundwater motion. *J. Geol,* 48: 785–944.

Kharaka, Y.K. and Barnes, I., 1973. *SOLMNEQ: Solution–Mineral Equilibrium Computations.* NTIS PB-215 899, 82 pp.

Kharaka, Y.K. and Berry, F.A.F., 1973. Simultaneous flow of water and solutes through geological membranes. I. Experimental investigation. *Geochim. Cosmochim. Acta,* 37: 2577–2603.

Kharaka, Y.K. and Berry, F.A.F., 1974. The influence of geological membranes on the geochemistry of subsurface waters from Miocene sediments at Kettleman North Dome in California. *Water Resour. Res.,* 10: 313–327.

Lebedev, L.M. and Nikitina, I.B., 1968. Chemical properties and ore content of hydrothermal solutions at Cheleken. *Dokl. Acad. Nauk USSR, Earth Sci. Sect.,* 183: 180–182.

Levinson, A.A., 1974. *Introduction to Exploration Geochemistry.* Applied Publ. Ltd., Calgary, Alta, 612 pp.

Magara, K., 1968. Compaction and migration of fluids in Miocene sandstone, Nagaoka plain, Japan. *Am. Assoc. Pet. Geol. Bull.,* 52: 2466–2501.

Magara, K., 1969. Upward and downward migrations of fluids in the subsurface. *Bull. Can. Pet. Geol.,* 17: 20–46.

Magara, K., 1973. Compaction and fluid migration in Cretaceous shales of western Canada. *Geol. Surv. Can., Pap.,* 72-18.

Penman, H.L., 1970. The water cycle. *Sci. Am.,* 223 (3): 98–108.

Roedder, E., 1972. Composition of fluid inclusions. *U.S. Geol. Surv., Prof. Pap.,* 440-JJ.

Rumeau, J.-L. and Sourisse, C., 1972. Compaction, diagenèse et migration dans les sédiments argileux. *Bull. Centre Rech. Pau-SNPA,* 6: 313–345.

Saxby, J.D., 1969. Metal-organic chemistry of the geochemical cycle. *Rev. Pure Appl. Chem.,* 19: 131–150.

Schwartz, F.W. and Domenico, P.A., 1973. Simulation of hydrochemical patterns in regional groundwater flow. *Water Resour. Res.,* 9: 707–720.

Skinner, B.J. and Barton, P.B., 1973. Genesis of mineral deposits. *Ann. Rev. Earth and Planet. Sci.,* 1: 183–211.

Tóth, J., 1962. A theory of groundwater motion in small drainage basins in central Alberta, Canada. *J. Geophys. Res.,* 67: 4375–4387.

Tóth, J., 1963. A theoretical analysis of groundwater flow in small drainage basins. *J. Geophys. Res.,* 68: 4795–4812.

Van Everdingen, R.O., 1968. Studies of formation waters in western Canada: geochemistry and hydrodynamics. *Can. J. Earth Sci.,* 5: 523–543.

White, D.E., 1957. Magmatic, connate, and metamorphic waters. *Geol. Soc. Am. Bull.,* 68: 1659–1682.

White, D.E., 1968. Environments of generation of some base-metal ore deposits. *Econ. Geol.,* 63: 301–335.

Wolf, K.H., 1976. The influence of compaction on ore genesis. In: G.V. Chilingar and K.H. Wolf (Editors), *Compaction of Coarse-grained Sediments, 2.* Elsevier, Amsterdam, in press.

Chapter 4

FLUID-INCLUSION EVIDENCE ON THE GENESIS OF ORES IN SEDIMENTARY AND VOLCANIC ROCKS

EDWIN ROEDDER

INTRODUCTION

Although this volume includes discussions of many types of deposits in sedimentary and volcanic rocks, fluid-inclusion evidence is most abundant and pertinent in the Mississippi Valley-type of deposit (and, to a lesser extent, in the Kuroko type); hence, data from these deposits are emphasized here. Only a small part of the inclusion literature deals with such deposits. Much of it deals with ore samples, either from vein deposits obviously associated with igneous intrusions ("magmatic hydrothermal"), or from quartz crystals in veins of metamorphic to quasi-igneous origin (particularly the large Russian literature — see Ermakov, 1950). In addition, particularly in recent years, studies of silicate melt inclusions in terrestrial, meteoritic, and lunar samples have provided data on the liquid lines of descent for various magmas. The principles are generally applicable, however, to any deposit in which the constituent minerals have grown from a fluid medium — gas, liquid, or melt. Because of the lack of suitable sample material (as described below), very few studies have been made of fluid inclusions in most of the other types of ore deposits covered in this volume. Usable material may eventually be found in at least some of these and could provide exceedingly important data on ore genesis, data that are in part otherwise simply unavailable.

"The origin" of the Mississippi Valley-type of ore deposit, using the type term in its most general meaning, has been the subject of an extensive, in part heated discussion, only a small part of which comes from problems of exactly what types of deposits are to be included. The range covered is well illustrated in the volume stemming from a symposium on the genesis of stratiform lead–zinc–barite–fluorite deposits held at the United Nations in New York in 1966 (Brown, 1967). In an excellent review and sequel to this volume, Brown (1970, p. 104) summarized eleven distinctive characteristics of these deposits, as they are found in the three main districts in the Mississippi Valley proper (see Sangster, Chapter 9, Vol. 6).

Other apparently similar deposits in the United States and throughout the world differ most drastically from these characteristics in terms of the isotopic composition of the lead. Brown therefore suggested (1970, p. 117) that the broad class of Mississippi Valley-type deposits might be divided into three categories, based on the presence of normal type, B-type, or J-type lead. These three leads yield model ages that are, respectively, approximately correct, "too old", and "too young".

These differences are particularly significant in discussions of "the origin" of these deposits, in that they are one of the most obvious indications that there can hardly be only one mechanism of origin. Indeed, one of the most important causes for differences of opinion as to genesis is that there *must* have been a multiplicity of mechanisms operating in the various specific deposits. One does not need to go to isotopic studies to see this, however; the tremendous range in gross composition of the individual deposits, from essentially pure Pb with only traces of Zn, F, and Ba, to those with essentially pure Zn, to mixtures containing major amounts of all four (and other) elements, suggests that there had to be significant differences in the chemistry and perhaps in the mechanism(s) operating.

Most ore deposits represent the results of one ore or more processes in which an ore element is dissolved from a (presumably) dispersed or dilute source area, transported to the future site of the ore body, and deposited there in a relatively concentrated form. The many theories of origin of the Mississippi Valley-type ores differ among themselves in part as to the source or sources chosen for the ore elements. The term "ore element" used here includes not only the obvious metals Pb and Zn (and to a minor extent, Cu) but also the F and Ba that are even more highly concentrated. Thus, the production of the Illinois—Kentucky district, reported by Grogan and Bradbury (1968), is Pb, 60,000; Zn, 200,000; and CaF_2, 10,000,000 tons (ratio 1/3/170). Other districts are equally biased, but toward high barite or high Pb. The sources to be considered are listed in Table I. Each of these eight has been put forward as the major source or as a significant contributing source.

It is very important to realize that as long as the fluids can move through distances measured in kilometers (e.g., 1 km equals 30 years flow at 1 μm/sec), during geologic time, the source need *not* be enriched. As a good approximation, each part per million of an element in a rock corresponds to about 10,000 tons of that element per cubic mile of

TABLE I

Proposed sources for the ore elements in Mississippi Valley-type deposits

1. Seawater (and hence originally from weathering of rocks)
2. Leached from sedimentary rocks or muds in place
3. Expelled from sedimentary minerals, particularly from shales, during diagenesis
4. Expelled from recrystallizing metamorphic minerals
5. Leached from metamorphic rocks in place
6. Leached from igneous rocks in place
7. Expelled directly from crystallizing magma
8. Volcanic exhalations (via fumaroles, hot springs, etc.)

rock (or 2400 tons/km^3). Thus the extraction process can remove enough metal to make a big ore deposit from geologically reasonable amounts of very dilute source materials; geology kindly provides such vast amounts of materials and time that the extraction (and concentration) processes can be (and probably are) exceedingly slow and inefficient.

Central to any understanding of the processes of dissolution, transport, and redeposition of the ore elements is the nature and origin of the responsible fluid — the ore-forming fluid. Because fluid inclusions provide us with *samples* of that fluid, the physical and chemical data obtained from their study supply useful information toward an understanding of the entire process.

A variety of sources has been suggested for the ore fluids forming the Mississippi Valley-type deposits, as listed in Table II. Also, some theories of origin invoke combinations of several sources for the fluids, i.e., mixing. This mechanism, involving separate sources for metal- and sulfide-bearing fluids, was first detailed, for the Pine Point deposits, by Beales and Jackson (1966; see also Jackson and Beales, 1967); Hoagland (1973) has since applied it to the Central Tennessee deposits. It is also important to note that the divisions in Table II, as in Table I, are based on rather indistinct boundaries. Thus, there may well be a continuum between sea, connate, compaction, and metamorphic water (White, 1957; Noble, 1963), and several varieties may be combined in the water evolved from a subduction zone.

Obviously there must have been a much larger volume and weight of ore fluid than the resultant ore itself; the ore fluid, therefore, must not only be transported from the site of dissolution to the ore body, but many successive volumes must pass through the site of the ore body in sequence. The volume of this ore fluid will bear a simple inverse relationship to the concentration change (input minus output). Thus, if this concentration change is only 1 ppm, 1 million times more fluid than ore is required, and at 100 ppm, 10,000 times more fluid. These volumes may seem large, but when examined in geological terms they are not unreasonable. Even when geologically very small segments of time are assumed for ore deposition (Roedder, 1960), the flow rates required are low (<1 μm/sec) and the quantity of flow is also reasonable (e.g., 10 gallons/min (= 38 l/min) to form the main ore body at Pine Point: Roedder, 1968a, p. 447).

TABLE II

Proposed sources for the fluids responsible for dissolution, transport, and deposition of the ore elements in Mississippi Valley-type deposits

1. Meteoric (surface) waters (± deep circulation and heating to form "hydatogenic" fluids)
2. Seawater (± evaporation to cause enrichment in salts)
3. Connate and compaction fluids in sediments (± dissolved evaporites)
4. Metamorphic waters, expelled during dehydration reactions
5. Magmatic waters, evolved from crystallizing magma ("magmatic–hydrothermal" or "telethermal")

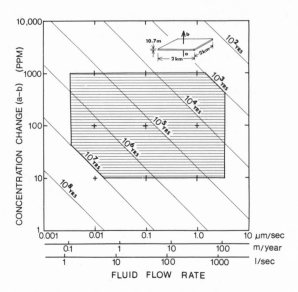

Fig. 1. Diagram showing the interrelations between time, flow rate, and concentration change (incoming minus outgoing fluid) needed for the deposition of 10^8 tons of 10% ore in a stratum 4 km² in area and 10.7 m thick. The rock is assumed to have a bulk density of 2.34 and a bulk porosity of 10 vol%, and the liquid a density of 1.0. The shaded area is the geologically most probable range of conditions under which most Mississippi Valley-type deposits formed. Adapted from Roedder (1960).

In Fig. 1, the interrelations of the three variables, concentration change, flow rate, and deposition time, are illustrated graphically on a log–log plot, for a very large hypothetical tabular deposit of a 10^8 tons of 10% ore (Pb, Zn, or any other element). The model has been set up for flow perpendicular to the beds for simplicity; in nature most of the flow has probably been parallel with the beds, but the numerical results will be identical.

The central crosshatched area on this diagram represents what the writer believes to be the geologically most reasonable range of these variables for the formation of most Mississippi Valley-type ore deposits. The right- and left-hand limits of this area are based on rather arbitrary guesses as to the maximum and minimum flow rates to be expected in such sedimentary terranes (see below under section "Types of data available: rate of movement"). The lower and upper limits of 10 and 1000 ppm concentration change (i.e., amount of precipitation) are based on data from fluid inclusions (Roedder, 1972) and brine analyses (White, 1967; Carpenter et al., 1974). The diagonal time limits are chosen somewhat arbitrarily at 1000 and 10 m.y. on the basis of several lines of evidence (Roedder, 1968a, p. 447; 1969, p. 807; 1971a, p. 786; Brown, 1971, p. 568; Lavery and Barnes, 1971; Leach, 1973; Skinner and Barton, 1973). Other data may result in these boundaries being moved one way or the other, but it seems unlikely that they can be moved very far. Thus, the single neutron activation analysis for zinc in inclusions in fluorite from southern Illinois giving the highest heavy metal content ever reported from inclusions (Czamanske et al., 1963) would move the upper boundary up to 10,000 ppm at a maximum, assuming that complete precipitation of all heavy metals occurs. Similarly, White (1971) has suggested that the ore fluids at White Pine, Michigan, flowed for 10^6-10^8 years. Regardless of the position chosen for the boundaries, the diagram gives the quantitative relationships between the variables.

In addition to the multiplicity of possible sources for the elements and the transporting fluids, the fluid may follow a wide variety of possible *pathways*, i.e., "structural control", both on a regional scale and within the space of the deposit itself. Thus, for long distances, major faults, unconformities, bedding planes, and porous horizons have been proposed; within the confines of the ore body itself, an even wider assortment of features may control the flow, including caverns, joints, solution collapse or tectonic breccias and breccia pipes, reefs, and fossil stream channels.

Some have suggested that diffusion in a static fluid might be involved in the localization of the ore (e.g., by precipitation at the ore bed setting up concentration gradients). In the model shown in Fig. 1, however, even if we assume a complete transfer of *all* zinc from the sediments above and below to the ore bed, and an original Zn content of 100 ppm in these sediments, the average distance that the Zn atoms would have to diffuse would be over 2.5 km. Thus diffusion might aid the transport of ore metal, but hardly seems adequate by itself in this example.

The final unknown in the process is the actual cause for deposition. Here also many possibilities have been proposed, including mixing of waters, changes in T or P, loss of gases, reaction with country rock, internal changes within the liquid itself (e.g., sulfate being reduced slowly by organic compounds in solution), and biological activity.

The possible combinations of these variables in the construction of a theory of origin for a given ore body are almost infinite, and the evidence to select or discard any given concept or model is frequently nebulous, almost always ambiguous, and sometimes completely lacking. Fluid inclusions provide one more line of evidence that has too frequently been ignored. Inclusions are no panacea for ore-deposition problems, but they do provide a surprising amount of evidence, sometimes otherwise unavailable, with a minimum of ambiguity. Thus they provide data on the density, temperature, rate of movement, pressure, salinity, and composition of the ore fluids.

The purpose of this chapter is to examine the evidence from fluid inclusions and how it is obtained, evaluate its significance and limitations, and show how it can be used to place limitations on the possible mechanisms of ore transport and deposition. The various data from fluid inclusions alone cannot determine the origin or origins of stratiform ores, but (regardless of this origin), the data place severe limitations on the mechanisms of ore deposition which could have been operative. Any theory of origin must be compatible with the inclusion data, or the data must be refuted, if the theory is to stand. It is important to note that the selection of possible mechanisms of origin, and the elimination of impossible ones, is of more than academic interest. These decisions can be all important in the selection, and in the success, of exploration procedures to be used for these deposits.

LITERATURE SOURCES

There is no single source book in English on fluid-inclusion research. A review of the extensive Russian work (Ermakov, 1950) is available in English translation but is now rather obsolete. A short semipopular article (Roedder, 1962b) provides a brief review, and a discussion of fluid-inclusion evidence concerning ore deposition in general is available (Roedder, 1967a). A summary of the world literature on all aspects of the composition of inclusions is also available (Roedder, 1972). Abstracts of the scattered world literature (including particularly the large volume of articles in Russian) have been published each year since 1968 in a series of privately printed volumes entitled *Fluid Inclusion Research − Proceedings of COFFI*, available at cost from the present writer.

Although perhaps half the world literature on fluid inclusions is in Russian, very few of these Soviet publications deal with deposits of the type covered by the present volume. Furthermore, many of them pertain to studies using a method (decrepitation) that has not found favor in the West.

ASSUMPTIONS ON WHICH FLUID-INCLUSION DATA ARE BASED

There has been little dispute since Sorby's classic paper (1858) about what assumptions must be made in fluid-inclusion research, but debates have been heated on the validity of these assumptions. The use of fluid inclusions as evidence of the nature of the ore-forming fluid, and the environment of ore deposition is based on the following major assumptions (for a more detailed discussion of these and other assumptions, see Roedder, 1967a):

(1) If a crystal of an ore mineral (or a gangue mineral *known* to be cogenetic with ore) grows from a fluid medium, irregularities in the growth process can, and usually do, cause the trapping of small amounts of this medium within the crystal.

(2) The fluid trapped is a representative sample of the ore fluid at the moment of trapping.

(3) Significant quantities of material are neither lost nor gained from the inclusions after this trapping.

There is little argument concerning the validity of the first assumption, as many natural crystals provide convincing proof of trapping by various types of irregularities; perhaps the most lucid examples come from the Mississippi Valley deposits (e.g., Roedder, 1965, 1972). Such inclusions are called *primary* because they form at the same time as the surrounding host crystal. There are many different criteria for establishing a primary origin for a given inclusion, and these vary widely with the individual sample or deposit. The specific criteria used are normally given in each published report (see also Table III).

Along with these primary inclusions, however, other inclusions may be found that were trapped at some later time, during the spontaneous healing of fractures. These *secondary* inclusions not only contain a later fluid, possibly millions of years later, which may have no connection with the ore fluid, but they are also very common, frequently far more common than primary inclusions. Inclusions along healed fractures (e.g., along cleavage directions in fluorite or sphalerite) are not necessarily secondary, however; evidence indicates that many if not most of them in Mississippi Valley-type deposits, although appearing to be secondary, are actually formed by fracturing and rehealing *during* the growth of the host crystal (e.g., Roedder, 1965, figs. 18, 19; Roedder et al., 1968, figs. 12—15). Such inclusions are termed *pseudosecondary*. In practice, data from pseudosecondary inclusions are found to be identical with those from equivalent primary inclusions (i.e., those trapped at the same time), but data from true secondary inclusions may be very different. The criteria for distinguishing the three types of origin are summarized in Table III.

Proof of the validity of the second assumption is a little more difficult, but supportive evidence in the literature on fluid inclusions is extensive. Certainly, there *are* differences between the bulk of any fluid and the fluid at its interface with any other phase. These equilibrium boundary layer effects are believed to extend out into the fluid too short a distance (tens or hundreds of Ångstroms) to be significant in present-day experimental inclusion studies. Nonequilibrium conditions, in which crystal growth is more rapid than diffusion in the fluid, may yield very much thicker nonrepresentative layers, but such rapid growth does not generally occur in geological environments. The only discordant notes here are some data reported by Barnes et al. (1969), indicating that the fluids trapped in some synthetic inclusions were slightly more dilute than the bulk of the fluid in the experiment. To my knowledge, no one has attempted corroboration of these results. One other situation which almost certainly invalidates the second assumption is the presence of an inhomogeneous fluid — liquid plus a gas, another liquid, or suspended solids. This very limitation turns out to be useful, as detailed beyond.

The third assumption, concerning leakage, was once thought to cause a major problem in the interpretation of all inclusion data (Skinner, 1953) but has since been shown to be of little consequence, except in a relatively few, generally recognizable situations, e.g., inclusions in barite (Roedder and Skinner, 1968). In addition to these three basic assumptions, each specific type of application of inclusion data involves additional assumptions that will be mentioned in the individual sections. One other general assumption should be mentioned, however, concerning the problem of possible reworking of the material in the deposit. Brown (1970) pointed out that although fluid-inclusion evidence on Mississippi Valley deposits was widely quoted and generally believed in North America, many students of this type of deposit in Western Europe, and particularly in France and Germany, tend to ignore the inclusion data, under the assumption that it gives no information on the *original* ore fluids. As will be shown later, the fluid-inclusion data, if true, preclude several of the low-temperature and syngenetic concepts of origin most generally held in Europe. Some small part of these problems may arise simply from different usage of terminology by various researchers, but there still remains a major difference

TABLE III

Criteria for primary vs. pseudosecondary vs. secondary origin for fluid inclusions, summarized in part from Roedder (1967a, pp. 519–523)[1]

───

I. Primary
A. Criteria based on occurrence in a single crystal with or without evidence of direction of growth or growth zonation.
    1. Occurrence as a single (or a small three-dimensional group) in an otherwise inclusion-free crystal, e.g., Roedder, 1965, fig. 10; Roedder, 1972, plate 6.
    2. Large size relative to that of the enclosing crystal, e.g., with a diameter $\sim \geqslant 0.1$ that of crystal. Occurrence as *multiple* relatively large size inclusions is an even more valid criterion. Large size *in three dimensions* is the most obviously valid criterion.
    3. Isolated occurrence, away from other inclusions, for a distance of $\sim \geqslant 5$ times the diameter of the inclusion.
    4. Occurrence as part of a random, three-dimensional distribution throughout the crystal, e.g., Roedder and Coombs, 1967, plate 4, figs. A, B.
    5. Disturbance of otherwise regular decorated dislocations surrounding the inclusion, particularly if they appear to radiate from it, e.g., Roedder and Weiblen, 1970, fig. 9.
    6. Include daughter crystals (or accidental solid inclusions) of the same phase(s) as occur as solid inclusions in the host crystal or as contemporaneous phases.
    7. Occurrence associated with a group of solid inclusions.
B. Criteria based on occurrence in a single crystal showing evidence of direction of growth.
    1. Occurrence beyond (in the direction of growth), and sometimes immediately before extraneous solids (the same or other phases) interfering with the growth, where the host crystal fails to close in completely. (Inclusion may be attached to the solid or at some distance beyond, from imperfect growth, e.g., Roedder, 1972, plate 1).
    2. Occurrence beyond a healed crack in an earlier growth stage, where new crystal growth has been imperfect (e.g., Roedder, 1965, figs. 18, 19; Roedder et al., 1968, fig. 15).
    3. Occurrence between subparallel units of a composite crystal, e.g., Roedder, 1972, frontispiece upper right.
    4. Occurrence at the intersection of several growth spirals, or at the center of a growth spiral visible on the outer surface.
    5. Occurrence, particularly as relatively large, flat inclusions, parallel to an external crystal face, and near its center (i.e., from "starvation" of the growth at the center of the crystal face), e.g., much "hopper salt".
    6. Occurrence in the core of a tubular crystal (e.g., beryl). This may be merely an extreme case of previous item.
    7. Occurrence, particularly as a row, along the edge from the intersection of two crystal faces.
C. Criteria based on occurrence in a single crystal showing evidence of growth zonation (as determined by color, clarity, composition, X-ray darkening, trapped solid inclusions, etch zones, exsolution phases, etc.).
    1. Occurrence in random, three-dimensional distribution, with different concentrations in adjacent zones (as from a surge of sudden, feathery or dendritic growth), e.g., Roedder, 1965, fig. 11.
    2. Occurrence as subparallel groups (outlining growth directions), particularly with different concentrations in adjacent zones, as in previous item.
    3. Multiple occurrence in planar array(s) parallel to a growth zone, e.g., Roedder and Coombs, 1967, plate 4E. (Note that if this is also a cleavage direction in this crystal, there is ambiguity.)
D. Criteria based on growth from a heterogeneous (i.e., two-phase), or a changing fluid.
    1. Planar arrays (as in C-3), or other occurrence in growth zones, in which the composition of inclusions in adjacent zones are different (e.g., gas inclusions in one and liquid in another), e.g., Roedder et al., 1968, fig. 9.
    2. Planar arrays (as in C-3) in which trapping of some of the growth medium has occurred at

TABLE III *(continued)*

points where the host crystal has overgrown and surrounded adhering globules of the immiscible, dispersed phase (e.g., oil droplets or steam bubbles).

3. Otherwise primary-appearing inclusions of a fluid phase that is unlikely to be the mineral-forming fluid, e.g., mercury in calcite, oil in fluorite (Roedder, 1972, plate 9, fig. 2) or air in sugar (Roedder, 1972, plate 9, fig. 4).

E. Criteria based on occurrence in hosts other than single crystals.

1. Occurrence on a compromise growth surface between two non-parallel crystals. (These inclusions have generally leaked, and could also be secondary).

2. Occurrence within polycrystalline hosts, e.g., as pores in fine grained dolomite, cavities within chalcedony-lined geodes ("enhydros"), or as crystal-lined vugs in metal deposits or pegmatites. (These latter are the largest "inclusions", and have almost always leaked.)

3. Occurrence in noncrystalline hosts (e.g., gas bubbles in amber; vesicles in pumice).

F. Criteria based on inclusion shape or size.

1. In a given sample, larger size and/or equant shape.

2. Negative crystal shape—valid only in certain specific samples (and a negative criterion in others).

G. Occurrence in euhedral crystals, projecting into vugs (suggestive, but far from positive – see Roedder, 1967a, p. 523).

II. Secondary

A. Occurrence as planar groups outlining healed fractures (cleavage or otherwise) that come to the surface of crystal (note that movement of inclusions with recrystallization can disperse – Roedder, 1971b, fig. 11).

B. Very thin and flat; in process of necking down.

C. Primary inclusions with filling representative of secondary conditions.

1. Located on secondary healed fracture, hence presumably refilled with later fluids (Kalyuzhnyi, 1971).

2. Decrepitated and rehealed following exposure to higher temperatures or lower external pressures than at time of trapping; new filling may have original composition but lower density.

III. Pseudosecondary

A. Occurrence as with secondary inclusions, but with fracture visibly terminating within crystal, e.g., Roedder, 1965, figs. 18 and 19; Roedder et al., 1968, figs. 12, 14 and 15.

B. Generally more apt to be equant and of negative crystal shape than secondary inclusions in same sample (use with care).

C. Occurrence as a result of the covering of etch pits cross-cutting growth zones (Roedder, 1972, plate 1, fig. 8).

---

[1] These criteria are in part only suggestive, and *must be applied with care,* as there is considerable chance for ambiguity.

between paleokarst, sabkha, lagoon, and similar environments at essentially *surface* temperatures (although perhaps sunheated) and the 150°C or higher reported for inclusions from many Mississippi Valley-type deposits. Although rarely discussed in any detail in the literature, the crux of the "reworking" problem lies in exactly which reincarnation of the ore deposit one is referring to as *the* origin. If an ore body is formed from fluids at one set of $P-T-X$ conditions (and presumably contains inclusions representative of such conditions), and then is *completely* reworked (i.e., all crystals are dissolved and reprecipitated via a different fluid medium), then the new inclusion data will give only the $P-T-X$ parameters of the reworking process (Bernard, 1973, p. 53–55). The necessary assumption here is that primary inclusions reveal the conditions of formation of the host crystal *where we see it now*, and obviously can tell us nothing of any previous deposit. Recognition of the possibility of such former deposits would then become a field problem rather than an inclusion problem.

In this connection, Bernard (1973, p. 54) has made the point that fluid-inclusion specialists sample only the late recrystallized idiomorphic ore minerals. As a result they get $P-T-X$ data corresponding only to the fluids present during the last recrystallization of the ores, under conditions of deep burial, whereas the original fine-grained ores formed from fresh waters under near-surface conditions. It is true that good primary inclusions are easier to find, more photogenic, and larger, in the coarser crystals from any deposit; hence, these have received the most attention. The new crystal containing the inclusions may have formed by growth at the expense of immediately adjacent crystals via an intergranular pore-fluid (i.e., typical recrystallization), or by direct free crystallization from a moving mass of liquid that had previously dissolved other crystals at some distance. The fuzzy line between "recrystallization" and "reworking" (or "remobilization") depends upon the interpretation of this distance. In either case, the inclusions represent samples of the fluid from which the present host crystal actually formed. If any fluid inclusions containing earlier fluids were present before, they would be effectively dispersed and lost by either process.

The writer believes that this argument is refuted by the inclusion evidence itself, as well as by the field data. Many examples in the literature on inclusions verify that the tiny inclusions in very fine-grained ores have salinities and homogenization temperatures as high as those of the coarse crystals (e.g., Roedder, 1967b, 1968a; 1971a). Also, the very tiny inclusions found in fine-grained ores, although frequently too small to use for homogenization studies, still show visually similar gas/liquid ratios; only the very tiniest inclusions may remain single phase, and these are believed to be metastable (Roedder, 1971b). If the hot saline inclusions are a result of later reworking, this would require that *every* sample, from *every* Mississippi Valley-type deposit studied, has been recrystallized in the presence of hot saline fluids; this would seem to require a very unlikely coincidence. On the other hand, the writer knows of no data to show that *any* primary inclusions in *any* Mississippi Valley-type deposit were either filled with fresh water or were formed at surface temperatures. The only possible exceptions are some barite deposits that may well have been reworked by surface waters.

CLASSIFICATION OF INCLUSIONS

In addition to the threefold classification of inclusions by mode of origin discussed above, into primary, pseudosecondary, and secondary, many attempts have been made, particularly in the Soviet literature, to set up an all-inclusive classification of inclusions. Thus, Ermakov (p. 18 in vol. 2 of *Fluid Inclusion Research – Proceedings of COFFI*, 1969) has set up 21 classes of inclusions based on composition and origin, and had earlier established another widely used classification based on the relative proportion of the phases present (Ermakov, 1950, p. 28, in translation). Classifications are only useful, however, if they result in elucidation of principles by organizing an otherwise confusing mass of data, reveal otherwise hidden relationships between categories, or provide insight into the nature of otherwise unknown samples. Unfortunately, most inclusion classifications are not very useful in practice, as they require more knowledge about the inclusions than one has, or consist of categories based on arbitrarily chosen numerical ratios of phases present (e.g., division at 25, 50 and 75% gas). These ratios are part of a continuum and cannot generally be measured accurately anyway. Most particularly, in spite of many explicit statements in the literature to the contrary, there is no real physical, chemical, or genetic significance to the division "line" at 50/50 gas/liquid at room temperature.

LABORATORY TECHNIQUE

As is all too common in most scientific literature, the important *practical* aspects of fluid inclusion research are usually omitted in the reports and have to be learned by each individual worker "by experience" (i.e., at a considerable waste of time). Each type of inclusion data requires a specific type of technique and has its own set of problems. Some of these are given under the individual headings in the succeeding section ("Types of data available"), and many others, particularly those dealing with various aspects of determining the composition of fluid inclusions, are detailed elsewhere (Roedder, 1972) and need not be repeated. A few of the more common and basic problems are covered in the following section, as there is no single general reference work on inclusion laboratory technique.

Sample selection

The selection of material to be studied is one of the most important aspects of fluid inclusion research. Optimum material will vary, of course, with the specific study, but for normal optical determinations, the optimum would be large, clear, euhedral zoned crystals of an ore mineral, such as sphalerite, protruding into vugs. If all the ore minerals are opaque, transparent gangue minerals growing *contemporaneously* with the ore minerals can be used, but obviously the significance of the data obtained is a direct function of the validity of this determination of contemporaneity. Far too often, mere association in the hand specimen is assumed to establish contemporaneity; this assumption is generally invalid. Even in intimate intergrowth and incrustation relationships, there are difficulties in establishing which minerals were in what type of equilibrium with each other (Barton et al., 1963). The common occurrence of complex paragenetic sequences of various ore and gangue minerals, clearly established by textural features of crustified ores, is proof that fluctuations took place in the ore fluid of such a nature or magnitude that the mineral assemblage being deposited changed with time. Under such a situation, it is hazardous to extrapolate from one mineral to another (e.g., from a transparent gangue mineral to an opaque ore mineral). Even the common occurrence of crystals of an ore mineral within a crystal of gangue mineral does *not* prove contemporaneous deposition of the two; when examined with care (e.g., Poty, 1969; Grigor'yev, 1965), such solid inclusions frequently are found to have grown by themselves, at a time when the enclosing host mineral was not precipitating. Truly contemporaneous growth of two crystals in contact with each other yields the characteristic ribbed "compromise-growth surface" (Grigor'yev, 1959).

Large crystal size merely increases the probability of finding adequate criteria for establishing fluid inclusion origin (Table III), as well as the probability of finding the larger inclusions that are easier to study. On occasion, crystals less than 1 mm are found to contain adequate inclusions, but these are difficult to prepare for study.

It is not unusual to find that the vagaries of inclusion formation have resulted in some samples having much better inclusions than others, and it is simply not possible to predict what material will be good. As there are usually far more samples than can be properly prepared in polished plates, time can be saved by careful hand-specimen study with a low-power binocular microscope to find those areas or individual crystals that are most likely candidates for further study. Cleavage plates are particularly useful. A coverglass stuck on which an index liquid that matches the mineral helps in looking into imperfect crystal faces or cleavage surfaces. The larger inclusions in clear crystals can be readily seen with a hand lens, but lack of such inclusions should not exclude the material from further study. Less transparent minerals can be crushed and examined in a matching index liquid. Small crystals are also best examined in a bath of matching index liquid; for this purpose the writer cements thin rings of glass (2–5 mm sections cut from glass tubing) onto ordinary glass slides with epoxy. The crystals to be examined are placed in the ring, and index liquid added until the reflection from the top surface shows a change from concave to very slightly convex. A coverglass is then placed on top and is held in place by capillarity. (Too little liquid results in a bubble which invariably lies over the grain; too much can also cause trouble.) These slides can be used also for handpicking grains under the binocular microscope for further preparation by filling with index liquid to the point where the meniscus is flat, eliminating the need for a coverglass. As only an approximate match is needed, several inexpensive liquids suffice: kerosene of $n = 1.45$ (for fluorite); "HB 40 oil" made by the Monsanto Chemical Co., Wilmington, Delaware, adjusted with kerosene to $n = 1.55$ (for quartz); and α-monobromonaphthalene of $n = 1.64$ (for olivine). Methylene iodide saturated with sulfur ($n = 1.788$) is a rather inadequate liquid for sphalerite (there is no inexpensive matching liquid). For groups of tiny grains, ordinary liquid mounts with a coverglass suffice. When a single tiny grain is to be examined, however, it is easiest to put it in a tiny three-sided "corral" of appropriately thin wire under the coverglass. It can then be manipulated for observation from various directions using a probe tipped with a single tapering camel's hair, inserted under the coverglass in the open side of the corral (Roedder and Weiblen, 1970, p. 808).

Cutting and polishing

The cutting and polishing of doubly polished plates suitable for microscopy require more care than is generally given to these steps. Optical examination of inclusions, particularly under the less-than-optimum optical characteristics of heating and cooling stages, is sufficiently difficult without being penalized by additional problems from poor sample polish, which causes serious image degradation. For ordinary microscopy (but not for heating or cooling); a coverglass with a closely matching index liquid is a poor substitute for a good polish.

Plates are usually prepared by cutting a slice in the optimum direction for the specific sample. This should be done with a thin continuous-rim diamond cutoff wheel and water coolant. These wheels can be obtained in a variety of grits and grades for cutting various materials. This type of wheel causes much less fracturing of the surface adjacent to the saw cut than occurs with the standard notched-edge diamond blade used in thin-section preparation. The thinner the blade the less pressure is needed in the cutting and the less sample is lost, but the more chance there is of blade breakage. For small crystals, and for cutting wedges out of prepared slides, the writer uses a blade 0.015 inch thick and 5 inch in diameter. The quality of the polish, especially the top surface, is particularly critical. (For details on polishing techniques, see Cameron, 1961; Saager, 1967; Allman and Lawrence, 1972.)

Thin crusts of crystal druses and fragile or porous samples are best prepared by coating or embedding in epoxy (under vacuum if necessary) and curing, grinding down to the point where a maximum amount of the desired crystals are intersected, and then polishing. This polished surface is cemented to a slide with a soluble cement and a second parallel saw cut is made. After the grinding and polishing, the cement can be dissolved to provide a doubly polished plate suitable for heating or freezing. Before reducing the thickness of any section by grinding, it is best to examine it with a "temporary polish" in the form of a coverglass and matching index liquid. In this way, one can frequently recognize a large inclusion that is near the surface and that can be saved only by grinding and polishing from the opposite side. A fine matt surface on the lower side of the plate can be tolerated with good inclusions but makes difficult inclusions impossible.

All impregnating and cementing materials must be chosen with the nature of the sample and its final use in mind. Thus, many epoxies can stand both the heating and cooling operations necessary for samples from low-temperature deposits, and hence, if the stage permits, the sample plate can be left on the glass slide during these operations. This is particularly important with very dark sphalerites that must be cut into very thin (and, hence, fragile) plates. If at all possible, however, the sample plates should be removed from the glass slides before heating or cooling runs. If balsam or Lakeside is present during heating-runs at ~150°C, even just as films in cracks, it will creep out over the surface, spoiling the optics, and will fog the stage windows. For several reasons, the writer prefers to complete all normal microscopy with the plates still cemented, mark the parts on which runs are to be made, and then remove them by cutting out a wedge of sample (and glass slide) with the diamond saw, holding the slide in the hand rather than in a clamp. The cuts are made so as to place the desired inclusions near the point of the wedge, to permit several chips to be placed in the heating stage simultaneously, still mounted on the glass wedges, all within a radius of several millimeters. To keep the thermal mass to a minimum, a coverglass can be substituted for the glass slide (pers. comm., R.E. Bennet, Jr.).

A record should be kept of what impregnating and cementing materials were used, and the maximum actual sample temperature attained during preparation. Some curing procedures on some epoxies form thousands of tiny spherical immiscible liquid inclusions in the cured resin, many of which are complete with a moving vapor bubble (Roedder and Weiblen, 1970, p. 801). When these occur in the excess epoxy at the edge of the sample, they are harmless, but when they occur in nearly invisible films of epoxy within cracks in the plate, they can appear to be indigenous to the sample. Also, as the hot plate temperatures needed (and even more so, the higher temperatures generally used) for the thermoplastic cements such as Canada balsam and Lakeside 70 are well above the homogenization temperature of the low-temperature inclusions characteristic of some Mississippi Valley-type deposits, such inclusions will be either lost through decrepitation, or will have become stretched and hence yield grossly misleading results (Larson et al., 1973). Bubbles in the cementing materials should be carefully avoided, as they generally spoil the plate for photography, and if numerous, they confuse the eye in searching for suitable inclusions.

The thickness of the doubly polished plate must be appropriate for the stage to be used and for the particular sample. Plates that are too thick prevent adequate illumination and usually result also in degraded images from birefringence or in the superposition of other inclusions, cracks, etc., upon the image being examined. They also are slower to equilibrate in the heating stage. On the other hand, plates that are too thin present less actual volume of mineral for examination, for a given amount of sample preparation. Average material is usually suitable at ~1 mm thickness, but if a dark and a clear mineral are present together (e.g., sphalerite and quartz), the quartz should be scanned when the slide is still thick, or the slide can be polished and cut in half at this stage and only half ground down for the darker material. Very dark sphalerite may require doubly polished sections of less than standard thin-section thickness (<0.030 mm); warpage in the cement can be a serious problem here. Standard thin sections are satisfactory for examination for very tiny inclusions only, as all larger inclusions are necessarily lost in preparation. Further, as inclusions even in the center of the section are at maximum only 15 μm from a normally rather roughly prepared surface, they are apt to either have leaked or to leak during heating (Roedder and Skinner, 1968).

Choice of inclusions and problems in microscopy

This is the single most critical step in all inclusion studies and, hence, must be handled with utmost care. The difficult problem here is generally that of verifying the primary origin of the inclusions. Pseudosecondary and secondary inclusions are frequently run along with the primaries, but they should be specifically selected, *in advance*, as examples of these types (see Table III for criteria for distinguishing).

An additional problem is presented by occasional inclusions that differ visually in their phase ratios from the bulk of the inclusions, either in gas/liquid ratio or in the presence or nature of "daughter crystals" precipitated from the liquid after trapping. Generally, such visually divergent inclusions will be found to have grossly different homogenization temperatures. The problem lies in deciding whether they are valid samples of pre-existing fluids (e.g., formed from rapidly changing or inhomogeneous fluids), or whether they have been made divergent by some secondary later process and, hence, should be eliminated from the group to be run. If they are valid but are eliminated under the assumption that they have been altered, valuable information is lost; conversely, good data from valid inclusions are confused and diluted with grossly misleading numbers. The main secondary processes that must be considered when one is presented with such divergent inclusions are: (1) leakage (natural or laboratory induced); (2) later refilling; and (3) necking down.

Leakage occurs in nature for a variety of reasons but is usually related to some deformation of the host crystal, causing cracks. These cracks are not always visible, as they must be an appreciable fraction of a wavelength of the light to affect it. Leakage from such fracturing is fairly evident when a series of primary-appearing, but divergent, inclusions (often empty of all liquid) occur as a planar array through a group of otherwise uniform primary inclusions. In thin sections, inclusions within a few μm of the top or bottom surfaces are frequently empty, particularly if their contents had been under high pressure. Overheating of inclusions beyond their homogenization temperatures, whether from later dikes (Lokerman, 1965) or in the laboratory (Roedder and Skinner, 1968; Larson et al., 1973) can build up high pressures and cause decrepitation or stretching, particularly of the larger inclusions. Freezing can also cause similar damage if the vapor bubble is small. "Later refilling" is a special case of leakage in which fracturing and rehealing of the inclusions occurs in the presence of a later fluid, under a different set of conditions. The opened inclusions will thus retain their original shape and spatial arrangement but will yield a different homogenization temperature.

Necking down is the process whereby a long thin or flat inclusion spontaneously reduces its total interfacial energy by selective solution and redeposition, to yield two or more separate, smaller, more equant inclusions of the same total volume (Roedder, 1972, plate 10). Nothing is lost or gained from the system as a whole, but if a phase separation has occurred previous to the necking down, the separated inclusions will differ in phase ratios and homogenization temperatures (Roedder, 1962b, p. 45 in original). The writer believes that necking down occurs commonly in nature, but that it generally takes place near the original trapping temperature. As such, it is probably responsible for much of the scatter of results that is apparent in any careful thermometric studies (Roedder, 1967a, pp. 534 and 558; Roedder et al., 1968, fig. 13). In the laboratory one must be alert to recognize the existence of necking down. When it has been interrupted while in progress, the long tails between inclusions are easy to see. After sufficient recrystallization, however, it may only be recognized by the adjacent occurrence of two or more inclusions that are divergent *in opposite directions* (too much and too little vapor relative to other nearby inclusions), and sometimes by abandoned daughter crystals, left surrounded by host crystal.

It is important to keep in mind that the *apparent* gas/liquid ratio is strongly affected by inclusion shape, so inclusions that appear to be different may have identical homogenization temperatures. A visual estimate of the gas/liquid ratio is normally used to guide the heating-stage operation. Thus, if the inclusion with the smallest ratio looks as though it would homogenize at 150°C, one may be tempted to set the heating stage to level out at 125°C and to creep up from there. However, the coefficient of expansion of water solutions varies inversely with the salinity, so an inclusion filled with pure water at 150°C can have a bubble at room temperature that is double the volume of one that was filled with saline brine at 150°C.

It is particularly important to spend enough time on microscopy before any runs are made. Not only will additional time frequently yield inclusion data not recognized at first, but a more careful search will almost always result in finding bigger, clearer, or more obviously primary inclusions than were found at first. This is particularly important because the optical images with the heating stages are so much poorer than with normal microscopy. It is especially important with freezing work, as ice is much more difficult to see than a bubble. Although the temptation is great to start immediately with high magnification, this is very wasteful of time. The entire plate should be scanned at low power, then at intermediate power before "homing in" on individual inclusions at high power. The location of particularly good inclusions seen during the scanning should be noted, of course, but the assumption should always be: "There are probably other, better, inclusions". It is also imperative to view at low power before assigning a primary or secondary origin to a given inclusion, as many of the critical features (faint traces of fractures, alignments with other inclusions, vague growth banding, etc.) are only visible at low power. On the other hand, high magnification is absolutely essential for all small inclusions and for small daughter phases in larger inclusions. In this connection, it is important to note that optical data from inclusions too small to run on heating or cooling stages may still provide valuable corroboration of the meager quantitative data from a few larger inclusions, and in some samples, such optical data may be all that are available from certain zones. The writer uses a 100X oil immersion objective (plus 12.5X oculars) regularly in his own microscopy and would recommend it as the most important single special tool for inclusion work, second only to a good microscope. If its depth of focus is inadequate to reach a desired inclusion, try turning the sample plate over and focusing through the bottom.

Lighting is an important consideration in searching for good inclusions. If the plate is very clear and transparent, and contains rare small inclusions, these are best found by closing the substage diaphragm down to a very small diameter; this shows up small inclusions over a considerable depth of focus. When searching for good inclusions in samples that are crowded with inclusions, cracks, or solid debris (and for close examination of any given inclusion), the diaphragm should generally be wide open. The eliminates much of the troublesome superposition of images.

Birefringent minerals should always be examined with one polarizer in place, set parallel to a vibration direction of the sample. This eliminates the annoying double images seen when looking deep into a mineral of low birefringence, and visible at almost any depth in calcite. The polarizer should be set parallel to the *ordinary* ray on highly birefringent, uniaxial negative minerals such as the rhombohedral carbonates, unless the inclusions are very close to the surface. This is because the extraordinary ray image is severely distorted and fuzzy. (As the ordinary ray in these minerals has a much higher index of refraction, this also permits one to focus significantly deeper into the section and hence brings more of the sample within range of the oil immersion objective.) If coverglasses have to be used, it is best to use the thinnest possible grade ("00"), to obtain the maximum depth of focus into the sample.

Faint growth-banding in colored minerals can sometimes be enhanced by the use of a wedge interference filter, adjusted to give maximum visual contrast between adjacent bands. Growth-banding and other planar features with which inclusions may be associated are best viewed first at very low power on a binocular microscope, where the working distance is large enough to permit tipping the plate up at high angles. Samples containing solid opaque inclusions outlining growth-bands should be viewed alternately in transmitted and reflected light; for this purpose two foot switches, one for each light source, are particularly convenient, as they leave the hands free (P.M. Bethke, personal communication, 1971). Growth-banding that is otherwise invisible may sometimes be revealed vividly when the plate is viewed under cathodoluminescence. Poty (1969) made effective use of radiation coloration from intense X-ray dosage to reveal very fine growth-banding, otherwise invisible, in quartz plates.

Relocating an inclusion, particularly a small one in a large plate crowded with inclusions, can be very frustrating. The writer uses a combination of rough sketches (starting at actual size) to show the approximate location, plus photographs or sketches of the area as seen through the microscope at one or more stages of magnification. It is important to record at which point in any such series of sketches one switches from the erect image as seen by the naked eye to that of the inverted image seen through the microscope, as this difference can cause much confusion on returning to the sample at a later time. Mechanical-stage vernier coordinates can be used if the section is still mounted on a glass slide. Poor reproducibility in the position of the slide when seated against the stops in the stage, and backlash in the stage mechanism, may make such coordinates useless at high magnification. To avoid loss of valuable data, each stage in disaggregation of a plate (sawing, breakage, etc.) should be well documented, so that interrelationships of inclusions to each other and to growth zones can be reconstructed. Not infrequently a sketch can be more useful than a photograph, because it can readily show features such as inclusions at slightly different levels that cannot be simultaneously photographed and can delete the clutter of unwanted and confusing detail that the camera insists on seeing. Record photographs of special inclusions should be made before running in case they decrepitate and are lost.

Heating stages and their operation

It is comparatively simple to set up a homemade heating stage on the microscope, and many different models have been described (Roedder, 1972, pp. JJ27–JJ28). Surprising as it may seem, however, a really satisfactory solution has not yet been achieved. It is very difficult to obtain an adequately high, known, and controlled temperature in a sample under such conditions that an objective of adequately high magnification and resolution, and adequately convergent transmitted illumination, may be used and still permit flexibility and speed in operation. The major problem with most heating stages is the working distance needed for the objective lens on the microscope. In order to keep the objective out of the heat, long-working-distance lenses must be used. Most commonly these are objectives designed for use with the universal stage, and as they must be used here without the hemispheres, the effective magnification is considerably below their normal rating. The optical quality under such conditions is also far from optimum. It is frequently better optically to use an ordinary objective of lower magnification, with higher magnification oculars, than the universal-stage objectives. The writer seldom uses oculars less than 12.5X, and frequently goes to 20X oculars, in order to get magnification and still use long-working-distance objectives. The resolution is poor, but this is not too serious with moving black bubbles.

At the time of writing, two commercially available heating stages appear to be the most appropriate for inclusion studies. The Leitz 350 stage is the most commonly used; some of its limitations have been discussed elsewhere (Roedder et al., 1968, p. 341). The other is a new combination heating and freezing stage, based on a design by B. Poty of Nancy, France, and manufactured by R. Chaix, Nancy, France; the writer has not had the opportunity to try this stage.

The operation of any heating stage must be done with care and constant consideration of the possible sources of error, as it is surprisingly easy to get beautifully consistent, reproducible, but incorrect numbers. One of the major problems comes from the difficulty in minimizing thermal gradients within the sample chamber (they can never be eliminated). The necessarily large temperature gradients between the small hot chamber and the cold microscope stage, objective, and condenser, at most only tens of millimeters away, make significant gradients within the sample chamber almost inevitable. These gradients should (and do) change with sample size, shape, and placement, as they are a function of the thermal conductivities of the metal, glass and sample, and of the effectiveness of hot-air convection in the chamber. If thermocouples are used, heat flow along thermocouple lead wires can be a serious source of error, particularly if the wires are thick (Larson et al., 1973).

Perhaps the most difficult problem is that of calibration. Normally the writer uses tiny crystals of organic compounds of known melting point, provided commercially for this purpose (Arthur H. Thomas Company, Philadelphia, Pa.). These standards may be individually sealed into short segments of thin-walled capillary glass tubing and placed on the stage between or beside the sample chips. Because most impurities lower the melting point and cause it to occur over a range of temperature, the melting of the *last* crystal should be used. Additional crystals of a given melting point standard, sprinkled beneath and on top of the sample plates, may yield some surprising data on thermal gradients. All calibrations should be made with a stage and sample geometry (and heating schedule) as close as possible to that used with the unknown samples. As the standards cannot experience the identical gradients that exist in the samples, errors in the measurements will still take place that will vary greatly with the equipment and technique.

Although dynamic methods are commonly used to obtain homogenization temperatures (i.e., using a continuously increasing temperature — sometimes as much as 3–5°/min), this technique probably yields very large errors. It is *far* preferable to effectively level off the temperature of the stage after each temperature increment, as the thermal gradients under static heat-flow conditions will generally be much less than those under dynamic conditions.

To check on the possibility that inclusion leakage during the run may cause erroneously high homogenization temperatures, it is best to measure the diameters of the gas bubbles before and after the run, using a graduated ocular in the microscope. These measurements may be in arbitrary units, as only changes are of importance. A useful substitute check on leakage is the duplication of homogenization temperature on a second run. Such duplication is a necessary but not sufficient check, as overheating of inclusions above their homogenization temperatures, particularly the low-temperature inclusions in soft minerals characteristic of the Mississippi Valley-type deposits, can cause permanent stretching of the inclusion walls. They then yield *duplicable* but seriously erroneous homogenization temperatures (Larson et al., 1973). This source of error can be avoided by making all desired determinations on the low-temperature inclusions first, before attempting to homogenize the higher-temperature inclusions in the same plate.

The point of homogenization is taken as the temperature at which the bubble disappears. There is a semantic problem here that can cause difficulty. Most fluid inclusions consist of only liquid plus vapor, and so the temperature at which the bubble disappears is the temperature of homogenization. If daughter crystals are also present, they may not have dissolved completely at this temperature and hence, it is not a true temperature of homogenization for the inclusion. Confusion can be eliminated by the use of "L–V homogenization" (or "liquid–vapor homogenization"), and "complete homogenization" whenever ambiguity might occur.

Homogenization usually occurs suddenly, because of surface tension, when the bubble has been reduced to 1–2 μm diameter, introducing a minor error that can be ignored (Roedder, 1971b, p. 328). The behavior of any daughter crystals should also be noted.

On cooling after homogenization, the temperature at which heterogenization (i.e., forma-tion of a bubble) occurs should be noted, particularly if the temperature at homogeniza-tion had been rising, as the true temperature of homogenization, under equilibrium static conditions, should fall between these two. If the homogenization temperature was low (i.e., the bubble at room temperature was very small), heterogenization may not occur even on cooling to room temperature because of metastable stretching of the inclusion liquid (Roedder, 1971b). If many different inclusions are being followed simultaneously, or if there are rapid changes, as in critical phenomena or in the solution of multiple daughter crystals, a dictaphone can be used during the observations, to avoid loss of continuity of observation.

Probably the most frustrating aspect of all fluid-inclusion study is the problem of illumination in the heating or cooling stages. Large inclusions that appear clear and have only thin black lines at their boundaries will usually be usable even in poorly designed heating stages that severely limit the angle of convergence possible for the light. Most inclusions, however, that require the use of strongly convergent light (i.e., the high power condenser) for ordinary microscopy will appear opaque or almost so in the nearly col-limated lighting that results from the common heating-stage design. The more collimated the lighting, the broader the dark borders around inclusions (from total reflection at the inclusion—mineral interface). This effect is most bothersome in high index minerals such as sphalerite, in which it is not uncommon to find that perhaps only one inclusion out of ten selected for running can actually be used in the stage. Even if the black border is not very broad, the small bubble will frequently disappear in these dark areas. Such inclusions should be checked at higher temperatures, however, as the bubble generally moves about much more actively when it gets very small, and hence it may reappear. (Stretching may yield similar behavior — see Larson et al., 1973). The most effective remedy (other than finding better inclusions) is the use of a flexible fiber optics illuminator. By bringing the light in from beside the objective (above the stage) or from below the stage, it is generally possible to find some position where the bubble can still be seen. The 1/8 inch by 24 inch light guide is best, because of its flexibility (American Optical Co., Space Defense Divi-sion, Southbrige, Mass.). It is often most effective if the normal transmitted light beam is blocked.

Cooling stages

The problems in design and operation of a cooling stage are more difficult than those for a heating stage. Most commercial cooling stages, except the newly designed Chaix stage mentioned above, are for metallurgical use, in reflected light. Several different stages for inclusion use have been described in the literature (Roedder, 1972, p. JJ28), and some of the problems in their use have been discussed (Roedder, 1962a). Mississippi Valley-type inclusions commonly show gross supercooling and other metastable phenomena that can lead to serious errors if not recognized (Roedder, 1967c, 1971b). As freezing is less apt to damage inclusions than homogenization (unless they contain low-salinity fluids and small bubbles), it is generally best to make freezing runs on a given sample before heating runs. If only a few determinations need be made, the writer recommends a manually controlled freezing cell (Roedder, 1962a, p. 1051).

TYPES OF DATA AVAILABLE – MISSISSIPPI VALLEY-TYPE DEPOSITS

Inclusions have yielded at least some data on each of the following features of Mississippi Valley ore fluids: density, rate of movement, pressure, temperature, gross salinity, pH, noncondensable gases, isotopic composition, and nonvolatile ions in solution. When inclusions are studied from various parts of zoned crystals, or from various minerals in a previously established paragenetic sequence, it is frequently found that changes took place in the ore fluid with time, particularly in composition and temperature (e.g., Roedder et al., 1968; Roedder, 1971a). For this reason it is important to view the data from any given deposit in this larger context, to see if there is a correlation of the occurrence of changes in the mineralogy (e.g., mineral assemblage, or the size, crystal habit, or color of a given mineral) with changes in the fluid-inclusion data. Note that such correlations do not necessitate a cause and effect relationship. Many studies of magmatic hydrothermal deposits have been made showing such correlations (e.g., Ermakov, 1950). Although obvious mineralogical changes have taken place during the deposition of the Mississippi Valley-type ores (e.g., the district-wide changes in calcite crystal habit in the Upper Mississippi Valley district, Heyl et al., 1959, p. 100), the correlative changes in the fluid inclusions are generally less abrupt or may not even be recognized. One should always be alert to the possibility of such changes, however, and hence to the hazards of assuming equivalence of data from different parts of a given deposit.

Density

Unlike the estimates of pressure and temperature, fluid inclusions permit reasonably accurate and unambiguous estimates of the density of the ore-forming fluids, and there does not appear to be any other source for such information. If the relative volumes of the liquid and gas phases are determined at room temperature (by using geometrically regular inclusions), and if the salinity is known (from freezing data), the density of the originally homogeneous fluid can be calculated. Thus, simple linear intercepts of the phase boundaries on long tubular inclusions give an approximate phase ratio, and thin flat inclusions can be photographed or sketched onto paper, using a camera lucida and the relative volumes of liquid and vapor obtained with a pantograph or (much more easily) by cutting out and weighing the two "phases" on a good balance. Although these methods can be highly precise, their accuracy can be very poor when the inclusion walls are not truly parallel, as is commonly found. Inclusions with an appreciable third dimension can also be measured, but the errors involved become much larger unless a universal stage is used. At first glance, it appears that the volume of the spherical gas bubble could be obtained accurately, from its diameter, but that the volume of the liquid phase (total inclusion volume minus gas-bubble volume) would be inexact, except in very geometrically regular inclusions whose volumes can be assumed to consist of the sum of measurable cylinders, cones, pyramids, or prisms. Unfortunately, however, although it is true that the inclusion volumes are inexact, the volume of the spherical gas bubble may also be inexact, because of the negative lens effects of a curved inclusion wall. This may cause an error of as much as 50% in the estimate of the bubble volume in minerals of high index of refraction, such as sphalerite (Roedder, 1972, plate 11, figs. 7, 8). Volume changes from external or internal pressure, and from thermal expansion of the host, are negligible compared with the effects of composition and inaccuracies in phase-volume measurement. With rare exceptions, the overall inclusion fluid densities (i.e., the density of the fluids at the time of trapping) for Mississippi Valley-type deposits are found to be greater than 1.0, and frequently are as high as 1.1 g/cm^3. This is because the room-temperature solutions have densities reaching a maximum of about 1.2, and there is about 10% of vapor of effectively zero density. The most important point here is that this overall density (i.e., the density of the hot fluid that was trapped) is greater than 1.0, as this controls the direction of hydraulic gradients of the hot ore-forming fluids against fresh cold meteoric and ground waters at density near 1.0. The room-temperature density of the liquid phase must be estimated from the salinity (as determined on the freezing stage), or if chemical analyses have been made, it can be measured experimentally on synthetic fluids of such composition. Brown (1942) has shown that even very small compositional differences may result in density differences that can be significant in surface-water circulation patterns. However, it should be noted that the density increase from dissolving 1 wt.% salt would be cancelled out by a 30°C rise in temperature.

Rate of movement

Except for possible surges from large-scale solution collapse of the wall rock (as in East Tennessee, Roedder, 1967b, pp. 352–353), the ore fluids are believed to have moved very slowly at the site of deposition. Several lines of evidence support this, the first two of which are based on metastable equilibria:

(1) All published freezing studies of fluid inclusions in Mississippi Valley ore and gangue minerals report gross supercooling on the microscope freezing stage. This is taken to signify a complete freedom from such solid nuclei in suspension as are present in surface waters; these nuclei in surface waters normally preclude more than 10°C supercooling, whereas many inclusions in Mississippi Valley-type samples require supercooling of 20–40°C below their equilibrium freezing temperature before they will freeze (Roedder, 1962a, 1963, 1968a, c). (Daughter crystals of NaCl and other phases generally do not act as nuclei for ice.)

(2) If the expansion on freezing in the microscope cold-stage eliminates the vapor bubble, it should renucleate as the ice melts to form more dense water, but in many Mississippi Valley-type samples, metastable superheated ice occurs, at high negative pressures — as much as 1000 atm negative pressure. At such negative pressures, ice is in equilibrium (metastable) with water at temperatures as high as $+6°C$ (Roedder, 1967c), because of the failure to nucleate a vapor bubble. This is a much larger negative pressure on a liquid than has ever been recorded before in extensive studies of laboratory liquids, and just as with item (1), this metastability requires that the inclusion liquid be exceptionally clean and free of suspended solid nuclei.

(3) Crystals of ore and gangue minerals growing in open fractures or vugs may contain crystals of other phases that nucleated and grew simultaneously, but they are very clean and free of inclusions of clay or other debris that might have been carried in suspension, as are found in some magmatic–hydrothermal veins (Barton et al., 1971).

(4) The inclusions and other microscopic features reveal exceedingly minute regular oscillatory growth bands in several Mississippi Valley ores and related types of occurrences. If these bands are truly annual varves as suggested (Roedder, 1968a, b; 1969; Leach, 1973), they indicate crystal-growth rates in the range of only 10 μm/year and hence suggest very quiet conditions, fluid flows being in the range of perhaps one μm/sec over tens of thousands of years, unless the concentration change (amount precipitated) is very low (<1 ppm, see fig. 1).

Although they do not concern inclusions, several features observed in the field also suggest relatively slow flow rates. Included here would be the following:

(1) Although some crystal growth anisotropy from fluid movement has been found (Stoiber, 1946; Kesler et al., 1972) such anisotropy is generally minor or absent.

(2) The large size and perfection of the crystals in some deposits.

(3) The obvious vertical settling of most nuclei in open fracture fillings, as in the East Tennessee district where the sphalerite crystals are found almost entirely on the lower sides of even steeply inclined fractures, apparently as a result of sphalerite nuclei forming in the fluid and settling to the bottom. Bastin (1931, p. 39) presented similar evidence for the vuggy Illinois fluorite ores.

(4) The relatively great distance between individual crystals of a given phase in some mines (this indicates only that little spontaneous nucleation took place, hence very little supersaturation, hence very slow growth, and hence very slow flow rates; Roedder, 1971a, p. 786).

Pressure

The information available from inclusions as to the pressure of deposition is meager but may still be useful. Several features permit the placing of rough lower limits. Although boiling has occurred in some magmatic–hydrothermal deposits, as evidenced by the presence of low-density primary inclusions that had been filled with steam as well as others filled with hot liquid, such evidence is not found in Mississippi Valley-type inclusions. Thus, we know that the hydrostatic pressure was always greater than the vapor pressure of the solutions present at the inclusion-homogenization temperatures. At a homogenization temperature of 150°C, pure water has a vapor pressure of 3.7 bars over atmospheric and hence would require a minimum of 37 m of hydrostatic head of cold water (at 1.0 g/cm^3) to keep it from boiling. Gases, such as methane in solution, would of course raise this minimum. If the water over it were heated to the boiling point throughout, the column would be lighter and this minimum head would have to be higher (40.9 m); if the solution were saline, its vapor pressure would be slightly lower, requiring a lower head (28.5 m for 20% NaCl; Haas, 1971).

The inclusion evidence places no upper limit on the pressure at the time. However, if another, independent (and correct) thermometer were available, the difference between this temperature and the homogenization temperature would be a true "pressure correction" and would permit a valid estimate of the pressure. Such an independent thermometer would have to be very accurate. Thus the compressibility of 20% NaCl solution, at 150°C, is so small that even 3,300 m of hydrostatic head on it would only cause a pressure correction of 25°C (Klevtsov and Lemmlein, 1959). Bethke and Barton (1971, p. 160) have shown, for example, that the distribution of Mn between galena and sphalerite can be combined with $P-V-T$ data on fluid-inclusion fluids to give *both* pressure and temperature within surprisingly narrow limits.

Another rough lower limit is established by the presence of gases under pressure (mainly methane), dissolved in the brines of inclusions in many Mississippi Valley-type ores. Thus, when these are opened on the crushing stage, the gas bubble expands approximately twenty fold (Roedder, 1970b, fig. 4). If this represents strictly the expansion of the gas in the bubble itself, the twenty fold volume change would correspond to approximately 20 atm pressure. The true pressure in the bubble is undoubtedly lower than this, as much gas can be contributed to the bubble from the liquid. The solubility of methane in these brines at geologically reasonable pressures is probably about one volume of gas/volume of liquid, and because distances for diffusion are small, only 10—100 μm, degassing can occur very rapidly. On the other hand, the pressure necessary to keep this gas in solution at the elevated temperature of trapping, although unknown, would be much higher.

Another potential source of data on pressure is present in the many deposits that contain inclusions of liquid petroleum as well as brine. There is abundant inclusion evidence that the ore-fluid brines frequently contained suspended immiscible globules of petroleum-like liquids and that these oil globules stuck to growing fluorite surfaces and, hence, were trapped preferentially. On cooling after trapping, these "oil" inclusions form a shrinkage bubble just as do the brine inclusions, and homogenization temperatures can be determined on a heating stage (see next section). They usually give lower homogenization temperatures than adjacent presumably coeval brine inclusions (Freas, 1961; Roedder, 1963, p. 175; 1971a, p. 785). This is a result of the much higher compressibility of oils than brines (approximately four fold), causing a much larger pressure correction for the oil inclusion than for the brine. If independent data on the relative compressibilities of the two fluids were available, both pressure and temperature of trapping could be obtained. In any such application, care should be taken to avoid material in which spontaneous decomposition reactions have taken place within the oil inclusions, as has occurred in some inclusions in Illinois fluorite (Roedder, 1962b, p. 40 in original).

Temperature

The temperature at the time of fluid trapping, and hence the temperature or a particular ore fluid, can best be obtained from optical determination of homogenization temperatures of primary two-phase, gas-liquid inclusions in ore or in gangue minerals *known* to be coeval with ore. Although there are many serious problems in the use of homogenization temperatures for geological thermometry, particularly for deposits formed at high temperature and pressure, most of these problems are minor for samples from low-temperature stratiform deposits. In addition, the temperature corrections needed for the effects of hydrostatic pressure are small and fairly well known. It should be noted, however, that the added basic assumption here — that the reported measurements are correct — is not always valid because of various problems in sample selection and measurement technique. When care has been used, the values obtained are probably the most precise and accurate in the field of geologic thermometry. Large numbers of such homogenization determinations have been made, particularly on Mississippi Valley-type deposits in the United States (Table IV). In summary, they show that during the ore-forming stage, temperatures were generally 100–150°C and rarely were as high as 200°C. Temperatures of late-stage, calcite-depositing fluids in many deposits were less than 100°C.

As inclusions are heated, the internal pressure builds up, and for those homogenizing in the liquid phase, dP/dT increases abruptly at the point of homogenization. Further heating causes the internal pressure eventually to become greater than the tensile strength of the host, and it bursts. This is the basis for the *decrepitation* technique of geothermometry. The coarsely crushed sample is heated at a specified rate and the frequency (and/or magnitude) of decrepitation is detected with a microphone and suitable electronic amplification and averaging techniques, and plotted vs. sample temperature to form a decrepigram. Although there are many serious problems in both theory and practice with this technique, the temperature of decrepitation of any given inclusion *is* a function, in part, of the temperature of homogenization. It is also a function of a series of other factors that in part cannot be controlled. As a result, the difference between these two temperatures can be large and either positive or negative. However, as a result of the relative incompressibility of water solutions in the low-temperature range, decrepitation in Mississippi Valley-type samples tends to occur only a small temperature increment above the homogenization temperature. The amount of this *overshoot* varies appreciably with the nature of the sample, but is commonly at least 20°C. As decrepitation can be

TABLE IV

Homogenization and freezing temperatures on inclusions in major Mississippi Valley-type deposits and possibly related occurrences

District, mine, or occurrence[1]	Homogenization temperature, °C			Freezing data[14]			Reference
	minimum[2]	maximum	accepted best range or value[3]	minimum salinity[2]	maximum salinity[4]	accepted best salinity[4] range or value[3]	
Aachen, West Germany, Zn	125	140					Roedder, unpublished data
Alabama, Ba	73	91					Hughes and Lynch (1973)
Bytom, Poland, Zn			100–120[11]				Roedder, unpublished data
Cartagena, Spain, Pb–Zn	60	140	110	−7.9	−9.4	−9.2	Roedder (1963)
Central Kentucky, Zn	80	110	110	−2.5	−35	−15–20	Roedder (1971a)
Central Missouri, Ba–Pb–Zn	50	140	110	−10	−24	−22	Leach (1973)
Central Tennessee, Zn	64	151	90–120	−7.5	−30	−20	Roedder (1971a)
Central Tennessee, Zn	70	140		−15	−22		De Groodt (1973)
Derbyshire, England, F				−13	−23	−15–20	Ford (1969)
East Tennessee, Zn	100	280[6]	194[6]				Roedder (1967b)
East Tennessee, Zn	60	175[6]	100–150	−7.5	−32	−10–30	Miller (1969)
East Tennessee, Zn	89	135	114	−12	−26		Roedder (1963, 1971a)
Friedensville, Pa., Zn	50	150					Larson et al. (1973)
Guatemala, Pb–Zn							Roedder (1967b)
Hansonburg, N. Mex., Pb–Ba–F	140	195	160–180	−7	−13	−10	Kesler and Ascarrunz-K (1972)
Ireland, Keel mine, Pb–Zn	176	185	175	−7.3	−8.0		Roedder et al. (1968)
Ireland, Keel mine, Pb–Zn							Roedder, unpublished data
Ireland, Mogul, Pb	150	275	225				Patterson (1970, in Morrissey et al., 1971)
Ireland, Silver mines, Pb–Zn	190	250					Greig et al. (1971)
Ireland, Silver mines, Pb–Zn							Greig et al. (1971)
Laisvall, Sweden, Pb	83	223[6]	100–150	−19.7	−1.05[7]	−23–27	Roedder (1968c)
Missouri, Ba–Zn	<50[13]	115		(5–10)[15]	(23)		Leach (1971)
Missouri, Pb–Zn–Ba	<40[15]	110		(4–10)[15]	(>22)		Leach (1973)
New Lead Belt, Missouri, Pb	82	145	90–120	−10.6	−28.2	−22	Roedder, unpublished data

Deposit							Reference
Northern Arkansas, Zn	83					(>22)	Leach (1973)
Northern Arkansas, Ferndale limestone*16		132					Nelson (1973)
Northern Arkansas, Rush Creek, Zn	85	170	110–150	(5)	(25)		Potter (1970)
North Pennines, England, Pb–Zn–F	114	169*17		−17.2	−21		Harker (1971)
North Pennines, England, Pb–Zn–F	88	150		−17	−24	−20	Sawkins (1966)
North Pennines, England, Pb–Zn–F	99	205	130–160	−23.0	−23.4		Roedder (1967b)
Picher (Tri-State) Pb–Zn	115	135					Roedder (1963)
Picher (Tri-State) Pb–Zn	83	120					Newhouse (1933)
Picher (Tri-State) Pb–Zn	51	97					Schmidt (1962)
Pine Point N.W.T., Can., Pb–Zn	52	131	52–105	−12	−35*8	−25	Roedder (1968a)
Polaris deposit, N.W.T., Can., Pb–Zn				−12.5	−12.9		Jowett (1972)
Santander, Spain, Pb–Zn							Roedder (1963)
Sardinia, Litopone Ba			<40*12				Roedder, unpublished data
Southeast Missouri Pb belt				−6.5*5	−23.5*5		Roedder (1967b)
Southern Illinois, F–Pb–Zn	84*10	172	140				Grogan and Shrode (1952)
Southern Illinois, F–Pb–Zn	122	144	130				Pinckney and Rye (1972)
Southern Illinois, F–Pb–Zn	70*10	129					Freas (1961)
Southern Illinois, F–Pb–Zn	139	153	142–145				Roedder et al. (1963)
Southern Illinois, F–Pb–Zn				−15.6	−19.0		Hall and Friedman (1963)
Southern Illinois, F–Pb–Zn				−14.5	−23.4	−15–20	Roedder (1963, 1967b)
Sweetwater, Tenn., Ba*9	109	198		−5	−11		Nadeau (1967)
Sweetwater, Tenn., Ba*9	52	130	96	−13	−23		Roedder (1967b, 1971a)
Timberville, Va., Zn	140	150					Roedder (1967b, 1971a)
Tri-State Pb–Zn	83	120				(>22)	Leach (1973)
Upper Miss. Valley, Pb–Zn	80	105					Newhouse (1933)
Upper Miss. Valley, Pb–Zn	75	121					Bailey and Cameron (1951)
Upper Miss. Valley, Pb–Zn							Hall and Friedman (1963)
Upper Miss. Valley, Pb–Zn				−2.9*5	−3.1*5		Erickson (1965)
Upper Miss. Valley, Pb–Zn	46*5	74*5				−20	Roedder (1967b)

*1 See original reference for details on samples, locality and geology; only data on sphalerite, fluorite, or barite are listed unless otherwise stated; no data on decrepitation listed.

*2 Data on *known* secondary inclusions are excluded.

*3 Given only where appropriate.

*4 As −21.1° is the minimum liquidus temperature (eutectic) in the system $NaCl–H_2O$, temperatures below this must have salts other than NaCl.

TABLE IV (continued)

*5   Late calcite only.

*6   This has been shown to be incorrect – see Larson et al. (1973).

*7   This inclusion had a NaCl · $2H_2O$ liquidus, hence, it was very strongly saline.

*8   Some as high as –10° had a NaCl · $2H_2O$ liquidus and hence were very strongly saline.

*9   Data only on fluorite cogenetic with barite.

*10   Oil inclusions.

*11   Estimated, from tiny oil inclusions in sphalerite and brine inclusions in associated carbonates.

*12   Estimated, as all inclusions are full of liquid at room temperature.

*13   This temperature obtained only on calcite and barite; sphalerite temperatures were 108–115°.

*14   Values in parentheses are salinities (NaCl % equiv.).

*15   Low values obtained on barite.

*16   Possibly the same brine that formed the northern Arkansas zinc district?

*17   These homogenization temperatures are believed to be approximately 30°C too high, because of thermal gradients in the heating stage (R.W. Potter, personal communication, 1974).

performed in the field and on opaque materials as well, however, it is useful as a *qualitative* field tool, particularly on low-temperature deposits. The writer has serious doubts that it can ever be calibrated to yield accurate data, free of subjectivity, on the temperatures of homogenization (or formation).

A necessary corollary of the slow flow and deposition rates and the large volumes of liquid passing any given point in the ore is that a considerable volume of rock in the vicinity of the ore body must have been heated to the temperatures recorded in the inclusions. The heat capacity of water is so high that the fluid from six complete changes of the water in the pores of a rock with only 10% porosity has a total heat capacity greater than that of the rock. As such, any temperature differences will soon be levelled, and hence the inclusion temperatures cannot be considered to represent merely some local late-stage recrystallization environment, as some have argued.

Gross salinity

The salinity of the ore fluid is most easily determined from the equilibrium freezing temperature of fluid inclusions (Roedder, 1962a, 1963). This is the depression of the freezing point for the trapped solution and is so determined that metastable phenomena such as superheating or supercooling are not involved. There is obviously some uncertainty in estimating the salinity from the depression of the freezing point for mixed salt solutions. However, the gross composition of these salts is fairly well known and uniform (see below). The conversion of freezing points to salinities (in weight percent mixed salts) can be determined experimentally by freezing a synthetic fluid simulating the inclusion fluids. Such data have been verified by actual quantitative analyses of inclusions for salts and water, as mentioned below.

In summary, the salinity of Mississippi Valley ore fluids usually exceeds 15 wt.% salts, and frequently exceeds 20% (see Table IV), yet daughter crystals of NaCl are almost never found, implying appreciable amounts of ions other than Na and Cl. It is important to note also that several details of the freezing work prove the presence of at least some salts other than just NaCl, even though this is usually the main salt. One such detail is that pure NaCl can only lower the freezing point to $-21.1°C$ (Roedder, 1962a, fig. 4), yet many inclusions have a lower freezing temperature.

pH

It would be very desirable to know the pH of the fluids in inclusions at the time of trapping. If these values could be obtained at room temperature, together with inclusion composition, extrapolations to the temperature and pressure of formation might be possible. The pH at room temperature can be obtained by calculation, e.g., by using analytical results for CO_3^{2-} and HCO_3^-. Direct measurement is possible only with large inclusions, which can be broken into and absorbed onto indicator papers under a low-power microscope. If gases are evolved when the inclusion is opened, gross changes in pH can occur almost instantly. An extensive literature in Russian concerns such procedures (Roedder, 1972, p. JJ39), but only one modern attempt has been made in the Western world (Erickson, 1965; pH about 7.5, on samples from Upper Mississippi Valley lead—zinc deposits). Another procedure for determination of inclusion pH is that of crushing the host mineral in water and measuring the pH of the slurry. Hundreds of such measurements will be found in the Soviet inclusion literature (Roedder, 1970a), but for several reasons, in particular the effects of various mineral surfaces, loss of gases, and the several-thousandfold dilution, the writer believes that they are practically worthless (Roedder, 1972, pp. JJ38—JJ41).

Noncondensable gases and organic matter

Although very little information is available on noncondensable gases and organic matter, it could be of considerable import to problems of the chemistry of ore deposition. Thus, unless the mixing of two fluids is assumed to take place, one is faced with the difficult task of possible simultaneous transport of sulfur and metals in a given fluid. Organic matter should, in general, reduce sulfate to sulfide, although these reactions are apparently very slow. Barton (1967) has proposed that the Mississippi Valley-type ore metals have been transported by a fluid bearing organic matter, and in which the sulfur was present as sulfate; slow reduction of the sulfate in this nonequilibrium solution avoids the problems of simultaneous transport of sulfide and metal in the same fluid. The solubility of methane alone in 50,000-ppm brines is adequate to provide, at saturation under pressures equivalent to a depth of only 1,000 ft. (Duffy et al., 1961), reducing capacity to form about 2 g S^{2-}/l of brine. In addition, many subsurface waters are known to be saturated with respect to methane (Buckley et al., 1958), and studies with the crushing stage (Roedder, 1970b) reveal appreciable gas in solution, presumably mainly methane. In contrast, magmatic hydrothermal fluids also contain methane, but usually in much smaller amounts, which could well be from simple inorganic equilibrium reactions in the system C—H—O.

In addition to the obvious methane, a surprisingly high percentage of Mississippi Valley-type ores contain small amounts of a yellow-brown, fluorescent oily fluid, which, when a sample is crushed, smells strongly of petroleum. This is present as separate fluid inclusions, as immiscible globules in primary and secondary brine inclusions, or in larger quantities as seeps in the mine workings. In some samples, this fluid has degraded spontaneously, inside the inclusions, into a dark-brown viscous phase and a colorless low-viscosity liquid (and sometimes a birefringent crystal as well). The two liquid phases are immiscible; the darker one has an amazingly selective preference for only certain crystallographically oriented surfaces on the walls of the approximately spherical inclusions in fluorite (see Roedder, 1962b, p. 40 in original; 1972, plate 9). The colorless fluid flashes into vapor when the inclusion is opened and hence probably consists mainly of light hydrocarbons.

Organic materials in solution may be of importance to several aspects of the origin of Mississippi Valley ores. In addition to the reducing capacity of the methane mentioned above, the oily phase, present as immiscible globules in the brine and the rather surprising amounts in solution (Price, 1973), would have been a continuing internal source for reductants of sulfate to form sulfide. In some deposits some semisolid organic debris may also have been present in the solutions (e.g. Hansonburg, New Mexico; Roedder et al., 1968, figs. 9–11), and it is important to remember that the most effective reductants are probably used up as a result of this fact, and hence are not found on analysis. Many other organic compounds may be present. Thus, the oxalate ion was present in at least some connate brines, as the mineral whewellite ($CaC_2O_4 \cdot H_2O$) has been recognized in several occurrences. Veitch and McLeroy (1972) found that amino acids in solution greatly increased the solubility of heavy metals in carbonate environments, and Miller et al. (1972) found fatty acids in bedded barites that they believe are from sulfate-reducing bacteria. (This would explain the common occurrence of free hydrogen sulfide in the inclusions in these "fetid" barites.)

Isotopic ratios

Practically all the isotopic studies of fluid inclusions have been made in the last decade, and most of these have been on magmatic–hydrothermal-type deposits. Roedder et al. (1963) reported a procedure for obtaining the amount of water, its isotopic ratio of deuterium to hydrogen (D/H), and the chemical analysis of the seven major ions in solution, all from a sample containing several milligrams of water, and they list seven such analyses from Mississippi Valley-type deposits. Hall and Friedman (1963) then used this technique on 33 samples of known paragenesis, from the southern Illinois fluorite–lead–zinc district, and the Upper Mississippi Valley lead–zinc district. The D/H ratios (and the concentration and gross chemical composition of the salts present) are all similar to present connate waters in those formations for the main ore stage, but some late minerals were deposited from fluids of different composition. It may seem that "several milli-grams" is a very small sample, but actually this amount of inclusion liquid of *verified* singular origin is only possible with special types of material, and usually requires much work to obtain. Samples are needed containing large single inclusions, in the 1-mm range, or large numbers of smaller primary inclusions, cut to eliminate most secondaries. The sample size requirements can be expected to decline as techniques improve, but as in all inclusion analyses, the single most important step is that of verifying the origin of the selected inclusions (Roedder, 1972).

The original isotopic signature of the water (e.g., sea or meteoric) can be changed by isotopic exchange with the rocks through which the water percolates only if the exchange is sufficiently rapid, and if there is a significant reservoir present. Oxygen will exchange in geological environments at as low as 150°C (Clayton et al., 1968; Pinckney and Rye, 1972), and rocks contain a large reservoir of oxygen. Hydrogen is probably exchanged more readily, but the reservoir of hydrogen in the rocks is so very small that it will be controlled by the water composition, rather than the reverse. Thus, even a normal shale contains less hydrogen than would be present in the water filling its pores, but it contains about seven times as much oxygen as the water. Several changes of water could thus exchange the hydrogen of the rock completely, but far more changes would be needed for the oxygen. No oxygen-isotopic determinations have yet been reported for Mississippi Valley-type inclusion liquids, but as the improvements in this technique have been rapid, we may expect more work of this kind. Although it would be of considerable interest, the isotopic composition of the sulfur in the inclusion fluids is not within reach of present experimental methods, considering the available sample size.

Nonvolatile ions (i.e., "salts") in solution

The composition of the salts in inclusion fluids can be determined by semimicroanalysis of water leaches prepared from samples crushed under conditions carefully controlled to avoid loss and contamination (Roedder, 1972). The salts are surprisingly uniform among the various districts examined and consist mainly of sodium and calcium chlorides. A typical analysis is shown in Table V. Almost without exception, such analyses show the descending weight percent sequence, chlorine—sodium—calcium—potassium—magnesium—boron. Bicarbonate is probably low. Total sulfur, stated as sulfate, seldom exceeds a few thousand ppm. The value given for total sulfur as sulfate in Table V is a maximum, because an appreciable but unknown amount of sulfur is also contributed by oxidation of the sphalerite during the leaching. Unfortunately the determination of microgram quantities of sulfur is one of the most difficult analytical problems and at present does not permit distinction between the various valence states. Some evidence, however, indicates

TABLE V

Typical fluid-inclusion analysis: sphalerite, Tri-State district, Oklahoma (Roedder, 1967b)

	Parts per million	Moles per liter
Na^+	57,100	2.47
K^+	2,700	0.07
Ca^{2+}	18,000	0.45
Mg^{2+}	2,400	0.10
Cl^-	124,600	3.51
SO_4^{2-}	<3,300	<0.03
$B_4O_7^{2-}$	107	0.0007
HCO_3^-	?	?
Total salts	208,000 (does not include heavy metals)	
H_2O	792,000 (probably includes approximately 800 ppm CH_4)	

that sulfide sulfur must be very low. Most significant, however, is the concentration of salts. The freezing data indicating very strong brines are corroborated by those few analyses in which actual concentrations were determined (rather than just ratios). Analyses of this type are reported for the southern Illinois fluorspar–lead–zinc district, the Tri-State lead–zinc district (Kansas–Oklahoma–Missouri), and Santander, Spain (Roedder et al., 1963), and for the southern Illinois fluorite–lead–zinc district and the Upper Mississippi Valley lead–zinc district (Illinois–Iowa–Wisconsin; Hall and Friedman, 1963). Additional studies of this sort are needed on other deposits, but most deposits simply do not provide suitable sample material.

A few preliminary data on inclusions in a sample of fluorite from southern Illinois, obtained by neutron activation (Czamanske et al., 1963), showed unexpectedly high heavy-metal contents. The fluid contained approximately 1% each of copper and zinc, and 0.4% manganese. Pinckney and Haffty (1970) analyzed some other samples from the same locality by atomic absorption spectroscopy and found much lower concentrations (maxima in ppm: Zn 1040; Cu 350; Cl 152,000); the cause for the difference is not known, and it should be resolved.

As sodium and potassium are among the most abundant constituents present in inclusions and may be determined by flame photometry with relative ease, precision, and accuracy, the Na/K ratio is one of the most useful parameters. The fluids that formed the Mississippi Valley-type ore deposits seem to be characterized by much higher Na/K ratios (about 17, by weight) than those having magmatic–hydrothermal affiliations. Even though the Mississippi Valley fluids have very high Na/K ratios and are very similar in many respects to "normal" connate and oil-field waters, the most striking difference is that the inclusion fluids have *lower* Na/K ratios than the lowest ratio reported in oil-field waters (Roedder et al., 1963). Hall and Friedman (1963) and Sawkins (1968) suggest that the extra potassium may represent a magmatic contribution. Similarly, the Mississippi Valley fluids have lower Cl/SO$_4$ ratios than subsurface waters (Sawkins, 1968).

With few exceptions, calcium exceeds magnesium, frequently by a large factor. Although little attention has been given it, the Ca/Mg ratio in the ore-forming fluid is of considerable significance in controlling dolomitization. Care is needed to obtain valid determinations of this ratio in inclusions, however, because of contamination from embedded carbonate crystals exposed during crushing. Mississippi Valley-type ores formed from fluids with a rather uniform Ca/Mg weight ratio between 4 and 8, but other types of deposits deviate widely at both ends of this range.

Comparison of fluid-inclusion data from Mississippi Valley-type deposits with those from magmatic hydrothermal deposits

Inclusions in Mississippi Valley-type deposits show surprisingly little variation, whereas those from magmatic–hydrothermal deposits vary widely, although they do have some features that are characteristically in certain ranges for specific ore types.

In density, most Mississippi Valley-type fluids had densities very close to or slightly >1.0, whereas the magmatic fluids, with a few notable exceptions, were significantly <1.0. In this connection, Hanor (1973) has shown that subsurface brines in many areas have *in situ* densities for which the increasing salinity and temperature with depth just compensate, to yield a gravitationally stable column.

In terms of rate of movement, evidence is fairly abundant for very low rates for the Mississippi Valley-type fluids and is rather scanty for possibly much faster flows in magmatic systems (Barton et al., 1971).

The maximum pressure at the time of deposition cannot be determined in most inclusions, but nothing suggests a high pressure for the Mississippi Valley-type (in keeping with the generally low estimates of depth to cover during formation). In contrast, inclusion evidence can be found for the formation of magmatic deposits over a wide range of pressure, from near surface to >1000 atm.

The temperature ranges of formation for the two types overlap, in that Mississippi Valley inclusions generally have homogenization temperatures between 100 and 150°C and rarely as high as 200°C, whereas magmatic deposits cover the entire range from 100 to >500°C and frequently show a range of >100°C for ore minerals in a single deposit. The largest differences found in Mississippi Valley-type deposits are between inclusions in the main ore-stage and those in late calcite, which are always at the low-temperature end, as well as being much lower in salinity.

The salinity, as determined by the freezing-stage technique, is high in almost all inclusions from Mississippi Valley-type ore deposits. It is seldom < 15% NaCl equivalent and frequently exceeds 20%. All examples with <10% NaCl equivalent were from secondary inclusions or very late stage minerals (see Roedder, 1967b, for details on many of the individual deposits or related occurrences listed in Table IV). In contrast, magmatic—hydrothermal deposits have inclusion salinities generally <10 wt% NaCl equivalent, with a few notable exceptions such as the porphyry copper deposits, where boiling may have occurred, leaving brines of >50 wt% actual NaCl (i.e., with large NaCl daughter crystals in the inclusions).

The differences between the inclusions in the two classes of ore deposits are most notable in terms of composition. Mississippi Valley-type ore deposits have formed from fluids that were essentially sodium—calcium—chloride brines, generally containing appreciable methane (perhaps ~800 ppm) in solution and frequently containing droplets of a brownish immiscible oily phase (Roedder, 1967b, 1972). Magmatic—hydrothermal deposits, on the other hand, have formed from a wide range of solution types containing little or no organic matter, but to my knowledge, these never approach the ratios found in the Mississippi Valley-type. Among the major constituents, the ratios Na/K, Na/Ca, and Na/Cl are always much higher in the Mississippi Valley-type fluids. (It is necessary to use ratios for comparison, as the overall concentrations are so different.)

DATA FROM OTHER THAN MISSISSIPPI VALLEY-TYPE DEPOSITS

Many types of ore deposits are included in the subject matter of this volume other than the Mississippi Valley-type deposits, but relatively few studies have been made of their fluid inclusions. The fifteen most pertinent examples are listed in Table VI. The major hurdle in most of these is the problem of finding suitable host material for inclusions in these generally fine-grained ores; when coarse-grained material is found, the relation of its inclusions to the ore is frequently ambiguous. Thus, the high temperatures found for inclusions in Soviet taconites and jaspilites (Sivoronov, 1968) probably refer to a later metamorphic event, and similar explanations might be appropriate for the Witwatersrand (Krendelev et al., 1970).

Fluid inclusions in detrital grains in sediments can give much information on the nature of the source rocks (e.g., Clocchiatti, 1970; Lofoli, 1970, 1971), and hence aid in the solution of structure and correlation problems. Furthermore, recognizable special fluid inclusion types that are characteristic of a specific type of ore deposit, e.g., the very high-temperature, high-salinity inclusions found in porphyry copper deposits, can be used as an exploration tool, since they will persist through the gossan stage of weathering and even into stream sediments (Nash, 1971; Roedder, 1971c).

The origin of the Colorado Plateau uranium ores has been a particularly controversial subject, and if fluid inclusions could be found in these ores, they might help to place some constraints on the possible conditions of origin. Unfortunately, the primary ores are very difficult subjects for inclusion investigation, as they consist mainly of exceedingly fine-grained opaque minerals. One occurrence, however, a pipelike deposit in Permian sediments (Gornitz, 1969), had usable inclusions in calcite, with homogenization temperatures of $60-110°C$. The close relation of the calcite to the primary ore minerals is based on the presence in it of solid inclusions of chalcopyrite and hematite.

The Kuroko-type deposits have been the subject of rather intensive geological study, particularly by the Japanese. Ten of these studies involve inclusion work, the data from which show that the fluids responsible for the Kuroko deposits were quite different from those forming the Mississippi Valley-type deposits in that they were similar to seawater in concentration of salts and were much hotter (as high as $310°C$; Table VI). The data are very consistent with the model of a marine volcanic environment generally proposed for these ores (e.g., Motegi, 1968; Otagaki et al., 1968; Horikoshi, 1969; Tatsumi and Watanabe, 1971; Kajiwara, 1973), probably including mixing of seawater and meteoric water (Sato, 1968; Sakai and Matsubaya, 1973).

SUGGESTIONS FOR FUTURE WORK

Many of the more obvious avenues for further work on inclusions are implicit in the preceding discussion. In particular, a concerted effort should be made to develop microscope heating and cooling stages that combine better optical characteristics with good control and measurement. The writer is currently attempting to devise a stage using flowing heated (or cooled) gas as the heat-exchange fluid surrounding the sample plate, but many different design avenues should be explored. Refinement of the available criteria for assignment of a primary origin, and the development of new criteria, are both needed and presumably will come from more detailed microscopy.

Fluid-inclusion studies may be expected to contribute increasingly to an understanding of the genesis of several types of very fine-grained ore deposits in sediments, possibly formed during diagenesis, about which there has been much controversy. The Colorado Plateau-type of U–V–Cu ore deposits may seem at first to be an unlikely candidate for inclusion studies. Recent detailed studies, however, of the inclusions in cements in limestones deposited during diagenesis (Nelson, 1973) have shown that careful microscopy on the inclusions in such difficult materials, including the use of cathodoluminescence, can provide valuable information. An important distinction must be made, however, between

TABLE VI

Homogenization data on ores in sediments other than Mississippi Valley-type

Locality	Type	Mineral examined	Homogenization temperature C°	Reference
Kamoto, western Katanga	Co–Cu in shale	dolomite	140, 240*[1]	Bartholomé and Pirmolin (1970, 1971)
Kamoto, western Katanga	Co–Cu in shale	dolomite	55, 120, 200*[2]	Pirmolin (1970)
USSR – variou	taconites and jaspilites of iron ore deposits	quartz	60–530*[7]	Sivoronov (1968)
Orphan mine, Grand Canyon, Arizona	uranium deposit as a pipe in Permian sediments	calcite	60–110	Gornitz (1969)
Witwatersrand, Blind River and others	uranium deposits in Precambrian conglomerates throughout the world	quartz	60–560*[3]	Krendelev et al. (1970)
Japan	cpy.-bearing veins in silicified zone of Kuroko deposits	quartz	*[5]	Takenouchi and Imai (1968)

TABLE VI (*continued*)

Locality	Type	Mineral examined	Homogenization temperature C°	Reference
Kosaka mine, Japan	Kuroko-type	quartz of siliceous ores	200–295	Tokunaga (1968)
Furutobe mine, Japan	Kuroko-type	barite (Keiko) quartz (Keiko) sphalerite (Ohko) fluorite (Ohko) barite (Kuroko)	99–211 117–190 245 159 116–212	Homma and Miyazawa (1969)
Kosaka mine, Japan	Kuroko-type	quartz sphalerite barite	225–310*4 190–245 120–300	Lu (1969)
Kosaka and other mines, Japan	Kuroko-type	siliceous ore yellow ore black ore	180–290 220–290 80–200	Tokunaga et al. (1970)
Northern Japan	various Kuroko-type deposits	quartz, barite and sphalerite	100–265	Ohmoto et al. (1970)
Kosaka mine, Japan	Kuroko-type	Upper Kuroko (gn.–sp.–barite) Lower Kuroko (sp.–py.–cpy.–barite) Oko (py.–cpy.) Keiko (py.–cpy.–qtz.)	100–150 150–200 ~200 200–300	Sato (1970)
Shakanai mine, Japan	Kuroko-type	quartz	132–274*6	Enjoji (1972)
Japan	unspecified Kuroko-type deposits		150–300	Ohmoto (1973)

*1 Bubble homogenized at 140°; last daughter crystals at 240°. A leach yielded a solution with Na/K about 1.4 and (Na + K)/Cl about 0.4.

*2 These are the temperatures of disappearance of three different daughter crystals.

*3 Includes study of the gases evolved on crushing.

*4 Freezing temperatures of −1.5 to −5.3 indicate salinities of <10% (NaCl equivalent). The range of temperature for quartz is interpreted as due to trapping of primary gas bubbles and that of barite to leakage.

*5 Only freezing data given in abstract: −2.7 to −3.4°, corresponding to 5–6% salinity (NaCl equivalent).

*6 Freezing studies show salinity to be 2.4–5.2% salinity (NaCl equivalent) (his page 117).

*7 Temperature range related to degree of metamorphism.

Abbreviations: cpy = chalcopyrite; gn = galena; sp = sphalerite; py = pyrite; qtz = quartz.

what is theoretically possible to achieve with inclusion study and the practical problems presented by available sample materials. In theory, a large number of important questions in sedimentary petrology and ore research can be answered by studies of inclusions, such as primary vs. diagenetic vs. hydrothermal chert (or dolomite), but until new techniques are developed to observe, analyze, and interpret data from inclusions of <2 μm diameter, most such samples are simply not usable.

Another aspect of inclusion study may eventually aid in determining the origin of the heavy metals in the U–V–Cu deposits in sediments. The chemical alteration and leaching of pyroclastic volcanic glass in the sediments during diagenesis has frequently been proposed as the source for the uranium (and perhaps the vanadium) in these ores, but a good chemical mass balance calculation cannot be made because the compositions of the original unaltered glass can only be guessed. If, however, detrital quartz phenocrysts from this volcanic component remain in the altered sediments, these will almost always contain *un*altered inclusions of volcanic glass, complete with its original complement of heavy metals. Analysis of such glass inclusions for uranium, particularly, is possible with presently available alpha and fission track techniques, as well as by ion probe. The bulk composition of such inclusions, which is easily obtained by electron microprobe (see Roedder and Weiblen, 1970, and later papers in that series of Proceedings volumes), may also be compared with the altered composition to determine what amounts of each major component must have been added to the circulating brines.

Imminent new developments in analytical techniques offer exciting possibilities for the future of inclusion analysis. Neutron-activation analysis is a powerful technique particularly suited to heavy-metal studies, if the problems of contamination and loss in sample preparation can be overcome. It should also be particularly applicable to determining the important ratio Br/Cl, which has genetic significance. New developments in gas chromatography, alone or combined with mass spectrometry, show great promise in studies of the small but important amounts of gases and organic compounds present in so many inclusions. New and more precise analyses for K, Ar, Rb, and Sr may prove of value in understanding some oı the problems in using these elements in dating some types of samples (e.g., Rama et al., 1965).

The identification of daughter minerals is important in the study of fluid inclusions, as these minerals represent compounds with which the inclusion fluid has become saturated on cooling. Application of modifications of the new methods of small-particle handling, plus the electron microprobe and the scanning electron microscope, should help greatly. The main problem is the technique of extracting the daughter mineral from the inclusion.

One of the most fertile fields for inclusion study will be the hopefully inevitable refinement in analytical technique to permit the determination of isotopic ratios on much smaller samples than can now be run, as the major hurdle in many isotopic studies is that of preparing samples that contain an adequate amount of inclusion fluid of *known* origin. It can now be done with relative ease for H, but not for C, O, and S. This is not intended to minimize the significance of the major contributions already made by the study of the isotopes of hydrogen in fluid inclusions (and of all four elements in minerals), particularly from Mississippi Valley-type deposits. However, these have generally been on rather special types of material from which relatively large quantities of inclusion fluid (i.e., >1 mg) could be obtained. Reasonably accurate ratios for all four elements, determined on the *same* small inclusion sample, would provide a much more powerful tool for understanding the source and mixing of ore-forming fluids and the chemistry of ore precipitation than determinations of hydrogen alone.

Several problems on the analysis of inclusions remain which cause serious gaps in our knowledge of the chemistry of ore transport and deposition, and for which no new breakthrough appears on the horizon. These are pH, Eh, and the sulfur species present in solution. Although many determinations of the pH of inclusion fluids have been published, the writer believes that most are invalid. Determination of inclusion Eh is even more difficult, and very little is known about it, except via the identification of daughter phases and the gases in solution (Roedder, 1972, p. JJ40). The third problem, that of the sulfur species present, is a purely analytical problem, in that methods have not been devised to determine submicrogram amounts of sulfate and/or sulfide sulfur. Actually, even the procedures for total sulfur in submicrogram amounts are far from satisfactory. A good method for determining sulfide and sulfate sulfur would provide a measure of Eh, if equilibrium had been obtained.

All three of these determinations would presumably have to be made at room temperature and then extrapolated to the temperature of trapping, using known ionization constants. The possible value of the results to understanding ore transport and deposition are such that a concerted effort is warranted.

ACKNOWLEDGEMENTS

The writer is indebted to many colleagues for the numerous discussions that have contributed much to this paper. In particular, he is indebted to those who have reviewed the manuscript: P.M. Bethke, R.W. Potter, II, I. Haapala, H.E. Belkin, H. Wedow, Jr., J.S. Brown, B. Poty, W.C. Kelly, and of course, K.H. Wolf.

REFERENCES

Allman, M. and Lawrence, D.F., 1972. *Geological Laboratory Techniques.* New York, Arco Publ. Co., New York, N.Y., 335 pp.
Bailey, S.W. and Cameron, E.N., 1951. Temperatures of mineral formation in bottom-run lead–zinc deposits of the upper Mississippi Valley, as indicated by liquid inclusions. *Econ. Geol.*, 48: 626–651.
Barnes, H.L., Lusk, J. and Potter, R.W., 1969. Composition of fluid inclusions. Abstracts of papers presented at Third International COFFI Symposium on Fluid Inclusions. In: *Fluid Inclusion Research. Proc. COFFI*, 2: 13.

Bartholomé, P. and Pirmolin, J., 1970. Fluid inclusions and stratiform mineralization at Kamoto, Western Katanga, Republic of the Congo. In: *Collected Abstracts, IMA–IAGOD Meet., '70, Tokyo*. Science Council of Japan, Tokyo, p. 249.

Bartholomé, P. and Pirmolin, J., 1971. Fluid inclusions and stratiform mineralization at Kamoto, Western Katanga. *Proc. IMA–IAGOD Meet., 1970*, IAGOD Vol. – *Soc. Min. Geol. Jap.*, Spec. Issue, 3: 355 (Abstract).

Barton Jr., P.B., 1967. Possible role of organic matter in the precipitation of the Mississippi Valley ores. *Econ. Geol. Monogr.*, 3: 371–378.

Barton Jr., P.B., Bethke, P.M. and Toulmin, Priestley III, 1963. Equilibrium in ore deposits. *Mineral. Soc. Am., Spec. Pap.*, 1: 171–185.

Barton Jr., P.B., Bethke, P.M. and Toulmin, M.S., 1971. An attempt to determine the vertical component of flow rate of ore-forming solutions in the OH vein, Creede, Colorado. *Proc. IMA–IAGOD Meet., 1970*, Joint Symp. Vol. – *Soc. Min. Geol. Jap.*, Spec. Issue, 2: 132–136.

Bastin, E.S., 1931. The fluorspar deposits of Hardin and Pope Counties, Illinois. *Ill. Geol. Surv. Bull.*, 58: 116 pp.

Beales, F.W. and Jackson, S.A., 1966. Precipitation of lead–zinc ores in carbonate reservoirs as illustrated by Pine Point ore field, Canada. *Inst. Min. Metall. Trans.*, 75: B278–B285.

Bernard, A.J., 1973. Metallogenic processes of intra-karstic sedimentation. In: G.C. Amstutz and A.J. Bernard (Editors), *Ores in Sediments*. Springer, Berlin, pp. 43–57.

Bethke, P.M. and Barton Jr., P.B., 1971. Distribution of some minor elements between coexisting sulfide minerals. *Econ. Geol.*, 66: 140–163.

Brown, A.C., 1971. Zoning in the White Pine copper deposit, Ontonagon County, Michigan. *Econ. Geol.*, 66: 543–573.

Brown, J.S., 1942. Differential density of ground water as a factor in circulation, oxidation and ore deposition. *Econ. Geol.*, 37: 310–317.

Brown, J.S. (Editor), 1967. Genesis of stratiform lead–zinc–barite–fluorite deposits (Mississippi Valley-type deposits), a symposium. *Econ. Geol. Monogr.* 3: 443 pp.

Brown, J.S., 1970. Mississippi Valley-type lead–zinc ores. *Miner. Deposita*, 5: 103–119.

Buckley, S.E., Hocott, C.R. and Taggart Jr., M.S., 1958. Distribution of dissolved hydrocarbons in subsurface waters. In: L.G. Weeks (Editor), *Habitat of Oil*. Am. Assoc. Petrol. Geol., Tulsa, Okla., pp. 850–882.

Cameron, E.N., 1961. *Ore Microscopy*. Wiley, New York, N.Y., 293 pp.

Carpenter, A.B., Trout, M.L. and Pickett, E.E., 1974. Preliminary report on the origin and chemical evolution of lead- and zinc-rich oil-field brines in central Mississippi. *Econ. Geol.*, in press.

Clayton, R.N., Muffler, L.J.P. and White, D.E., 1968. Oxygen isotope study of calcite and silicates of the River Ranch no. 1 well, Salton Sea geothermal field, California. *Am. J. Sci.*, 266: 968–979.

Clocchiatti, R., 1970. Study of glass inclusions and their alteration, a regional example from the Dolomites (Bolzano, Italy). *Schweiz. Mineral. Petrol. Mitt.*, 50 (1): 159–166 (in French).

Czamanske, G.K., Roedder, E. and Burns, F.C., 1963. Neutron activation analysis of fluid inclusions for copper, manganese and zinc. *Science*, 140: 401–403.

DeGroodt Jr., J.H., 1973. Determination of temperatures of fluorite formation by fluid inclusion thermometry, Central Tennessee. *Geol. Soc. Am., Progr. Abstr.*, 5 (5): 394.

Duffy, J.R., Smith, N.O. and Nagy, B., 1961. Solubility of natural gases in aqueous salt solutions. I. Liquidus surfaces in the system CH_4–H_2O–$NaCl$–$CaCl_2$ at room temperatures and at pressures below 1,000 psia. *Geochim. Cosmochim. Acta*, 24: 23–31.

Enjoji, M., 1972. Studies on fluid inclusions as media of the ore formation. *Tokyo Kyoiku Daigaku, Sci. Rept., Sect. C*, 11 (106): 79–126.

Erickson Jr., A.J., 1965. Temperatures of calcite deposition in the Upper Mississippi Valley lead–zinc deposits. *Econ. Geol.*, 60: 506–528.

Ermakov, N.P., 1950. *Research on Mineral-Forming Solutions*. Kharkov Univ. Press., Kharkov, 460 pp (in Russian). (Most of this book, and two later volumes that were, in effect, supplements to it, have been translated in: Yermakov, N.P., [sic] and others, 1965. *Research on the Nature of Mineral-*

Forming Solutions, with Special Reference to Data from Fluid Inclusions. Int. Ser. Monogr. Earth Sci. – Pergamon, Oxford, 22: 743 pp.)

Ford, T.D., 1969. The stratiform ore deposits of Derbyshire. In: C.H. James (Editor), *Inter-Univ. Geol. Congr., 15th, 1967, Univ. Leicester*, pp. 73–96.

Freas, D.H., 1961. Temperatures of mineralization by liquid inclusions, Cave-in-Rock fluorspar district, Illinois. *Econ. Geol.*, 56: 542–556.

Gornitz, V.M., 1969. *Mineralization, Alteration and Mechanism of Emplacement, Orphan Ore Deposit, Grand Canyon, Arizona.* Diss., Columbia Univ., 196 pp. (Abstract). (Also published in *Econ. Geol.*, 65: 751–768 (1970), and 66: 983 (1971)).

Greig, J.A., Baadsgaard, H., Cumming, G.L., Folinsbee, R.E., Krouse, H.R., Ohmoto, H., Sasaki, A. and Smejkal, V., 1971. Lead and sulphur isotopes of the Irish base metal mines in Carboniferous carbonate host rocks. *Proc. IMA–IAGOD Meet., 1970*, Joint Symp. Vol. – *Soc. Min. Geol. Jap.*, Spec. Issue, 2: 84–92.

Grigor'yev, D.P., 1959. Rate of crystallization of minerals. *Vses. Mineral. Obshch., Zapiski, 2nd Ser.*, 88 (5): 497–511 (in Russian; translated in *Int. Geol. Rev.*, 3: 694–705, 1961).

Grigor'yev, D.P., 1965. *Ontogeny of Minerals.* Isr. Program Sci. Transl., Jerusalem, 250 pp. (Translated from the Russian).

Grogan, R.M. and Bradbury, J.C., 1968. Fluorite–lead–zinc deposits of the Illinois–Kentucky mining district. In: J.D. Ridge (Editor), *Ore Deposits in The United States, 1933–1967*. A.I.M.E., New York, N.Y., pp. 370–399.

Grogan, R.M. and Shrode, R.S., 1952. Formation temperatures of Southern Illinois bedded fluorite as determined from fluid inclusions. *Am. Mineral.*, 37: 555–556.

Haas Jr., J.L., 1971. The effect of salinity on the maximum thermal gradient of a hydrothermal system at hydrostatic pressure. *Econ. Geol.*, 66: 940–946.

Hall, W.E. and Friedman, I., 1963. Composition of fluid inclusions, Cave-in-Rock fluorite district, Illinois, and Upper Mississippi Valley zinc–lead district. *Econ. Geol.*, 58: 886–911.

Hanor, J.S., 1973. The role of in situ densities in the migration of subsurface brines. *Geol. Soc. Am., Abstr. Progr.*, 5 (7): 651–652.

Harker, R.S., 1971. *Some Aspects of Fluid Inclusion Geothermometry with Particular Reference to British Fluorites.* Diss., Univ. Leicester, Leicester, 155 pp.

Heyl, A.V., Agnew, A.F., Lyons, E.J. and Behre Jr., C.H., 1959. The geology of the upper Mississippi Valley zinc–lead district. *U.S. Geol. Surv., Prof. Pap.*, 309: 310 pp.

Hoagland, A.D., 1973. Appalachian zinc–lead and the deposits of middle Tennessee. *Geol. Soc. Am., Abstr. Progr.*, 5 (5): 404.

Homma, H. and Miyazawa, T., 1969. Formation temperatures of Kuroko (black ore) deposit in Japan. (1) Yunosawa ore body, Furutobe mine. *J. Soc. Min. Geol. Jap.*, 19: 73 (in Japanese; English abstract in *Fluid Inclusion Research – Proc. COFFI*, 2: 46–47, 1969).

Horikoshi, Ei, 1969. Volcanic activity related to the formation of the Kuroko-type deposits in the Kosaka district, Japan. *Miner. Deposita*, 4: 321–345.

Hughes, T.H. and Lynch Jr., R.E., 1973. Barite in Alabama. *Ala. Geol. Surv. Circ.*, 85: 21–23.

Jackson, S.A. and Beales, F.W., 1967. An aspect of sedimentary basin evolution: the concentration of Mississippi Valley-type ores during late stages of diagenesis. *Bull. Can. Pet. Geol.*, 15 (4): 383–433.

Jowett, E.C., 1972. *Nature of the Ore-Forming Fluids of the Polaris Lead–Zinc Deposits, Little Cornwallis Island, N.W.T., from Fluid Inclusion Studies.* Thesis, Univ. Toronto, Toronto, 35 pp.

Kajiwara, Y., 1973. Chemical composition of ore-forming solution responsible for the Kuroko-type mineralization in Japan. *Geochem. J.* (Japan) 6: 141–149 (in English).

Kalyuzhnyi, V.A., 1971. The refilling of liquid inclusions in minerals and its genetic significance. *L'vov. Gos. Univ. Mineral. Sb.*, 25: 124–131 (in Russian).

Kesler, S.E. and Ascarrunz-K, R.E., 1972. Guatemalan lead–zinc mineralization: magmatic–hydrothermal or Mississippi Valley-type? *Geol. Soc. Am., Abstr. Progr.*, 4 (7): 561.

Kesler, S.E., Stoiber, R.E. and Billings, G.K., 1972. Direction of flow of mineralizing solutions at Pine Point, N.W.T. *Econ. Geol.*, 67: 19–24.

Klevtsov, P.V. and Lemmlein, G.G., 1959. Pressure corrections for the homogenization temperatures of aqueous NaCl solutions. *Dokl. Akad. Nauk SSSR*, 128: 1250–1253 (in Russian; translated in *Acad. Sci. U.S.S.R. Dokl.*, 128 (1960): 995–997).

Krendelev, F.P., Zozulenko, L.B. and Orlova, L.M., 1970. Temperature of homogenization and composition of gases in gas–fluid inclusions in pebbles from ancient conglomerates of sulfide type. In: *Fluid inclusion Research – Proc. COFFI*, 3: 101–107 (in English).

Larson, L.T., Miller, J.D., Nadeau, J.E. and Roedder, E., 1973. Two sources of error in low-temperature inclusion homogenization determination and corrections on published temperatures for the East Tennessee and Laisvall deposits. *Econ. Geol.*, 68: 113–116.

Lavery, N.G. and Barnes, H.L., 1971. Zinc dispersion in the Wisconsin zinc–lead district. *Econ. Geol.*, 66: 226–242.

Leach Jr., D.L., 1971. *Investigation of Fluid Inclusions in Missouri Barites*. M.A. Thesis, Univ. Missouri, Columbia, Mo.

Leach, D., 1973. *A study of the Barite–Lead–Zinc Deposits of Central Missouri and Related Mineral Deposits in the Ozark Region*. Ph.D. Diss. Univ. Missouri, Columbia, Missouri, 186 pp (see also abstract by Leach, et al., 1972, in: *Fluid Inclusion Research – Proc. COFFI*. 3: 40).

Lofoli, P., 1970. Internal characteristics of detrital minerals as exemplified by the deposits of the Congo Basin. *Schweiz. Mineral. Petrol. Mitt.*, 50 (1): 37–40 (in French).

Lofoli, P., 1971. Electron microscopy of sand grains. *C.R. Acad. Sci., Paris, Ser. D*, 273: 462–465 (in French).

Lokerman, A.A., 1965. Mineral thermometry in connection with the problem "Ores and Dikes". In: *Mineralogical Thermometry and Barometry*. "Nauka", Moscow, pp. 288–290 (in Russian).

Lu, K.I., 1969. *Geology and Ore Deposits of the Uchinotai–Higashi Ore Deposits, Kosaka Mine, Akita Prefecture, with Special Reference to the Environments of Ore Deposition*. Diss., Univ. Tokyo, 1969, 116 pp. (in English) (Also referenced as 1970).

Miller, J.D., 1969. Fluid inclusion temperature measurements in the East Tennessee zinc district. *Econ. Geol.*, 64: 109–110.

Miller, R.E., Brobst, D.A. and Beck, P.C., 1972. Fatty acids as a key to the genesis and economic potential of black bedded barites in Arkansas and Nevada. *Geol. Soc. Am., Abstr. Progr.*, 4 (7): 596.

Miyazawa, T., Tokunaga, M., Okamura, S. and Enjoji, M., 1971. Formation temperatures of veins in Japan. *Proc. IMA–IAGOD Meet. 1970*, IAGOD Vol. – *Soc. Min. Geol. Jap.*, Spec. Issue, 3: 340–344.

Morrissey, C.J., Davis, G.R. and Steed, G.M., 1971. Mineralization in the Lower Carboniferous of central Ireland. *Inst. Min. Metall. Trans.*, 80, Sect. B: B-174–B-185.

Motegi, M., 1968. Geology and sedimentary environments of Iwami Kuroko mine, Shimane Prefecture. *J. Soc. Min. Geol. Jap.*, 18 (4): 200–205 (in Japanese, with English abstract).

Nadeau, J.E., 1967. *Temperatures of Fluorite Mineralization by Fluid Inclusion Thermometry, Sweetwater Barite District, East Tennessee*. M.S. Thesis, Univ. Tennessee, 36 pp.

Nash, J.T., 1971. Composition of fluids in porphyry-type deposits. *U.S. Geol. Surv., Prof. Pap.*, 750 A; *Geol. Surv. Res.*, p. A-2 (Abstract).

Nelson, R.C., 1973. Fluid inclusions as a clue to diagenesis of carbonate rocks. *Geol. Soc. Am., Abstr. Progr.*, 5 (7): 748.

Newhouse, W.H., 1933. The temperature of formation of the Mississippi Valley lead–zinc deposits. *Econ. Geol.*, 28: 744–750.

Noble, E.A., 1963. Formation of ore deposits by water of compaction. *Econ. Geol.*, 58: 1145–1156.

Ohmoto, H., 1973. Origin of hydrothermal fluids responsible for the Kuroko deposits in Japan. *Econ. Geol.*, 68: 139 (Abstract).

Ohmoto, H., Kajiwara, Y. and Date, J., 1970. The Kuroko ores in Japan: product of sea water? *Econ. Geol.*, 65: 738–739 (Abstract).

Otagaki, T., Tsukada, Y., Osada, T. and Suzuki, H., 1968. Geology and ore deposits of the Shakanai mine. I. On the mode of occurrence of Kuroko (black ore) in the Daiichi ore deposit. *Min. Geol. Jap.*, 18 (1): 1–10 (in Japanese, with English abstract).

Pinckney, D.M. and Haffty, J., 1970. Content of zinc and copper in some fluid inclusions from the Cave-in-Rock district, southern Illinois. *Econ. Geol.*, 65: 451–458.

Pinckney, D.M. and Rye, R.O., 1972. Variation of $^{18}O/^{16}O$, $^{13}C/^{12}C$, texture, and mineralogy in altered limestone in the Hill mine, Cave-in-(Rock) District, Illinois. *Econ. Geol.*, 67: 1–18.

Pirmolin, J., 1970. Inclusions fluides dans la dolomite du gisement stratiform de Kamoto (Katanga Occidental). *Ann. Soc. Géol. Belg.*, 93 (II): 397–406.

Potter, R.W., 1970. *Geochemical, Geothermetric* [sic.] *and Petrographic Investigation of the Rush Creek Mining District, Arkansas.* M.S. Thesis, Univ. Arkansas, Fayetteville, Ark., 115 pp.

Poty, B., 1969. The growth of quartz crystals in veins, exemplified by "La Gardette" vein (Bourg d'Oisans) and the clefts in the Mont-Blanc Massif. *Sci. Terre, Mem.*, 17: 162 pp (in French with English abstract).

Price, L.C., 1973. *The Solubility of Hydrocarbons and Petroleum in Water as Applied to the Primary Migration of Petroleum.* Diss., Univ. California, Riverside.

Rama, S.N.I., Hart, S.R. and Roedder, E., 1965. Excess radiogenic argon in fluid inclusions. *J. Geophys. Res.*, 70: 509–511.

Roedder, E., 1960. Fluid inclusions as samples of the ore-forming fluids. *Int. Geol. Congr, 21st, Proc.*, Sect. 16: 218–229.

Roedder, E., 1962a. Studies of fluid inclusions. I: Low-temperature application of a dual-purpose freezing and heating stage. *Econ. Geol.*, 57: 1045–1061.

Roedder, E., 1962b. Ancient fluids in crystals. *Sci. Am.*, 207: 38–47.

Roedder, E., 1963. Studies of fluid inclusions. II: Freezing data and their interpretation. *Econ. Geol.*, 58: 167–211.

Roedder, E., 1965. Evidence from fluid inclusions as to the nature of the ore-forming fluids. In: *Symposium-Problems of Postmagmatic Ore Deposition, Prague, 1963.* Czech. Geol. Surv., Prague, 2: 375–384, 13 plates.

Roedder, E., 1967a. Fluid inclusions as samples of ore fluids. In: H.L. Barnes (Editor), *Geochemistry of Hydrothermal Ore Deposits.* Holt, Rinehart and Winston, New York, N.Y., pp. 515–574.

Roedder, E., 1967b. Environment of deposition of stratiform (Mississippi Valley-type) ore deposits, from studies of fluid inclusions. *Econ. Geol., Monogr.*, 3: 349–362.

Roedder, E., 1967c. Metastable superheated ice in liquid–water inclusions under high negative pressure. *Science*, 155: 1413–1417.

Roedder, E., 1968a. Temperature, salinity, and origin of the ore-forming fluids at Pine Point, North-west Territories, Canada, from fluid inclusion studies. *Econ. Geol.*, 63: 439–450.

Roedder, E., 1968b. The noncolloidal origin of "colloform" textures in sphalerite ores. *Econ. Geol.*, 63: 451–471.

Roedder, E., 1968c. Environment of deposition of the disseminated lead ores at Laisvall, Sweden, as indicated by fluid inclusions. *Int. Geol. Congr., 23d, Prague, 1968, Rep., Sect., 7, Endogenous Ore Deposits, Proc.*, pp. 389–401.

Roedder, E., 1969. Varvelike banding of possible annual origin in celestite crystals from Clay Center, Ohio, and in other minerals. *Am. Mineral.*, 54: 796–810.

Roedder, E., (Editor), 1970a, *Fluid Inclusion Research, Proc. COFFI, 1970*, vol. 3 – Wash., D.C. (privately published).

Roedder, E., 1970b. Application of an improved crushing microscope stage to studies of the gases in fluid inclusions. *Schweiz. Mineral. Petrogr. Mitt.*, 50 (1): 41–58.

Roedder, E., 1971a. Fluid inclusion evidence on the environment of formation of mineral deposits of the southern Appalachian Valley. *Econ. Geol.*, 66: 777–791.

Roedder, E., 1971b. Metastability in fluid inclusions. *Proc. IMA–IAGOD Meet. 1970*, IAGOD Vol. – *Soc. Min. Geol. Jap.*, Spec. Issue, 3: 327–334.

Roedder, E., 1972. Composition of fluid inclusions. Chapter JJ. In M. Fleischer (Editor), *Data of Geochemistry*, 6th ed. *U.S. Geol. Surv., Prof. Pap.*, 440JJ, 164 pp.

Roedder, E. and Coombs, D.S., 1967. Immiscibility in granitic melts, indicated by fluid inclusions in ejected granitic blocks from Ascension Island. *J. Petrol.*, 8 (3): 417–451.

Roedder, E., and Skinner, B.J., 1968. Experimental evidence that fluid inclusions do not leak. *Econ. Geol.*, 63: 715–730.

Roedder, E., and Weiblen, P.W., 1970. Lunar petrology of silicate melt inclusions, Apollo 11 rocks. *Proc. Apollo 11 Lunar Sci. Conf. – Geochim. Cosmochim. Acta*, Suppl. 1 (1): 801–837.

Roedder, E., Ingram, B. and Hall, W.E., 1963. Studies of fluid inclusions. III: Extraction and quantitative analysis of inclusions in the milligram range. *Econ. Geol.*, 58: 353–374.

Roedder, E., Heyl, A.V. and Creel, J.P., 1968. Environment of ore deposition at the Mex–Tex deposits, Hansonburg district, New Mexico, from studies of fluid inclusions. *Econ. Geol.*, 63: 336–348.

Saager, R., 1967. New techniques for polishing ore minerals. *Univ. Witwatersrand Econ. Geol. Res. Unit Info., Circ.*, 39: 13 pp. plus 5 plates.

Sakai, H. and Matsubaya, O., 1973. Isotope geochemistry of the thermal waters of Japan and its implication to Kuroko ore deposits. *Geol. Soc. Am., Abstr. Progr.*, 5 (7): 792.

Sato, T., 1968. Ore deposit and mechanism of its formation of Uchinotai Western ore body, Kosaka mine, Akita Prefecture, Japan. *J. Min. Geol. Jap.*, 18 (5): 241–256 (in Japanese, with English abstract).

Sato, T., 1970. Physiochemical environments of "Kuroko" mineralization at Uchinotai deposit of Kosaka mine, Akita prefecture, In: *Collected Abstracts, IMA–IAGOD Meetings '70: Tokyo*. Science Council of Japan, Tokyo, p. 143.

Sawkins, F.J., 1966. Ore genesis in the North Pennine orefield, in the light of fluid inclusion studies. *Econ. Geol.*, 61: 385–401.

Sawkins, F.J., 1968. The significance of Na/K and Cl/SO₄ ratios in fluid inclusions and subsurface water, with respect to the genesis of Mississippi Valley-type ore deposits. *Econ. Geol.*, 63: 935–942.

Schmidt, R.A., 1962. Temperatures of mineral formation in the Miami–Picher district as indicated by liquid inclusions. *Econ. Geol.*, 57: 1–20.

Sivoronov, A.A., 1968. The use of included solutions in study of metamorphic rocks in deposits of iron ores. *Mineral. Thermom. Barom.*, 1: 181–184 (in Russian; English abstract in: *Fluid Inclusion Res. – Proc. of COFFI*, 1, 19, 1968).

Skinner, B.J., 1953. Some considerations regarding liquid inclusions as geologic thermometers. *Econ. Geol.*, 48: 541–550.

Skinner, B.J. and Barton Jr., P.B., 1973. Genesis of mineral deposits. *Ann. Rev. Earth Planet. Sci.*, 1: 183–211.

Sorby, H.C., 1858. On the microscopic structure of crystals, indicating the origin of minerals and rocks. *Q. J. Geol. Soc. Lond.*, 14 (1): 453–500.

Stoiber, R.E., 1946. Movement of mineralizing solutions in the Picher field, Oklahoma–Kansas. *Econ. Geol.*, 41: 800–812.

Takenouchi, S. and Imai, H., 1968. On the salinity of liquid inclusions in quartz crystals. *J. Geochem. Soc. Jap.*, 2 (1): 41–42 (in Japanese; English abstract in *Fluid Inclusion Res. – Proc. COFFI*, 2, 73, 1969).

Tatsumi, Tatsuo and Watanabe, Takeo, 1971. Geological environment of formation of the Kuroko-type deposits. *Proc. IMA–IAGOD Meet. 1970*, IAGOD Vol. – *Soc. Min. Geol. Jap.*, Spec. Issue, 3: 216–220.

Tokunaga, M., 1968. The formation temperature of the Uchinotai deposits, Kosaka mine. *Min. Geol., Jap.*, 18: 54 (in Japanese).

Tokunaga, M., Miyazawa, T., Homma, H. and Park, H., 1970. Formation temperature of some black ore deposits in Akita Prefecture, Japan. In: *Collected Abstracts, IMA–IAGOD Meetings 1970: Tokyo*. Science Council of Japan, Tokyo, p. 246.

Veitch, J.D. and McLeroy, D.G., 1972. Organic mobilization of ore metals in low-temperature carbonate environments. *Geol. Soc. Am., Abstr. Progr.*, 4 (7): 110–111.

White, D.E., 1957. Magmatic, connate, and metamorphic waters. *Geol. Soc. Am. Bull.*, 68: 1659–1682.

White, D.E., 1967. Mercury and base-metal deposits with associated thermal and mineral waters. In: H.L. Barnes (Editor), *Geochemistry of Hydrothermal Ore Deposits*. Holt, Rinehart and Winston, New York, N.Y., pp. 575–631.

White, W.S., 1971. A paleohydrologic model for mineralization of the White Pine copper deposit, Northern Michigan. *Econ. Geol.*, 66: 1–13.

Chapter 5

THE SIGNIFICANCE OF ORGANIC MATTER IN ORE GENESIS

J.D. SAXBY

INTRODUCTION

Processes involving organic material have often been suggested as contributing to the formation of a variety of ore bodies (Dunham, 1961, 1971). Unfortunately these mechanisms are often not well defined or understood. This has led Krauskopf (1955) to suggest that organic matter is often the "principal stumbling block" to reconstructing the environment of metal deposition.

Observed or hypothetical associations between ores and organic carbon have been reported for many metals including: copper (Haranczyk, 1961; Barghoorn et al., 1965; Yui, 1966; Hamilton, 1967; Szczepkowska-Mamczarczyk, 1971; Tokarska, 1971; Wedepohl, 1971; Bartholomé et al., 1973; Van Eden, 1974), silver, lead and zinc (Metsger et al., 1958; Frolovskaya et al., 1966; Croxford and Jephcott, 1972; Asanaliyev, 1973; Pering, 1973), uranium and thorium (Gruner, 1956; Breger and Deul, 1959; Granger et al., 1961; Kornfeld, 1964; Schmidt-Collerus, 1969; Calvo, 1971; Haji-Vassiliou and Kerr, 1972, 1973), gold (Merwin, 1968; Ong and Swanson, 1969; Baskakova, 1970; Radtke and Scheiner, 1970), iron (Lepp and Goldich, 1964; Cloud et al., 1965; Becker and Clayton, 1972), and manganese (Doyen, 1973). Ore genesis usually requires abnormal accumulations of a particular ion, together with a suitable anion, and mechanisms involving both living and dead organic matter have been postulated to achieve the required concentrations. Even after initial deposition or emplacement of the ore body, organic material can contribute significantly to the course of subsequent diagenesis or metamorphism (Taylor, 1971; Vassoevich, 1971).

In this chapter the role of carbonaceous material in ore genesis is briefly surveyed, with particular reference to metal-organic compounds. Data are also presented on organic matter isolated from a range of metalliferous deposits. Organic processes leading to deposits of metal carbonates and phosphates (Degens, 1965; Trudinger, this Vol.) have not been included in the discussion which is mainly concerned with sulphide mineralization. As well as being of geological and geochemical interest, the presence of organic matter is of importance in minerals exploration, particularly where interference is caused to geophysical techniques such as induced polarization (Schrage, 1956; Liddy, 1974) and in ore treatment when flotation efficiency may be adversely affected by carbonaceous material (Sullivan et al., 1969; Lyon and Fewings, 1971).

CONCENTRATION OF METALS

Consideration of the approximate average metal content of various sediments (Table I) suggests that certain elements tend to occur preferentially in organic-rich rocks. Metals which show the greatest tendency to be concentrated in bituminous shales or enriched in organic fractions are Ag, As, Co, Cu, Ge, Mn, Mo, Ni, Pb, Sn, U, V and Zn (Krauskopf, 1955; Degens et al., 1957; Brongersma-Sanders, 1970). In general, such an association of metals and carbonaceous materials can be attributed to metal-organic compounds or to adsorption and precipitation mechanisms during sedimentation (Ahrens, 1966; Saxby, 1969).

Metal-organic compounds

Living organisms. Biological studies have shown that metals bonded to certain organic ligands are present in all living matter and unusual accumulations are possible (Vinogradov, 1963; Peterson, 1971). Boychenko et al. (1968) have stated that the total content of Co, Cu, Mn, Mo, Ni and Zn in plants alone is greater than that in ore bodies. Microorganisms can concentrate elements from sea water with concentration factors up to 10^6 (Trudinger and Bubela, 1967).

When organisms die, at least some of these metal-organic compounds can resist the action of bacteria and give rise to an association of biogenically derived metals and carbonaceous matter in the consolidated sediment. In biological systems the nature of the metal-organic interaction is often uncertain, although predictions can be made on the basis of stabilities of known complexes. The preference of metals for certain coordinating groups is shown in the following list, in which metals are arranged in approximately decreasing order of biological occurrence (Bowen, 1966; Williams, 1967):

Fe :   porphyrin, imidazole, $-NH_2$, $=NH$, R_2S, $-S^-$, $-COO^-$, $-O^-$, $=PO_4^-$
Mg: $-COO^-$, $=PO_4^-$, porphyrin, imidazole
Ca: $-COO^-$, $=PO_4^-$, porphyrin, imidazole
Cu: $-NH_2$, $=NH$, R_2S, $-S^-$
Zn: $-NH_2$, $=NH$, R_2S, $-S^-$
Mn: $-COO^-$, $=PO_4^-$, imidazole
Co: $-NH_2$, $=NH$, corrin, benzimidazole, carbanion of a sugar
Mo: $-NH_2$, $=NH$, R_2S, $-S^-$
Cd: R_2S, $-S^-$.
Cr: $-COO^-$, $=PO_4^-$
Ni: not known
V :   not known

In general, carboxylate groupings form their strongest bonds with "hard" metal ions

TABLE I

Approximate concentrations of common ore-forming metals in sediments and organisms (p.p.m.)*

Metal	Earth's crust	Shales	Sand-stones	Lime-stones	Coals	Soils	Plants	Animals	Decomposers
Ag	0.07	0.07	0.05	0.05	<0.5	0.1	0.06	0.006	0.15
As	1.8	13	1	1	3	6	0.2	0.2	
Au	0.004	0.005	0.005	0.005	≤0.125		<0.00045	<0.009	
Bi	0.17	1	0.3		1		0.06		4
Cd	0.2	0.3	0.05	0.035	0.25	0.06	0.64	0.3	0.5
Co	25	19	0.3	0.1	4	8	0.48	<0.3	1.5
Cr	100	90	35	11	6	100	0.23		
Cu	55	45	5	4	15	20	14	2.4	15
Fe	56300	47200	9800	3800		38000	140	160	130
Ge	5.4	1.6	0.8	0.2	6	1			
Hg	0.08	0.4	0.03	0.04		0.03	0.015	0.05	25
Mn	950	850	50	1100	150	850	630	0.2	
Mo	1.5	2.6	0.2	0.4	1.5	2	0.9	<1	1.5
Ni	75	68	2	20	15	40	27	<1	1.5
Pb	12.5	20	7	9	10	10	2.7	4	50
Sb	0.2	1.5	0.05	0.2	<30	5	0.06	0.14	
Sn	2	6	0.5	0.5	<3	10	<0.3	<0.16	5
Ti	5700	4600	1500	400	900	5000	1	<0.7	
U	2.7	3.7	0.45	2.2	0.5	1	0.038	0.023	0.25
V	135	130	20	20	20	100	1.6	<0.4	0.67
Zn	70	95	16	20	<100	50	160	160	150
Zr	165	160	220	19	100	300	0.64		5

*Values are taken (or estimated) from Clark and Swain (1962), and Bowen (1966).

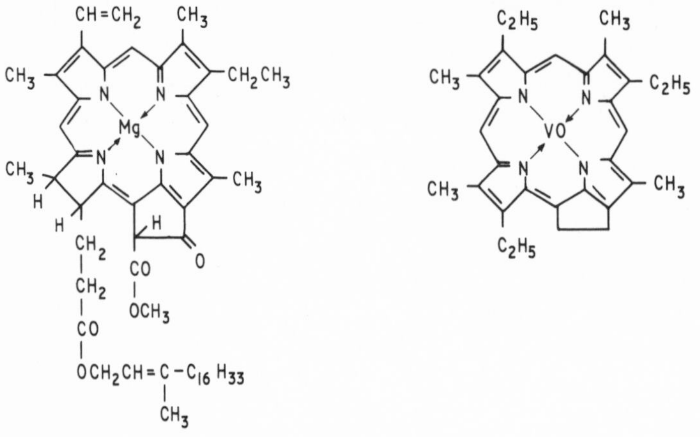

Fig. 1. A. Chlorophyll-a (a common green plant pigment). B. Vanadyl deoxophylloerythroetio-porphyrin (a common metal-porphyrin complex in petroleum).

such as Ca^{2+} and Mg^{2+}, nitrogen atoms favour "hard/soft" metal ions such as Fe^{2+}, Co^{2+}, Cu^{2+} and Zn^{2+}, while sulphur ligands prefer "soft" metal ions like Cu^+ (Livingstone, 1965). A complexed metal may sometimes assist in the coupling of different ligands, such as porphyrins and amino acids (Hodgson et al., 1970).

During diagenesis of the sediment most metal-organic complexes will decompose. If organic matter is lost as carbon dioxide, volatile hydrocarbons or soluble organic compounds, the metal so released will be even more concentrated in the sediment than in the original living material. Experimental data on diagenesis of metal-amino acid complexes is given later in this chapter. The great stability of the porphyrin nucleus prevents decomposition at normal sediment temperatures and hence porphyrins can be extracted from a wide range of carbonaceous rocks (Saxby, 1969). Usually, however, the metal originally present has been replaced by either vanadium or nickel to give a more stable complex. In the case of chlorophyll, various ring substituents are lost or hydrogenated and the magnesium atom is replaced (Fig. 1) but the resonating ring is stable until extreme conditions are encountered (e.g., temperatures $> 500°C$).

Complex formation. Metals dissolved in natural waters can be concentrated by reaction with suitable organic ligands in the sedimentary basin. The resulting metal-organic compounds may be incorporated directly into the sedimentary pile or be adsorbed on clays and other minerals which have a large surface area (Weiss, 1969). A wide range of organic molecules, which are capable of binding metals through coordinate bonds involving nitrogen, oxygen and sulphur atoms, are present in various waters. These compounds include amino acids, other organic acids and certain alcohols, as well as heterocyclic compounds containing nitrogen and sulphur. The presence of metal-organic compounds in hydrothermal ore-forming solutions has also been proposed and supporting evidence comes from fluid inclusion studies (Kranz, 1968). Warm fluids generated within a sedimentary basin are particularly important in this regard. Clearly, however, at the temperatures usually suggested for igneous hydrothermal solutions only a limited number of complexes are sufficiently stable to persist for any appreciable time.

Another mechanism for the formation of metal complexes in solution involves selective attack on a particular mineral by percolating waters containing a specific organic ligand (Baker, 1973). Ligands which can participate in such solubilization or weathering mechanisms include salicylic acid, 2-ketogluconic acid and adenosinetriphosphate. Ong et al. (1970) have found that natural organic acids ("humic acids") are effective agents in the weathering and transport of metals such as Cu, Al, Fe, Zn and Pb. Rachid (1972) and Rachid and Leonard (1973) have examined the complexing role of amino acids in humic materials and found that dissolution of large quantities of metals (up to 682 mg/g of organic matter) from their insoluble salts can be achieved. Once a metal-organic complex has been formed and held in a sedimentary basin, it will undergo diagenesis in the same way as similar compounds introduced directly from living sources.

Physical adsorption

The attractive interaction of metallic ions with soluble, colloidal or particulate organic material may range from weak forces — which leave the ion easily replaceable (physical adsorption) — to strong forces — which are indistinguishable from chemical bonds (chemisorption) and lead to metal-organic compounds of the type just discussed. Although physical adsorption may be less specific than chemisorption in binding a particular metal, it is possible to concentrate large quantities of metals on certain naturally-occurring organic materials. Ferguson and Bubela (1974) have shown that green filamentous algae react with solutions of copper, lead and zinc by both metal complex formation in solution and physical adsorption on particulate organic matter. High enrichment factors are obtained.

With regard to both metal-organic compounds and metals adsorbed physically on organic matter, it is important to know whether the concentration of metal so achieved in a sedimentary basin is sufficient to give rise to an ore body. Clearly, if some type of post-depositional concentration is invoked, economic deposits can readily be achieved (Roberts, 1967). However, in the absence of such a secondary mechanism, a clearcut answer cannot be given.

On the evidence available, it seems unlikely that primary processes could concentrate sufficient quantities of metals such as copper, lead and zinc from present-day sea water (Krauskopf, 1971). However, since ore bodies are not "normal" occurrences, under exceptionally favourable conditions one cannot exclude the possibility of an organism, organic ligand or organic adsorbent selectively concentrating a metal such as copper in the proportions necessary to lead ultimately to a sedimentary ore body (Boychenko et al., 1968). It is perhaps significant that copper concentrations of 3—10% in a peat from southeastern New Brunswick have been attributed to chelation of copper (Fraser, 1961). This peat contains 60–80% organic matter having an elemental composition of 59% carbon, 6% hydrogen, 33% oxygen and 2% nitrogen.

Chemical precipitation

A third mechanism of metal concentration during sedimentation involves reaction of carbonaceous material with metals or metal complexes dissolved in percolating ground waters or hydrothermal fluids. In particular, if the metallic ion can be reduced and precipitated as an insoluble species, an association of metal with carbonaceous material will be achieved, at least until all the latter has been oxidized and hence solubilized. If a large quantity of reducible soluble species encounters an organic "indicator" bed over a period of time, all the carbonaceous material in the bed may be consumed by oxidation to carbon dioxide. The initial association of ore with carbonaceous material will thus be eliminated.

This type of process can account for many carbonaceous uranium deposits, where it has been suggested that soluble U(VI) complexes in ground waters are precipitated as a result of reduction to U(IV) by carbonaceous shales. Alternatively, where low-rank humic material is present, uranium may be bound as uranium humates, such as $[M_{2-2a}^{I}(UO_2)_a]_x[H_{1-x}(H_m)]_2$ (Martin and Bessot, 1966). Brief details on organic matter from some uranium deposits are given later in this chapter.

Similarly, reduction of soluble Au(I) or Au(III) complexes to Au(O) by carbonaceous beds may account for some types of gold mineralization. Ong and Swanson (1969) favour the transportation of gold as colloids, rendered hydrophilic by organic molecules, rather than as organo-gold chelate complexes. When such colloids encounter organic matter at low pH, colloidal gold is precipitated in intimate association with the carbonaceous matter. Deposition of gold from colloids, chloro-complexes or metal-organic compounds is potentially relevant to many areas including northeastern Nevada and the Witwatersrand in South Africa. (Editor's note: See Chapters 1 and 2 by Pretorius in Vol. 7.)

GENERATION OF SULPHIDE

Not only can organic matter contribute to the concentration of metals but its role in generating the sulphide ions, which are necessary for the formation of sedimentary sulphide ore bodies, is often more important. Even for metals which do not form sulphide deposits (such as uranium), sulphide ions generated by organic processes may be responsible for reduction and precipitation (Eargle and Weeks, 1973).

It is well known that at low temperatures sulphate-reducing bacteria can produce sufficient hydrogen sulphide for the formation of an ore body from sea water sulphate (Trudinger's Ch. 6). ZoBell (1963) has discussed limits for the existence of sulphate-reducing bacteria. Typical conditions for growth include Eh (+ 350 to −500 mV), pH (4.2−10.4), pressure (1−1000 atm), temperature (0−100°C), salinity (<1−30% NaCl) and pore spaces (> 5 μm in diameter), together with the absence of toxic substances and the presence of an energy source and trace elements (Ca, Mg, K, Fe, P, Cl, N). Consideration of the rates of sulphate reduction, organic matter production and ore deposition places limits on the feasibility of the formation of completely biogenic sulphide ores (Trudinger et al., 1972; Rickard, 1973).

Where bacterial sulphate reduction has occurred, remains of the organisms themselves and also of any organic energy source will, in all probability, be incorporated in the ore body. An initial, and possibly continuing, association of metal sulphide and organic matter will result (Oppenheimer, 1960). Of particular significance in many ore bodies and pyrite-rich sediments are the organic films which occur around and within pyrite framboids (Dunham, 1961; Love, 1969; Croxford and Jephcott, 1972; Sweeney and Kaplan, 1973).

In discussing Mississippi Valley-type ores, Jackson and Beales (1967) favour the generation of sulphide by sulphate-reducing bacteria in the presence of oil or bituminous sediments. On the other hand, in a model requiring higher temperatures, Barton (1967) has suggested that ore fluids, containing metal chloride complexes and metastable sulphate, encountered methane or associated carbonaceous material. A simplified reaction for the ensuing nonbiogenic reduction of sulphate to sulphide might be:

$$CH_4 + ZnCl_2 + SO_4{}^{2-} + Mg^{2+} + 3CaCO_3 \rightarrow ZnS + CaMg(CO_3)_2 + 2Ca^{2+} + 2Cl^- + H_2O + 2HCO_3{}^-$$

As with the reduction of gold and uranium complexes discussed earlier, organic matter will be completely oxidized to carbon dioxide in an extreme case. Germanov (1965) has found that up to 0.5% organic carbon is removed during hydrothermal sulphate reduction associated with ore deposition.

Organically bound sulphur may also be utilized in metal sulphide formation. The sulphur content of dry, dead organisms varies widely (commonly 0.2−2.0%), as does its distribution between organic and inorganic forms (ZoBell, 1963; Bowen, 1966). Part of the organic sulphur becomes available as hydrogen sulphide during bacterial attack before consolidation of the sediment, while the residue is gradually released during thermo-catalytic breakdown of the buried organic matter (Brooks, 1971). Although on the whole less important than bacterial sulphate reduction, this process can account for metal sulphide-organic matter associations in a subsiding basin where metal ions are available and large volumes of carbonaceous material are subject to increasing temperatures. In a slight variation on this mechanism, Skinner (1967) has proposed that Mississippi Valley-type ores result from hot brines carrying metal chloride complexes causing the release of hydrogen sulphide by thermal degradation of sulphur-containing organic compounds. Where sulphur becomes available from organic matter or other sources after deposition, the resulting metal sulphides may exhibit a variety of secondary textures and structures (Lambert, 1973).

The high content of organic matter in the Kupferschiefer base-metal deposits, as well as other evidence, suggests that bacterial sulphate reduction produced the sulphide necessary for ore formation. Some contribution from organic sulphur is also possible. Brongersma-Sanders (1969) has suggested that the metals in this stratabound deposit were derived initially from ocean water and concentrated by living plankton as metal complexes. However, other theories favour either the leaching of the red-bed country rocks or the presence of hydrothermal springs as a source for inorganic metal ions, which were then further concentrated by organic adsorption and precipitation as sulphides in an extensive euxinic basin (Dunham, 1961; Wedepohl, 1971; Renfro, 1974). Dunham (1961) and Stanton (1972a) have discussed many other deposits, such as the White Pine copper deposits of Michigan, the Rhodesian Copper Belt and Mount Isa in Queensland, where organic matter is relevant to the genesis of ores by submarine-exhalative and similar processes.

STAGES OF SEDIMENTARY ORE FORMATION

It is instructive at this point to consider the change occurring in organic matter which is incorporated in a sedimentary sulphide ore body. Initially this material will have an elementary composition characteristic of the organism concerned or of the biogenic source of the organic particles (Bordovskiy, 1965). In some cases "reworked" carbonaceous material or elemental carbon from other sediments or sources may also be incorporated as detrital grains (Darnell, 1967). Such material complicates the situation, particularly where bulk properties of all the organic matter are determined.

In general, plant or animal remains will pass through a type of coalification or maturation process depending on the properties of the basin and the sedimentary pile. In the case of an euxinic basin, bacterial attack on the carbonaceous material will release gases such as methane, hydrogen sulphide and ammonia, particularly in the top few centimetres. Hydrolysis of proteinaceous material may also take place and, if oxidizing conditions prevail, most types of organic matter will be extensively oxidized to soluble products. The amount of land- or marine-derived organic material remaining in a sediment depends, not only on biological and chemical factors, but also on the relative rates of supply and deposition of both inorganic and organic components (Sackett, 1964).

With increasing temperature and depth of burial, water will be progressively lost from the metal-rich sediment together with carbon dioxide by decarboxylation of organic acids. At the end of this stage, the organic material will chemically resemble a high volatile bituminous coal, if the original material was low in hydrogen (i.e., terrestrial), or oil shale kerogen, if the hydrogen content was high (i.e., marine). Continued subsidence and exposure to temperatures of $\sim 100°C$ or greater will result in thermal cracking of the organic matter possibly catalysed by clay minerals. Methane and/or other volatile hydrocarbons will be progressively released and organic remains, as their colour darkens, will be increasingly difficult to identify microscopically. Along with dehydration of clay minerals, slow recrystallization of metal sulphides may be occurring with the resulting increase in grain size being controlled to some extent by films of carbonaceous matter (particularly in framboidal pyrite). Organic matter in the Kupferschiefer, which is at this bituminous stage of alteration (Taylor, 1971), still exhibits recognizable plant structures. Germanov and Bannikova (1972) and Vershkovskaya et al. (1972) have used changes in aromatic hydrocarbons extracted from pyrite-, antimony- and mercury-rich rocks as an indication of thermal alteration. Hodgson et al. (1968) have used the concentration of compounds such as nickel and vanadyl porphyrins in soils, sediments and sedimentary rocks as an indicator of geochemical history.

As diagenesis merges into metamorphism the organic matter passes through an anthracite-like stage. Material of this type called shungite from the Precambrian of Karelia is believed to have formed from algal remains (Vologdin, 1970), although this is not certain (Krauskopf, 1955). If anhydrous reducing conditions prevail, loss of hydrogen (and to a lesser extent carbon) as methane will continue until fine graphite crystals are formed. The methane liberated at this point is sometimes encountered during exploratory drilling (Izawa, 1968; Taylor, 1971). Under more extreme metamorphic conditions of temperature and shear, mineralized coarse-grained graphitic slates and schists are produced. The intensity of graphitization is an indication of the conditions to which the coexisting mineralization has been subjected (Kromskaya, 1970).

Fig. 2 is a diagrammatic representation of the changes undergone by various coal macerals using atomic H/C and O/C ratios as the diagnostic parameters. In figures such as this, directions corresponding to loss of methane, carbon dioxide and water are dependent on the particular point in the diagram being considered. Macerals are more or less homogeneous microscopic constituents and progressive alterations in composition and other chemical and physical properties are observed with increasing diagenesis. Similar diagenesis-metamorphism paths might be expected for the relatively minor amount of organic matter in sedimentary ores.

When a carbonaceous sedimentary formation is intruded by an igneous body, progressive graphitization may be observed in the metamorphic aureole. Using X-ray diffraction, French (1964) has detected a trend from amorphous organic matter to well-crystallized graphite in samples within a few miles of the Duluth gabbro complex, which intrudes the Biwabik iron formation in northern Minnesota. A complete range of carbonaceous materials has also been detected in hydrothermal deposits of the Khibiny pluton, although in this case an inorganic origin has been suggested (Zezin and Sokolova, 1968).

Carbonaceous matter can exert a great influence on the pH and Eh of aqueous systems at temperatures above 200°C and participate in reactions which change the mineralogy of ores (Miyashiro, 1964; French, 1966; Yui, 1968; Taylor, 1971). Under metamorphic conditions graphite is not stable in the presence of pure water but can coexist with a gas phase rich in carbon dioxide. Thus, under oxidizing or aqueous conditions at high temperatures, graphite may be effectively eliminated from an ore, particularly if carbon dioxide

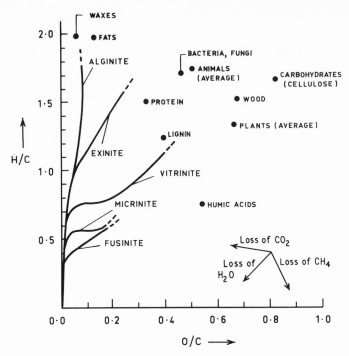

Fig. 2. Atomic H/C versus O/C diagram showing average development lines of coal macerals.

can be removed either as a gas or as a carbonate. Two of the many possible reactions are:

$$2H_2O + 2C \rightleftharpoons CH_4 + CO_2$$
$$2H_2O + C \rightleftharpoons 2H_2 + CO_2$$

Under these conditions the initial association of metals with organic carbon is either partially or completely destroyed. It is thus conceivable that at least some metamorphic ore bodies, which now show no evidence of organic carbon, were originally formed with carbonaceous matter playing a major role.

LABORATORY STUDIES

The ideas. summarized in the previous sections are of little value unless they are consistent with geological, chemical and biological data obtained from field and laboratory investigations. In particular, application of the techniques used by organic geochemists can contribute significantly in this area. The results of two types of experimental study recently carried out in the Minerals Research Laboratories of CSIRO illustrate this point.

Artificial diagenesis

The amino-acid cysteine and its oxidized form (cystine) are commonly found in geological systems and often constitute a major part of the organic sulphur in living organisms (Trudinger's Ch. 6). Through sulphur, nitrogen and oxygen functional groups, cysteine and cystine can coordinate with a wide range of metals (McAuliffe and Murray, 1972) and it is certain that some of these complexes have been incorporated in sediments by the mechanisms outlined earlier. Examples are given in Fig. 3. Hence the question arises as to the fate of such metal-organic compounds during subsequent diagenesis and metamorphism.

It is difficult to simulate the compositional and catalytic properties of a sediment and impossible to reproduce in a laboratory experiment the temperature/time relationships occurring naturally. However, by heating stoichiometric metal-cystine complexes at various temperatures under reducing conditions in sealed tubes, an indication is obtained on a molecular level of the reactions to be expected at lower temperatures in a sedimentary basin (Saxby, 1973). In a typical series of experiments, complexes of cystine (L = cystine $-2H = C_6H_{10}N_2S_2O_4$) were heated for 100 h at $200°C$ in stainless steel tubes and the following metal sulphides were readily detected by X-ray diffraction of the solid residue:

$FeL . \frac{5}{2}H_2O$	→	pyrite, pyrrhotite
$Fe_2L_3 . 6H_2O$	→	pyrite, marcasite
$CuL . H_2O$	→	covellite
ZnL	→	sphalerite
PbL	→	galena
$NiL . H_2O$	→	vaesite
$FeL . \frac{5}{2}H_2O + CuL . H_2O$	→	chalcopyrite, pyrite
$FeL . \frac{5}{2}H_2O + ZnL$	→	pyrite, sphalerite, marcasite
$FeL . \frac{5}{2}H_2O + PbL$	→	pyrite, galena
$FeL . \frac{5}{2}H_2O + NiL . H_2O$	→	pyrite, bravoite

It is noteworthy that metal sulphides are also formed when cystine is heated in the presence of metal salts either in the presence or absence of glycerol (Weil et al., 1954; Lambert, 1973).

These results, together with a consideration of the kinetic parameters applicable over long time intervals, suggest that metal sulphides can be formed in sediments at low temperatures ($< 100°C$) from the atoms already present in metal complexes of sulphur-containing ligands. If there is no sulphur in the ligand, a carbonate mineral may form or the released metal may migrate in the pore fluids. In the absence of coordinated metals, sulphur-containing organic matter will release hydrogen sulphide to the surrounding strata on thermal degradation.

In these experiments gaseous and oily products are formed as well as an intimate

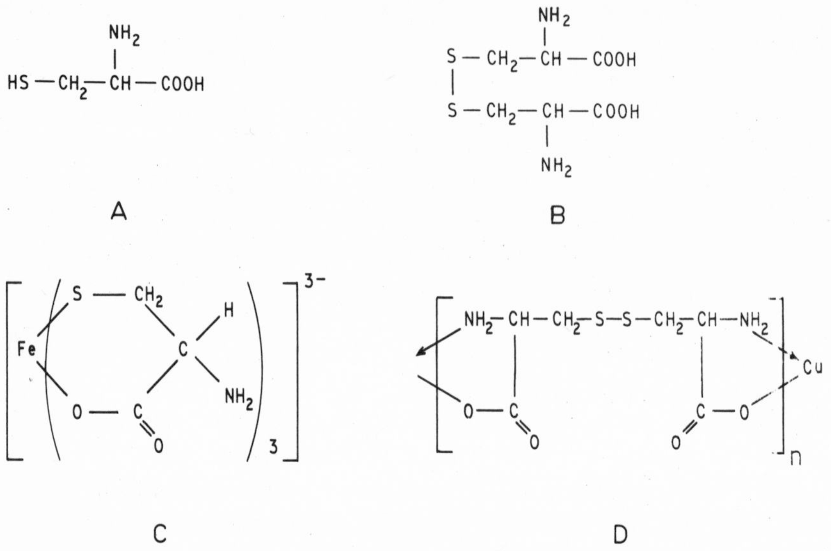

Fig. 3. A. Cysteine. B. Cystine. C. Iron(III)-cysteine complex ion. D. Copper(II)-cystine complex.

mixture of metal sulphides and insoluble organic matter. Data on the products formed from the iron(II)-cystine complex after 100 h at different temperatures are given in Table II. It is noteworthy that only a small proportion of the original organic carbon remains in the solid product. This leads to relatively high ratios of iron sulphides to organic matter, as in some ore-bearing sedimentary rocks. The rank of this organic matter, as revealed by H/C ratios, increases with temperature. This progressive increase in diagenesis and metamorphism is consistent with the theoretical considerations already given and the following data on actual ore bodies.

TABLE II

Artificial diagenesis of iron(II)-cystine complex

Temperature ($^\circ$C)	Gaseous product*	Oily product*	Insoluble product*		Insoluble organic matter	
			Fe-S compounds**	organic matter	atomic H/C	atomic N/C
100	20	38	38	3	2.4	0.3
200	41	25	34	<1	2.0	0.1
300	37	17	46	1	1.3	0.2
400	60	2	33	5	0.8	0.1

* Percentage of original metal-organic complex; ** mainly pyrite and pyrrhotite.

Isolation of organic matter in ores

Since the level of organic carbon is low (often < 1%) even in relatively carbonaceous ores, complete separation and characterization of such material can be a considerable problem. As a first step, soluble compounds may be extracted from crushed ores with common organic solvents such as chloroform or benzene/methanol. The distribution of such compounds (for example, normal alkanes) may be related to factors such as the source of the organic matter, diagenesis and selective bacterial alteration (Brooks, 1971; Pering, 1973). Recent contamination may obscure the situation, particularly if the extract is small.

Subsequent isolation of the insoluble organic material (kerogen) from the extracted ore depends on a somewhat tedious demineralizing procedure (Saxby, 1970). Treatment with dilute hydrochloric acid removes most carbonates and oxides, a mixture of hydrochloric and hydrofluoric acids dissolves most silicates, while HCl-resistant sulphides, particularly pyrite, are best removed by reaction with lithium aluminium hydride. Repeated treatments with these reagents are usually necessary to give kerogen containing a minimum of mineral matter. The exact procedure and number of treatments required vary with the physical properties and mineralogy of each ore. However, extreme conditions of temperature or reagent concentration, which are liable to alter the organic matter, should always be avoided. Physical methods rarely give quantitative yields of kerogen from ores, while oxidizing reagents such as nitric acid have limited use due to oxidation of organic functional groups during pyrite dissolution (Izawa, 1968).

Using this method of demineralization, kerogen has been recovered from various Australian ores and shales and from recent Red Sea sediments. Table III gives microanalytical data on the products remaining when no further mineral matter could be removed. The content of carbonaceous matter in most ores is highly variable. Values of organic carbon given in Table III are for a single sample and in most cases are higher than the average for the total ore. Fig. 4 shows how atomic ratios for representative kerogens are distributed on an expanded H/C versus O/C diagram. Many techniques such as functional group analysis, degradation by oxidation, reduction, hydrolysis or pyrolysis, electron spin resonance, X-ray diffraction, stable isotope measurement and infrared spectroscopy, as well as microscopic studies, can be used to characterize kerogen (Saxby, 1976). These and other physical and morphological properties are not considered in detail here.

Red Sea sediments. Core samples rich in zinc and copper sulphides, which have been recovered from the hot-brine area of the Red Sea, contain up to 1% organic matter (Saxby, 1972). Kerogens isolated from sediments from the Atlantis II Deep, the Discovery Deep and an apparently unmineralized area, have compositions typical of low-rank material (see Table III and Fig. 4). The nature and amount of kerogen in the Atlantis II Deep is consistent with the ore minerals being formed from metal- and sulphide-rich brines, which are injected periodically into the Deep from below. An apparent decrease

TABLE III

Carbonaceous matter isolated from metalliferous ores and associated rocks

Sample and economic metals	Organic carbon (%)	Elemental composition (%; dry, ash-free basis)					Ash (%)
		C	H	N	O	S	
Metalliferous Red Sea sediment; Cu, Zn	0.5	66.0	6.5	3.1	21.4	3.0	3.1
Julia Creek oil shale; V	15.0	74.6	8.0	2.2	9.8	5.4	2.4
Cobar ore; Cu, Pb, Zn	< 0.1	96.7	1.2	< 0.1	2.1	< 0.1	12.4
McArthur River ore; Pb, Zn	0.4	84.2	5.3	1.3	6.9	2.3	3.8
Mount Isa: Black Star ore; Pb, Zn	0.6	95.0	1.5	< 0.5	1.5	2.0	9.8
Black Rock ore; Pb, Zn	0.9	95.6	1.2	< 0.5	3.0	0.2	11.9
1100 ore; Cu	0.4	92.9	1.5	< 0.5	1.0	4.6	31.8
Hilton ore; Pb, Zn	0.3	93.6	1.8	0.4	2.4	1.8	30.1
Mount Oxide shale; (Cu)	1.5	91.0	1.7	0.2	3.6	3.5	11.7
Broken Hill shale; (Pb, Zn)	1.0	99.9	0.1	< 0.1	< 0.1	< 0.1	7.9
Woodcutters ore; Pb, Zn	1.2	96.9	0.7	0.1	2.3	< 0.1	8.7
Rum Jungle shale; (U)	2.2	98.9	0.1	< 0.1	0.6	0.4	6.3

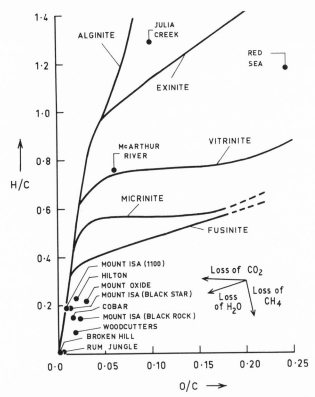

Fig. 4. Atomic H/C versus O/C diagram for kerogens from metalliferous ores and associated rocks.

with depth in H/C ratio of kerogens from Atlantis II (Saxby, 1972) suggests relatively rapid diagenesis of the organic matter at temperatures between 62 and 104°C or even higher (Degens and Ross, 1969; Ross, 1972). In the Discovery Deep and in the unmineralized area, larger amounts of kerogen associated with pyrite are consistent with most sulphide being derived from sulphate-reducing bacteria.

Julia Creek oil shale. This Cretaceous shale in northwest Queensland contains significant amounts of vanadium, which may prove to be recoverable from the spent shale residue after distillation. The mechanism of vanadium concentration is uncertain but, since at least part of the vanadium is organically bound, processes involving metal-organic complexes, either in the original marine organisms or as a result of later adsorption, seem to be involved. The composition of the kerogen is in the range expected for an oil shale, although the organic sulphur content is unusually high. Significantly, vanadium is also one of the metals showing a strong correlation with bituminous carbon in the Kupferschiefer of northwest Germany (Wedepohl, 1971).

Cobar. The copper and copper–lead–zinc ores at Cobar in northern N.S. Wales are located in shear zones of a Silurian carbonaceous slate. Although the organic matter may have influenced ore emplacement, only traces of well-graphitized material remain in the ore bodies. Most of the graphite consists of very fine crystals (thickness $\sim$ 50 Å) but where intense shearing has occurred a greater proportion of larger particles (thickness $\sim$ 400 Å) is present, probably as a result of small crystals coalescing or recrystallizing. Overall it appears that considerable amounts of organic carbon have been lost (as methane, carbon dioxide, etc.) during or after ore formation. In those micro-environments where the loss was not complete, the temperatures, pressures, and shearing effects have been sufficient to form graphite.

McArthur River and Mount Isa. These deposits, in the northeastern corner of the Northern Territory and northwestern Queensland, respectively, are Precambrian conformable base-metal deposits of similar age. Both have been studied extensively and are believed to have formed by syngenetic processes (Croxford and Jephcott, 1972; Stanton, 1972a; Lambert and Scott, 1973). The metals could well have been derived from metalliferous solutions ascending in nearby fault zones, possibly as a result of igneous activity. At the surface, injection of the hot solutions into a body of water could have been followed by syngenetic precipitation of metals. The sulphide may have been derived partly from the injected fluids and partly from bacterial sulphate reduction (Smith and Croxford, 1973). A general association of organic carbon with sulphides and particularly with fine-grained pyrite has been observed, especially for the McArthur River deposit.

Kerogen from the McArthur River ore resembles a high volatile bituminous coal in composition. This fact, together with the unmetamorphosed, fine-grained nature of the ore, suggests that temperatures above $\sim$ 100°C have never been reached. Carbonaceous material from an unmineralized, stratigraphically-equivalent shale has a similar composition (C 83.5%; H 6.4%; N 1.3%; O 7.1%; S 1.7%). Hence the thermal histories of the ore and of the equivalent shale must be virtually identical. Organic cellular structures have been observed microscopically in kerogen from McArthur River ore (Hamilton and Muir, 1974).

The Mount Isa ore bodies are at the greenschist stage of metamorphism, a fact sensitively shown by the anthracitic composition of the kerogen. Partial recrystallization of fine-grained metal sulphides has occurred, resulting in a change in the physical distribution of the organic matter, as well as in its concentration and chemical structure (Saxby and Stephens, 1973). Minor differences between ore bodies are probably present. For example, kerogen rank (based on H/C ratios) appears to decrease slightly in the Hilton and Mount Oxide deposits, which are 18 and 135 km north of Mount Isa, respectively.

Broken Hill. Organic matter is comparatively rare in the Precambrian silver–lead–zinc ores of Broken Hill in western New South Wales (Hamilton, 1965; King, 1969). This fact and the presence of graphite in shales of lower metamorphic grade to the north of the mines have been discussed by Taylor (1971). It is suggested that, where water was available under extreme metamorphic conditions, all solid carbonaceous material was reduced to methane or oxidized to carbon dioxide. The presence of carbonate minerals and occasional outflows of methane at Broken Hill is consistent with this hypothesis. The very high carbon content given in Table III is for a graphitic shale sample from north of Broken Hill.

It has often been suggested that the mineralizations at McArthur River, Mount Isa and Broken Hill represent three stages in the metamorphic history of somewhat similar original metalliferous sediments (Stanton, 1972a; Croxford and Jephcott, 1972). Certainly organic geochemical studies support such a model and provide quantitative data on the degree of alteration. It is tempting to suggest that the ores at present forming in the Red Sea represent the initial stage of the genetic sequence. Although there are similarities, the present water depth in the Red Sea (2000 m) is probably much deeper than that prevailing when the McArthur River and Mount Isa sediments were deposited (Croxford and Jephcott, 1972).

The presence at Broken Hill of thin beds of iron-rich sedimentary rock raises the controversial question of the origin of "banded iron formations" (Stanton, 1972b). Isotopic evidence has indicated that organic activity may have contributed to the derivation, transport and deposition of iron in the banded iron formations of Western Australia (Becker and Clayton, 1972). The absence of organic carbon in the latter deposits, except for traces of graphite in some shales, could be due to reactions (such as $6 Fe_2O_3 + C_{org} \rightarrow 4 Fe_3O_4 + CO_2$) at a temperature of $\sim 300°C$. This temperature is consistent with a burial depth of 14—22 km and geothermal gradient of $\sim 15°C/km$ (Ayres, 1972).

Woodcutters and Rum Jungle. Both the Woodcutters lead—zinc deposit and the Rum Jungle uranium mine are located in the Precambrian Golden Dyke Formation in the north of the Northern Territory. Kerogen from samples of Woodcutters ore and Rum Jungle slate are similarly high in rank and are approaching true graphite in composition (Table III). Roberts (1973) has discussed the genesis of the Woodcutters deposit and suggested that metal-organic complexes, formed from metal ions and the degradation products of algal material, could have been responsible for ore transport. Although shearing has taken place, the "graphite" appears to have formed at temperatures which have never exceeded about 200°C (Roberts, 1973).

As indicated earlier, association of organic matter with uranium (and gold) mineralization is well known, the most famous examples being thucolite from the Witwatersrand System in South Africa (Liebenberg, 1955) and the Colorado Plateau uranium ores (Breger and Deul, 1959). Snyman (1965) has suggested a possible role for organometallic complexes in uranium concentration and given analytical data on the organic matter in various thucolites. On a dry, ash-, sulphur-, and nitrogen-free basis, average values for the South African material are 85—88% C, 4% H, 8—11% O. These figures fall close to the micrinite curve in Fig. 4 but would be consistent with original algal material (alginite curve) that has lost hydrogen (by bombardment with alpha particles) or been oxidized at an appropriate stage of its history.

On the other hand, Colorado Plateau uranium ores are associated with a variety of carbonaceous materials including coalified wood. Most of the analytical data given by Breger and Deul (1959) for the Temple Mountain region in Utah fall near the vitrinite curve with compositions similar to lignite (75–81% C, 4–7% H, 12–19% O, N and S). Organic material richer in hydrogen (82.28% C, 8.13% H, 9.59% O) has been reported for the La Bajada deposit in New Mexico (Haji-Vassiliou and Kerr, 1972). The kerogen from Rum Jungle is much higher in rank than both the South African and Colorado Plateau samples and its genesis pathway cannot be determined.

CONCLUSION

An attempt has been made in this chapter to survey briefly the manner in which organic matter can control or direct the distribution of metals in sedimentary and meta-morphic rocks. Factors involved include the concentration of metals by organisms, the adsorption and precipitation of metal-organic compounds, the role of both living and dead bacteria, the interaction of metal- or sulphur-rich fluids with carbonaceous sediments and the gradual changes in consolidated organic matter caused principally by increasing temperature but also influenced by pressure, shearing forces, catalysis and time. There is a need in the future to escape from the rather vague phrases, which have sometimes been used regarding the interactions of metals and organic material, and to define more precisely the chemical, physical and biological processes occurring.

Laboratory studies, in which diagenesis and metamorphism are artificially simulated, can help to elucidate how associations of metals with organic material are formed or destroyed. Similarly, research involving a variety of techniques is needed to characterize both the soluble organic compounds and the insoluble organic matter in both ores and associated rocks. In constructing models for ore genesis, the presence (or even the absence) of small amounts of carbonaceous matter is a valuable parameter which has often been ignored in the past.

An understanding of the role of organic matter in ore formation requires the incorporation of chemical and biological principles into a geological framework. Extension of this multidisciplinary approach to problems in geochemical prospecting, geophysical exploration and ore processing may follow. It is important to relate these studies to the "mainstream" of organic geochemistry involving coals, oil shale, petroleum and natural gas. Future research may confirm that the chemical and biological reactions, which have given rise to much of the world's energy resources, have also been instrumental in the formation of many metalliferous deposits.

REFERENCES

Ahrens, L.H., 1966. Ionization potentials and metal-amino acid complex formation in the sedimentary cycle. *Geochim. Cosmochim. Acta,* 30: 1111–1119.

Asanaliyev, U., 1973. Prospecting criteria for stratiform lead-zinc mineralization in sedimentary formations (as in Central Tien Shan). *Int. Geol. Rev.,* 15: 1432–1439.

Ayres, D.E., 1972. Genesis of iron-bearing minerals in banded iron formation mesobands in the Dales Gorge Member, Hamersley Group, Western Australia. *Econ. Geol.,* 67: 1214–1233.

Baker, W.E., 1973. The role of humic acids from Tasmanian podzolic soils in mineral degradation and metal mobilization. *Geochim. Cosmochim. Acta,* 37: 269–281.

Barghoorn, E.S., Meinschein, W.G. and Schopf, J.W., 1965. Paleobiology of a Precambrian shale. *Science,* 148: 461–472.

Bartholome, P., Evrard, P., Katekesha, F., Lopez-Ruiz, J. and Ngongo, M., 1973. Diagenetic ore-forming processes at Kamoto, Katanga, Republic of the Congo. In: G.C. Amstutz and A.J. Bernard (Editors), *Ores in Sediments.* Springer, Berlin, pp. 21–41.

Barton, P.B., 1967. Possible role of organic matter in the precipitation of the Mississippi Valley ores. In: J.S. Brown (Editor), *Genesis of Stratiform Lead-Zinc-Barite-Fluorite Deposits in Carbonate Rocks.* Economic Geology Publishing Co., Lancaster, Pa., pp. 371–378.

Baskakova, M.P., 1970. Organic matter in gold-ore quartz veins in the Kyzyl-Kum sands. *Dokl. Akad. Nauk Uzb. S.S.R.,* 27(12): 30–32 (in Russian).

Becker, R.H. and Clayton, R.N., 1972. Carbon-isotopic evidence for the origin of a banded iron-formation in Western Australia. *Geochim. Cosmochim. Acta,* 36: 577–595.

Bordovskiy, O.K., 1965. Accumulation and transformation of organic substances in marine sediments. *Mar. Geol.,* 3: 3–114.

Bowen, H.J.M., 1966. *Trace Elements in Biochemistry.* Academic Press, London, 241 pp.

Boychenko, Ye.A., Sayenko, G.N. and Udel'nova, T.M., 1968. Behaviour of the concentrating action of plants in the biosphere. *Geochem. Int.,* 5: 1036.

Breger, I.A. and Deul, M., 1959. Association of uranium with carbonaceous materials, with special reference to Temple Mountain region. *U.S. Geol. Surv., Prof. Pap.,* 320(12): 139–149.

Brongersma-Sanders, M., 1969. Permian wind and the occurrence of fish and metals in the Kupferschiefer and marl slate. In: C.H. James (Editor), *Sedimentary Ores: Ancient and Modern* (revised) – *Proc. 15th Inter-Univ. Geol. Congr., Univ. Leicester,* pp. 61–71.

Brongersma-Sanders, M., 1970. Origin of trace metal enrichment in bituminous shales. *Advan. Org. Geochem., 1966 – Proc. Int. Congr., 3rd,* pp. 231–236.

Brooks, J.D., 1971. Organic matter in Archaean rocks. *Spec. Publ. Geol. Soc. Aust.,* 3: 413–418.

Calvo, M.M., 1971. Consideraciones en torno a la metalogenia de la asociación urano-orgánica natural. *Energ. Nucl. (Madrid),* 15: 381–387.

Clark, M.C. and Swaine, D.J., 1962. Trace elements in coal. *CSIRO Div. Coal Res., Tech. Commun.,* 45: 115 pp.

Cloud, P.E., Gruner, J.W. and Hagen, H., 1965. Carbonaceous rocks of the Soudan Iron Formation (early Precambrian). *Science,* 148: 1713–1716.

Croxford, N.J.W. and Jephcott, S., 1972. The McArthur lead-zinc-silver deposit, N.T. *Proc. Australasian Inst. Mining Metall.,* No. 243: 1–26.

Darnell, R.M., 1967. Organic detritus in relation to the estuarine ecosystem, *Publ. Am. Assoc. Advan. Sci.,* 83: 376–382.

Degens, E.T., 1965. *Geochemistry of Sediments.* Prentice-Hall, Englewood Cliffs, N.J., 342 pp.

Degens, E.T. and Ross, D.A. (Editors), 1969. *Hot Brines and Recent Heavy Metal Deposits in the Red Sea.* Springer, New York, N.Y., 600 pp.

Degens, E.T., Williams, E.G. and Keith, M.L., 1957. Environmental studies of carboniferous sediments, 1. Geochemical criteria for differentiating marine from fresh-water shales. *Bull. Am. Assoc. Pet. Geol.,* 41: 2427–2455.

Doyen, L., 1973. The manganese deposit of Kisenge-Kamata (western Katanga). Mineralogical and sedimentological aspects of the primary ore. In: G.C. Amstutz and A.J. Bernard (Editors), *Ores in Sediments*. Springer, Berlin, pp. 93–100.

Dunham, K., 1961. Black shale, oil and sulphide ore. *Advan. Sci.*, 18: 284–299.

Dunham, K., 1971. Introductory talk: rock association and genesis. *Soc. Mining Geol. Japan, Spec. Issue*, 3: 167–171.

Eargle, D.H. and Weeks, A.M.D., 1973. Geologic relations among uranium deposits, South Texas, Coastal Plain Region, U.S.A. In: G.C. Amstutz and A.J. Bernard (Editors), *Ores in Sediments*. Springer, Berlin, pp. 101–113.

Ferguson, J. and Bubela, B., 1974. The concentration of Cu(II), Pb(II) and Zn(II) from aqueous solutions by particulate algal matter. *Chem. Geol.*, 13 (3): 163–186.

Fraser, D.C., 1961. Organic sequestration of copper. *Econ. Geol.*, 56: 1063–1078.

French, B.M., 1964. Graphitization of organic matter in a progressively metamorphosed Precambrian iron formation. *Science*, 146: 917–918.

French, B.M., 1966. Some geological implications of equilibrium between graphite and a C-H-O gas phase at high temperatures and pressures. *Rev. Geophys.*, 4: 223–253.

Frolovskaya, V.N., Bocharova, G.I. and Ovchinnikova, L.I., 1966. Carbonaceous matter in ore of the Kurultykan polymetallic deposit. In: *Geology of Ore Deposits (Geologiya Rudnykh Mestorozhdeniy)*. Acad. Sci. U.S.S.R., 8 (3): 31–36 (in Russian).

Germanov, A.I., 1965. Geochemical significance of organic matter in the hydrothermal process. *Geochem. Int.*, 2: 643.

Germanov, A.I. and Bannikova, L.A., 1972. Change in the organic matter of sedimentary rocks in the hydrothermal process of sulfide concentration. *Dokl. Akad. Nauk S.S.S.R.*, 203(5): 1180–1182 (in Russian).

Granger, H.C., Santos, E.S., Dean, B.G. and Moore, F.B., 1961. Sandstone-type uranium deposits at Ambrosia Lake, New Mexico – an interim report. *Econ. Geol.*, 56: 1179–1210.

Gruner, J.W., 1956. Concentration of uranium in sediments by multiple migration-accretion. *Econ. Geol.*, 51: 495–520.

Haji-Vassiliou, A. and Kerr, P.F., 1972. Uranium-organic matter association at La Bajada, New Mexico. *Econ. Geol.*, 67: 41–54.

Haji-Vassiliou, A. and Kerr, P.F., 1973. Analytic data on nature of urano-organic deposits. *Am. Assoc. Pet. Geol., Bull.*, 57: 1291–1296.

Hamilton, L.H., 1965. Concepts of ore genesis applied to the Broken Hill lode, N.S.W. *J. Univ. N.S.W. Mining Geol. Soc.*, 3: 43–66.

Hamilton, L.H. and Muir, M.D., 1974. Precambrian microfossils from the McArthur River lead-zinc-silver deposit Northern Territory, Australia. *Miner. Deposita*, 9: 83–86.

Hamilton, S.K., 1967. Copper mineralization in the upper part of the Copper Harbor Conglomerate at White Pine, Michigan. *Econ. Geol.*, 62: 885–904 (see also 64: 462–464).

Haranczyk, C., 1961. Correlation between organic carbon, copper and silver content in Zechstein copper-bearing shales from the Lubin-Sieroszowice Region (Lower Silesia). *Bull. Acad. Polon. Sci., Sér. Sci. Géol. Géogr.*, 9(4): 183–189.

Hodgson, G.W., Hitchon, B., Taguchi, K., Baker, B.L. and Peake, E., 1968. Geochemistry of porphyrins, chlorins and polycyclic aromatics in soils, sediments and sedimentary rocks. *Geochim. Cosmochim. Acta*, 32: 737–772.

Hodgson, G.W., Holmes, M.A. and Halpern, B., 1970. Biogeochemistry of molecular complexes of amino acids with chlorins and porphyrins. *Geochim. Cosmochim. Acta*, 34: 1107–1119.

Izawa, E., 1968. Carbonaceous matter in some metamorphic rocks in Japan. *J. Geol. Soc. Japan*, 74: 427–432.

Jackson, S.A. and Beales, F.W., 1967. An aspect of sedimentary basin evolution: the concentration of Mississippi Valley-type ores during late stages of diagenesis. *Bull. Can. Pet. Geol.*, 15: 383–433.

King, H.F., 1969. Geological significance of stratiform ore deposits. In: C.H. James (Editor), *Sedimentary Ores: Ancient and Modern* (revised) – *Proc. 15th Inter-Univ. Geol. Congr., Univ. Leicester*, pp. 97–106.

Kornfeld, J.A., 1964. Geochemistry of uranyl oxides in Devonian marine black shales of North America. *Advan. Org. Geochem. 1962 – Proc. Int. Congr., 1st*, pp. 261–262.

Kranz, R.L., 1968. Participation of organic compounds in the transport of ore metals in hydrothermal solution. *Trans. Inst. Mining Metall.*, 77: B26–B36.

Krauskopf, K.B., 1955. Sedimentary deposits of rare metals. *Econ. Geol. 50th Anniv. Vol.*, 411–463.

Krauskopf, K.B., 1971. The source of ore metals. *Geochim. Cosmochim. Acta*, 35: 643–659.

Kromskaya, K.M., 1970. Sulfide mineralization in gabbroic rocks of the Bel'tau massif. *Zap. Uzb. Otdel. Vses. Mineral. Obshch.*, 21: 115–118 (in Russian).

Lambert, I.B., 1973. Post-depositional availability of sulphur and metals and formation of secondary textures and structures in stratiform sedimentary sulphide deposits. *J. Geol. Soc. Aust.*, 20: 205–215.

Lambert, I.B. and Scott, K.M., 1973. Implications of geochemical investigations of sedimentary rocks within and around the McArthur zinc-lead-silver deposit, Northern Territory. *J. Geochem. Explor.*, 2: 307–330.

Lepp, J. and Goldich, S.S., 1964. Origin of Precambrian iron formations. *Econ. Geol.*, 59: 1025–1060.

Liddy, J.C., 1974. Induced polarisation in the search for buried sulphide bodies. *Aust. Mining*, 66: 56–66.

Liebenberg, W.R., 1955. The occurrence and origin of gold and radioactive minerals in the Witwatersrand System, the Dominion Reef, the Ventersdorp Contact Reef and the Black Reef. *Trans. Geol. Surv. S. Afr.*, 58: 101–223.

Livingstone, S.E., 1965. Metal complexes of ligands containing sulphur, selenium, or tellurium as donor atoms. *Q. Rev.*, 19: 386–425.

Love, L.G., 1969. Sulphides of metals in recent sediments. In: C.H. James (Editor), *Sedimentary Ores: Ancient and Modern* (revised) – *Proc. 15th. Inter-Univ. Geol. Congr., Univ. Leicester*, pp. 31–60.

Lyon, G.C. and Fewings, J.H., 1971. Control of the flotation rate of carbonaceous material in Mount Isa chalcopyrite ore. *Proc. Australasian Inst. Mining Metall.*, No. 237: 41–46.

McAuliffe, C.A. and Murray, S.G., 1972. Metal complexes of sulphur-containing amino acids. *Inorg. Chim. Acta*, 6: 103–121.

Martin, J.A. and Bessot, P., 1966. Sur les acides humiques extraits d'une tourbe acide: étude de leurs propriétés, et de leur aptitude à fixer et transporter l'uranium. *Advan. Org. Geochem. 1964–Proc. Int. Congr., 2nd*, pp. 165–172.

Merwin, R.W., 1968. Gold resources in the oxidized ores and carbonaceous material in the sedimentary beds of northeastern Nevada. *U.S. Bur. Mines, Tech. Progress Rep.*, March.

Metsger, R.W., Tennant, C.B. and Rodda, J.L., 1958. Geochemistry of the Sterling Hill zinc deposit, Sussex County, New Jersey. *Bull. Geol. Soc. Am.*, 69: 775–788.

Miyashiro, A., 1964. Oxidation and reduction in the earth's crust with special reference to the role of graphite. *Geochim. Cosmochim. Acta*, 28: 717–729.

Ong, H.L. and Swanson, V.E., 1969. Natural organic acids in the transportation, deposition and concentration of gold. *Q. Colo. School Mines*, 64: 395–425.

Ong, H.L., Swanson, V.E. and Bisque, R.E., 1970. Natural organic acids as agents of chemical weathering. *U.S. Geol. Surv., Prof. Pap.*, 700-C: C130–C137.

Oppenheimer, C.H., 1960. Bacterial activity in sediments of shallow marine bays. *Geochim. Cosmochim. Acta*, 19: 244–260.

Pering, K.L., 1973. Bitumens associated with lead, zinc and fluorite ore minerals in North Derbyshire, England. *Geochim. Cosmochim. Acta*, 37: 401–417.

Peterson, P.J., 1971. Unusual accumulations of elements by plants and animals. *Sci. Progr. Oxford*, 59: 505–526.

Rachid, M.A., 1972. Amino acids associated with marine sediments and humic compounds and their role in solubility and complexing of metals. *24th Int. Geol. Congr.*, Sect. 10: 346–353.

Rachid, M.A. and Leonard, J.D., 1973. Modifications in the solubility and precipitation behavior of various metals as a result of their interaction with sedimentary humic acid. *Chem. Geol.*, 11: 89–97.

Radtke, A.S. and Scheiner, B.J., 1970. Studies of hydrothermal gold deposition, 1. Carlin Gold Deposit, Nevada: the role of carbonaceous materials in gold deposition. *Econ. Geol.*, 65: 87–102.

Renfro, A.R., 1974. Genesis of evaporite-associated stratiform metalliferous deposits – a sabkha process. *Econ. Geol.*, 69: 33–45.

Rickard, D.T., 1973. Limiting conditions for synsedimentary sulfide ore formation. *Econ. Geol.*, 68: 605–617.

Roberts, W.M.B., 1967. Sulphide synthesis and ore genesis. *Miner. Deposita*, 2: 188–199.

Roberts, W.M.B., 1973. Dolomitization and the genesis of the Woodcutters lead-zinc prospect, Northern Territory, Australia. *Miner. Deposita*, 8: 35–56.

Ross, D.A., 1972. Red Sea hot brine area: revisited. *Science*, 175: 1455–1457.

Sackett, W.M., 1964. The depositional history and isotopic organic carbon composition of marine sediments. *Mar. Geol.*, 2: 173–185.

Saxby, J.D., 1969. Metal-organic chemistry of the geochemical cycle. *Rev. Pure Appl. Chem.*, 19: 131–150.

Saxby, J.D., 1970. Technique for the isolation of kerogen from sulfide ores. *Geochim. Cosmochim. Acta*, 34: 1317–1326.

Saxby, J.D., 1972. Organic matter in Red Sea sediments. *Chem. Geol.*, 9: 233–240.

Saxby, J.D., 1973. Diagenesis of metal-organic complexes in sediments: formation of metal sulphides from cystine complexes. *Chem. Geol.*, 12: 241–248.

Saxby, J.D., 1976. Chemical separation and characterization of kerogen from oil shales. In: T.F. Yen and G.V. Chilingar (Editors), *Developments in Future Energy Sources, 1. Oil Shales*. Elsevier, Amsterdam (in press).

Saxby, J.D. and Stephens, J.F., 1973. Carbonaceous matter in sulphide ores from Mount Isa and McArthur River: an investigation using the electronprobe and the electron microscope. *Miner. Deposita*, 8: 127–137.

Schmidt-Collerus, J.J., 1969. Research in uranium geochemistry. Relation between organic matter and uranium deposits. *U.S. Atom. Energy Comm.*, GJO-933-2: 192 pp.

Schrage, I., 1956. Induced polarization of sulfide ores and graphite-containing rocks. *Freib. Forschungsh.*, 28C: 15–67.

Skinner, B.J., 1967. Precipitation of Mississippi Valley-type ores: a possible mechanism. In: J.S. Brown (Editor), *Genesis of Stratiform Lead-Zinc-Barite-Fluorite Deposits in Carbonate Rocks*. Economic Geology Publishing Co., Lancaster, Pa., pp. 363–370.

Smith, J.W. and Croxford, N.J.W., 1973. Sulphur isotope ratios in the McArthur lead-zinc-silver deposit. *Nature Phys. Sci.*, 245: 10–12.

Snyman, C.P., 1965. Possible biogenic structures in Witwatersrand thucolite. *Trans. Geol. Soc. S. Afr.*, 68: 225–235.

Stanton, R.L., 1972a. *Ore Petrology*. McGraw-Hill, New York, N.Y., 713 pp.

Stanton, R.L., 1972b. A preliminary account of chemical relationships between sulfide lode and "banded iron formation" at Broken Hill, New South Wales. *Econ. Geol.*, 67: 1128–1145.

Sullivan, G.V., Browning, J.S. and Sanders, S.J., 1969. Recovering gold from a graphitic schist from Tallapoosa County, Ala. *U.S. Bur. Mines*, RI 7251; 11 pp.

Sweeney, R.E. and Kaplan, I.R., 1973. Pyrite framboidal formation: laboratory synthesis and marine sediments. *Econ. Geol.*, 68: 618–634.

Szczepkowska-Mamczarczyk, I., 1971. Organic substance in the Zechstein copper shales of the Fore-Sudetic zone. *Kwart. Geol.*, 15(1): 41–55 (in Polish).

Taylor, G.H., 1971. Carbonaceous matter: a guide to the genesis and history of ores. *Soc. Mining Geol. Japan, Spec. Issue*, 3: 283–288.

Tokarska, K., 1971. Geochemical description of bitumen substance in the Zechstein copper shales. *Kwart. Geol.*, 15(1): 67–76 (in Polish).

Trudinger, P.A. and Bubela, B., 1967. Microorganisms and the natural environment. *Miner. Deposita*, 2: 147–157.

Trudinger, P.A., Lambert, I.B. and Skyring, G.W., 1972. Biogenic sulfide ores: a feasibility study. *Econ. Geol.*, 67: 1114–1127.

Vassoevich, N.B., 1971. Value of studying organic matter in recent and fossil sediments. *Org. Veshch. Sovrem. Iskop. Osadkov*, 1971: 5–11 (in Russian).

Van Eden, J.G., 1974. Depositional and diagenetic environment related to sulfide mineralization, Mufulira, Zambia, *Econ. Geol.*, 69: 59–79.

Vershkovskaya, O.V., Pikovskiy, Yu.I. and Solov'yev, A.A., 1972. Dispersed carbonaceous material in rocks and ores of the Plamennoye antimony-mercury deposit. *Dokl. Akad. Nauk S.S.S.R., Earth Sci. Sect.*, 205: 220–222.

Vinogradov, A.P., 1963. Biogeochemical provinces and their role in the organic evolution. *Geokhimiya*, 3: 199–213 (in Russian).

Vologdin, A.G., 1970. Organic remains from Shungites of the Precambrian of Karelia. *Dokl. Akad. Nauk S.S.S.R.*, 193: 258–261 (in Russian).

Wedepohl, K.H., 1971. "Kupferschiefer" as a prototype of syngenetic sedimentary ore deposits. *Soc. Mining Geol. Japan, Spec. Issue*, 3: 268–273.

Weil, R., Hocart, R. and Monier, J.C., 1954. Synthèses minérales en milieux organiques. *Bull. Soc. Fr. Minéral. Cristallogr.*, 77: 1084–1101.

Weiss, A., 1969. Organic derivatives of clay minerals, zeolites and related minerals. In: G. Eglinton and M.T.J. Murphy (Editors), *Organic Geochemistry*. Springer, Berlin, pp. 737–781.

Williams, R.J.P., 1967. Heavy metals in biological systems. *Endeavour*, 26: 96–100.

Yui, S., 1966. A thermochemical interpretation of the mode of occurrence of pyrrhotite in the Yaguki mine, Japan and the possible role of graphite. *J. Mining Coll. Akita Univ., Ser. A*, 4(1): 21–34.

Yui, S., 1968. Results of equilibrium calculations on the reaction between graphite and an H_2O-CO_2 fluid at 300–500°C and 100–2000 bars, and its implication to relict graphite in pyrometasomatic ore deposits, *J. Mining Coll. Akita Univ., Ser. A*, 4(3): 29–39.

Zezin, R.B. and Sokolova, M.N., 1968. Macroscopic occurrences of carbonaceous matter in hydrothermal deposits of the Khibiny pluton. *Dokl. Akad. Nauk S.S.S.R., Earth Sci. Sect.*, 177: 217–221.

ZoBell, C.E., 1963. Organic geochemistry of sulfur. In: I.A. Breger (Editor), *Organic Geochemistry*. Pergamon, Oxford, pp. 543–578.

Chapter 6

MICROBIOLOGICAL PROCESSES IN RELATION TO ORE GENESIS

P.A. TRUDINGER

INTRODUCTION

An ore body is a deposit containing an economically important mineral (or minerals) in a form and concentration which permit the commercial exploitation of that mineral. In theory, microorganisms could contribute in four main ways to the formation of mineral deposits, principally those arising by sedimentation at low temperatures in aqueous environments: (*1*) by accumulating mineral-forming elements; (*2*) by modifying physico-chemical conditions in the environment; (*3*) by generating organic matter; and (*4*) by transforming elements from one state to another by metabolism.

Accumulation of mineral-forming elements

Microorganisms in general have the ability to accumulate many elements from dilute solutions against enormous concentration gradients. As illustrated in Table I, element concentrations in the marine biota, which consists largely of algae and bacteria, may represent enrichments of up to 270,000 over the concentration in sea water.

A number of elements which are extracted from the surrounding environment have essential physiological functions, being components of cellular structures or of catalytic systems involved in energy production and synthetic reactions within the cell. In other instances, however, no physiological role has been reported (see Table I) and the accumulation of these elements may merely reflect the strong complexing character of the organic components of the living cell.

Specific organic complexing compounds may be elaborated by microorganisms to enable them to extract essential metals more efficiently when the latter become limiting. Low-molecular weight compounds containing hydroxamate or phenolate groups are excreted by certain microorganisms from media containing low levels of iron, and have been shown to function in iron uptake by these microorganisms. The iron complexes of siderochromes, which are organic molecules with hydroxamic acid centres, have stability constants in the order of $10^{29}-10^{32}$ and are thus amongst the strongest iron chelates known (Couglan, 1971).

Examples of extensive mineral deposition resulting from microbial accumulation of

TABLE I

Typical element concentrations in marine plants (from Bowen, 1966)

Element*	Concentration in sea water (ppm)	Concentration in marine plants (ppm dry wt.)	Enrichment factor
Ag	0.003	0.25	830
Al	0.01	60	6,000
As	0.003	30	10,000
Au	0.00001	0.012	1,200
B	5	120	24
Ba	0.03	30	1,000
Be	0.0000006	0.001	1,700
Bi	0.000017	0.06	3,530
C	28	345,000	12,300
Ca	400	10,000–300,000	25–750
Cd	0.0001	0.4	4,000
Co	0.003	0.7	2,300
Cr	0.00005	1	20,000
Cs	0.00005	0.07	1,400
Cu	0.003	11	3,700
F	1.3	4.5	3.5
Fe	0.01	700	70,000
Ga	0.00003	0.5	17,000
Hg	0.00003	0.03	1,000
I	0.06	30–1,500	500–25,000
K	380	52,000	140
Li	0.18	30	170
Mg	1,350	5,200	4
Mn	0.002	53	26,500
Mo	0.01	0.45	45
Na	10,500	33,000	3
Ni	0.005	3	600
P	0.07	3,500	50,000
Pb	0.00003	8	267,000
Rb	0.1	7	70
S	885	12,000	14
Se	0.00009	0.8	8,900
Si	3	1,500–20,000	500–6,700
Sn	0.003	1	330
Sr	8	260–1,400	33–175
Ti	0.001	12–80	12,000–80,000
V	0.002	2	1,000
W	0.0001	0.035	35
Zn	0.01	150	15,000

* Italicized elements are known or suspected to be physiologically important in microorganisms.

elements are the silica deposits arising from massive sedimentation of the siliceous skeletons of diatoms and radiolarians (Riedel, 1959; Lisitzin, 1972). It has been calculated that about $5 \cdot 10^{14}$ g of silica is removed annually from the oceans by biological activity (Harriss, 1972) which appears to be the main factor controlling the concentration of silica in sea water (Calvert, 1968).

In general, however, the concentration of economically important elements in microorganisms is low despite the large enrichment factors (Table I). Moreover, the protoplasm and soft structures of organisms are rapidly degraded after death and soluble elements are released into the aqueous environment. It is problematical, therefore, whether the amounts of economically important elements (e.g., base metals) normally present in microbial cells could contribute significantly to ore-grade accumulations of those elements in sediments. The special case of phosphorus is considered later.

Modification of physico-chemical environment

The various respiratory and fermentative reactions carried out by microorganisms lead to changes in the physical and chemical conditions of the environment which can modify mineralization processes occurring therein. Algae and bacteria, for example, are thought to exert a control on calcium carbonate deposition through their effects on environmental pH and CO_2 concentration (Chilingar et al., 1967). Suggestions have been made that alkalinity generated by bacterial sulphate reduction may be responsible for the formation of natron (hydrated Na_2CO_3) (Gubin and Tzechomaskaja, 1930; Verner and Orlovskii, 1948), a notion that has been particularly applied to the origin of the extensive natron deposits in Wadi Natrun in Egypt (Abd-el-Malek and Rizk, 1963). The creation of reducing conditions by microbial consumption of oxygen can clearly have profound indirect effects on the oxidation states of elements. For instance, Jensen (1958) and Viragh and Szolnoki (1970) consider that reduction of U^{6+} ions to the less soluble U^{4+} state by bacteriogenic H_2S may be a mechanism by which uranium becomes enriched in carbonaceous black shales and sandstones.

Generation of organic matter

About 50% dry weight of living organisms is carbon which is largely in the form of macromolecules such as proteins, lipids, carbohydrates and nucleic acids. The ultimate source of this organic matter is photosynthesis for which the marine phytoplankton is primarily responsible. Most organic matter is rapidly recycled to CO_2 and water by the activities of heterotrophic (i.e., organic matter-requiring) microorganisms but some soluble organic compounds have a significant lifetime in aqueous environments (e.g., Duursma, 1965) and a proportion of organic matter is incorporated into sediments. Organic contents as high as 25% C have been recorded in sediments of reducing environments (Strøm, 1955). Because of its complexing properties organic matter exerts a profound influence on the mobilization, transport and fixation of metals (W. Berger, 1950; Krauskopf, 1955). Microorganisms may play an additional role in the overall process by metabolizing the organic moieties of soluble metal-organic complexes and thus causing the release and precipitation of the metal (Aristovskaya, 1961).

The geochemical significance of metal-organic complexes has been discussed recently by Manskaya and Drosdova (1968), and by Saxby (1969, and Chapter 5, this volume).

Transformation by metabolism

Many microorganisms contribute to the geochemical recycling of elements by catalysing the oxidation and reduction of inorganic compounds (Table II). A number of these

TABLE II

Metabolism of inorganic compounds by microorganisms (from Woolfolk and Whiteley, 1962; Silverman and Ehrlich, 1964)

Element	Presumed reaction
	Oxidations
As	$AsO_3^{3-} + \frac{1}{2}O_2 \rightarrow AsO_4^{3-}$
Fe	$Fe^{2+} - e^- \rightarrow Fe^{3+}$
Mn	$Mn^{2+} + \frac{1}{2}O_2 + 2OH^- \rightarrow MnO_2 + H_2O$
P	$HPO_3^{2-} + H_2O - 2e^- \rightarrow HPO_4^{2-} + 2H^+$
S	$H_2S + \frac{1}{2}O_2 \rightarrow S^0 + H_2O$
	$H_2S + 2O_2 \rightarrow SO_4^{2-} + 2H^+$
	$S^0 + H_2O + 1\frac{1}{2}O_2 \rightarrow SO_4^{2-} + 2H^+$
Se	$H_2Se + 2O_2 \rightarrow SeO_4^{2-} + 2H^+$
	Reductions
As	$AsO_4^{3-} + H_2 \rightarrow AsO_2^- + 2OH^-$
Cu	$Cu(OH)_2 + \frac{1}{2}H_2 \rightarrow Cu(OH) + H_2O$
Fe	$Fe(OH)_3 + \frac{1}{2}H_2 \rightarrow Fe(OH)_2 + H_2O$
Mo	$Mo_6O_{21}^{6-} + 3H_2 \rightarrow 6MoO_{2.5} + 6OH^-$
Mn	$MnO_2 + H_2 \rightarrow Mn^{2+} + 2OH^-$
P	$\cdot HPO_4^{2-} + H_2 \rightarrow HPO_3^{2-} + H_2O$
S	$SO_4^{2-} + 10H^+ + 8e^- \rightarrow H_2S + 4H_2O$
Se	$SeO_4^{2-} + 10H^+ + 8e^- \rightarrow H_2Se + 4H_2O$
U	$UO_2(OH_2) + H_2 \rightarrow U(OH)_4$
V	$VO_4^{3-} + 2H^+ + \frac{1}{2}H_2 \rightarrow VO(OH)_2 + OH^-$

reactions result, directly or indirectly, in the immobilization of economically important elements and thus are potentially significant with respect to ore genesis.

A comprehensive discussion of the diverse aspects of microbial mineral transformation outlined above is beyond the scope of a single chapter. The references already cited, and Baas Becking (1959), Kuznetsov et al. (1963), Silverman and Ehrlich (1964), Trudinger and Bubela (1967), Pauli (1968), Zajic (1969), Schwartz (1972) and Krumbein (1972) provide further information on this subject. This chapter deals primarily with *metabolic* processes which are relevant to the question of the origin of metal sulphide, sulphur, phosphorite, manganese and iron deposits.

SULPHIDE MINERALIZATION

One of the earliest suggestions for a biological factor in ore genesis was in relation to the Mississippi Valley-type lead–zinc ore which, Siebenthal (1915) proposed, may have arisen as the result of bacterial sulphate reduction. Similar conclusions were reached by Bastin (1926) and, since that time, biological activity has been invoked to account, at least partially, for a number of sulphide deposits containing economic quantities of base metals.

The reduction of sulphate to sulphide is carried out by most microorganisms and plants. In the majority of instances the sulphide is utilized for synthesis of essential organic sulphur compounds such as the amino-acids cysteine [$HSCH_2.CH_2(NH_2).COOH$] and methionine [$CH_3SCH_2.CH_2(NH_2).COOH$]. This process is termed "assimilatory sulphate reduction" and generally little or no free sulphide is produced. Exceptions to this generalization are the thermophilic blue-green alga *Synechoccus lividus* (Sheridan and Castenholz, 1968) and yeasts grown under certain nutrient deficiencies (e.g., Wainright, 1970), or adapted to copper tolerance (Ashida, 1965), all of which have been reported to generate significant quantities of hydrogen sulphide.

Massive hydrogen sulphide production, however, is commonly attributed to a group of specialized anaerobic bacteria belonging to the genera *Desulfovibrio* and *Desulfotomaculum.* These carry out *dissimilatory* sulphate reduction (or sulphate respiration), a process by which the oxidation of organic compounds is coupled to the reduction of sulphate to provide the energy requirements for growth. Under laboratory conditions up to 1.5 g of sulphur may be transformed by dissimilatory sulphate reducers to produce 1 g of cellular material (Senez, 1962) and it is possible that, under less ideal natural conditions, the production of sulphide per unit cell mass may be even higher. This contrasts with assimilatory sulphate reduction where the sulphur reduced amounts to about 0.5–1% of the cell mass. The high sulphide production in dissimilatory reduction is a consequence of the relatively low energy yields of the reduction reactions. For example, anaerobic oxidation of a typical growth substrate, lactate, by sulphate (eq. 1) yields about 1.77 Kcal per mole electron compared with values of 25–28 Kcal per mole electron for complete oxidation of organic matter by oxygen (McCarty, 1972):

$$4CH_3CHOHCOO^- + 2SO_4^{2-} + 3H^+ \rightarrow 4CH_3COO^- + H_2S + HS^- + 4H_2O + CO_2 \qquad (1)$$

Sulphite is an intermediate in both assimilatory and dissimilatory sulphate reduction. Aside from their different physiological functions the two pathways are distinguished mainly by the steps preceding sulphite formation. In dissimilatory reduction sulphate is "activated" by a reaction with adenosine-triphosphate (ATP) to form adenosine-5'-phosphosulphate (APS) which is reduced to sulphite, while in the assimilatory pathway APS is further phosphorylated to 3'-phosphoadenosine-5'-phosphosulphate (PAPS) prior to reduction. These pathways are summarized in the following scheme and have been discussed in detail in a number of recent publications (e.g., Peck, 1962; Nicholas, 1967; Trudinger, 1969; Roy and Trudinger, 1970).

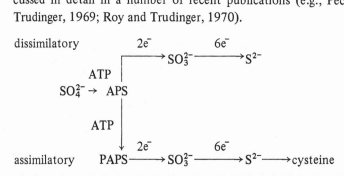

Cysteine is the principal product of assimilatory sulphate reduction and this compound, and its oxidized form cystine, readily undergo reactions with metallic ions and salts to form sulphides. Lambert (1973) synthesized sphalerite, galena, covellite, greigite, pyrrhotite, pyrite and marcasite from cystine and metallic salts in the dry state in experiments lasting 1–125 days at temperatures of 70–200°C. In general the rates of reaction increase in the presence of water, $CaCO_3$ and humic acids. Weil and his colleagues (Weil, 1955, 1958; Weil et al., 1954) found that sulphides of Fe, Ni, Mn, Co, Zn, Pb, Cu, Cd, Sn, Ag, As and Sb were formed from cystine and metallic salts or oxides in the presence of glycerol at temperatures below 100°C. Mixed sulphides such as chalcopyrite were also produced in Weil's experiments.

These reactions have been proposed as possible mechanisms by which metallic sulphides are produced in the biosphere (Weil, 1955). Quantitatively, however, they are of doubtful significance particularly where the main source of organic matter is fixation of CO_2 by planktonic photosynthesis. The Black Sea basin, for example, may be taken as representing a plausible environment in which biogenic sedimentary sulphides might arise (Trudinger et ål., 1972). The rate of primary organic matter production in this basin has been estimated at about 100 g C m^{-2} yr^{-1} (Sorokin, 1964) but it is probable that only about 4% of this primary organic matter is incorporated into the sediment (Deuser, 1971). Assuming an average C : S ratio of 100 : 1 for organic material (Alexander, 1965), the annual incorporation of organic sulphur into the sediments, therefore, would be expected to be in the order of 40 mg S m^{-2} yr^{-1}. The minimum rate of sedimentation in the euxinic zone of the Black Sea is about 0.01 cm yr^{-1} (Ross et al., 1970a,b). Thus, based on a density of 1.0 for fresh sediment the maximum expected concentration of organic sulphur which would be available for sulphide mineralization could be no more than about 0.04%. Since average sulphide concentrations of economic stratiform sulphide ores are 2–3 orders of magnitude greater than this figure (Trudinger et al., 1972), it appears unlikely that organic sulphur could be a major source of the sulphide in massive deposits although it may contribute to the development of secondary textures and structures during diagenesis (Lambert, 1973).

The dissimilatory sulphate-reducing bacteria are generally considered to be the most likely biological agents associated with large-scale sulphide mineralization. Their ecology and physiological characteristics have been summarized by ZoBell (1958) and by Postgate (1959, 1965). They are generally restricted to mildly acid to alkaline, reducing environments (Baas Becking et al., 1955, 1960) and as a group they appear to be able to withstand most of the extreme variations in pressure, temperature and salinity found in the natural environment (Table III). Perhaps understandably sulphate-reducing bacteria are remarkably tolerant to hydrogen sulphide, exposures to solutions of 2% or greater for up to 22 h having little effect on their viability (Miller, 1950a,b).

There is considerable evidence that in many natural situations the availability of organic matter may be the major factor controlling the extent of bacterial sulphate reduction (Sorokin, 1962; Koyama et al., 1965; Ivanov, 1968; Berner, 1970; Chukhrov, 1970). Numerous laboratory studies indicate that the sulphate-reducing bacteria are generally restricted to the utilization of simple organic molecules in the production of energy. Compounds whose oxidation by sulphate-reducing bacteria has been authenticated are listed in Table IV. Claims that more complex molecules, such as carbohydrates, hydrocarbons, and long-chain fatty acids and alcohols are metabolized have been queried by Postgate (1959) on the grounds that insufficient attention was paid to possible contamination by non sulphate-reducing organisms. One consequence of this apparently limited substrate metabolism is that the complex molecules of which primary organic

TABLE III

Some environmental limits of sulphate-reducing bacteria (from ZoBell, 1958)

Factor	Limits
Eh*	+ 350 to − 500 mV
pH*	4.2 − 10.4
Pressure	1 − 1,000 atmospheres
Temperature	0 − 100°C
Salinity	< 1 to 30% NaCl

* Not necessarily active over full range, see text.

TABLE IV

Organic compounds utilized in energy metabolism by sulphate-reducing bacteria (from Trudinger et al., 1972)

Compound	Structure
Fairly generally used	
Lactate	$CH_3.CHOH.COOH$
Pyruvate	$CH_3.CO.COOH$
Probably restricted to specific strains and species	
Formate	HCOOH
Choline	$(CH_3)_3.N(OH).CH_2CH_2OH$
Oxamate	$H_2N.CO.COOH$
Ethanol	$CH_3.CH_2OH$
Carbon monoxide	CO
Fumarate	$HOOC.CH = CH.COOH$
Malate	$HOOC.CH(OH).CH_2.COOH$

carbon is composed will most likely be extensively degraded by oxidative and fermentative reactions by other organisms before being available for sulphate respiration. The effects of environmental conditions on these associated reactions may, therefore, have a profound indirect influence on generation of hydrogen sulphide.

Dissimilatory sulphate-reducing bacteria are widely distributed on the earth's surface and they have been detected in a variety of soils, fresh and marine waters and ground waters. Their activities are particularly conspicuous in waters of euxinic basins (e.g., Sorokin, 1962, 1970) and in sediments where restricted circulation and depletion of oxygen by oxidation of organic matter result in ideal conditions for the proliferation of these organisms (Kaplan et al., 1963; Nissenbaum et al., 1972; Hallberg et al., 1972). Some examples of rates of sulphide formation in sediments are given in Table V.

Mineral synthesis by sulphate-reducing bacteria

The formation of metal sulphides in cultures of dissimilatory sulphate-reducing bacteria has been the subject of a number of investigations. Miller (1950c) reported that sulphides of Sb, Bi, Co, Cd, Fe, Pb, Ni and Zn were produced as the result of the activities of *Desulfovibrio desulfuricans* when relatively insoluble compounds of these metals were present in the nutrient medium. In Miller's experiments the mineral phases were not identified, but later Baas Becking and Moore (1961) synthesized many important minerals of Cu, Pb, Zn and Ag in crude biological systems (Table VI). Subsequently, Rickard (1969a) reported the formation of a variety of iron sulphide minerals by pure cultures of *D. desulfuricans* in the presence of goethite or a complex ferroso-ferric oxyhydroxide

TABLE V

Rates of sulphate reduction in sediments

Locality	Rate $(g\ S\ m^{-2}\ yr^{-1})$	Reference
Lake Belovod *	0.12–2.7	1
Lake Gel Gel*	0.55–83	1
Linsley Pond*	14.5–27	1
Black Sea	0.5–42	1
Santa Barbara Basin	3.1	1
Gulf of California	6.3–9.5	2
Black Sea	0.6–9.5	2
Santa Barbara Basin	9.5	2
Long Island Sound	3.15	2
Puget Sound	9.5	2
Pacific Sediments	1.9–44.2	2

* Fresh-water sediments.
References: 1 = Trudinger et al. (1972); 2 = Rickard (1973).

which formed on addition of ferrous iron to the culture medium. Rickard showed that the initial precipitate resulting from the reaction between ferrous iron and sulphide is a fine-grained black ferrous sulphide which he considered to be a mixture of amorphous FeS and poorly crystalline mackinawite (cf. also Sweeney and Kaplan, 1973). Based on the kinetics of mineral formation at different pH values he proposed the following sequence for the subsequent transformation of mackinawite in biological systems:

$$
\begin{array}{c}
\phantom{Fe^{2+} \to}\overset{\displaystyle \longrightarrow \text{ pyrite}}{\underset{\displaystyle S_n^{2-}}{\big|}} \\[4pt]
Fe^{2+} \to \text{mackinawite} \xrightarrow{\ S^{\circ}\ } \text{marcasite} \\[4pt]
\underset{\displaystyle \text{greigite} \longrightarrow \text{pyrrhotite}}{\big|_{S^{2-}}}
\end{array}
$$

Rickard also suggested that reactions between organic polysulphides and FeS (eq. 2) might be important in biological systems:

$$AS_n^{2-} + FeS \to FeS_2 + AS_{n-1}^{2-} \quad \text{(where A represents an organic molecule)} \tag{2}$$

Hallberg (1972a) demonstrated the formation of sphalerite and pyrite in continuous cultures (Hallberg, 1970) of *D. desulfuricans* which enabled relatively low concentrations

TABLE VI

"Biogenic" sulphide minerals*

Starting material	Mineral synthesized	Reference
Malachite [$CuCO_3.Cu(OH)$] or Chrysocolla [$CuSiO_3.2H_2O$]	covellite (CuS)	1
Cuprous oxide (Cu_2O)	digenite (approx. $Cu_{1.8}S$)	1
Lead carbonate [$PbCO_3$] or Lead hydroxy carbonate [$2PbCO_3.Pb(OH)_2$]	galena (PbS)	1
Smithsonite ($ZnCO_3$)	sphalerite (ZnS)	1
Silver chloride [AgCl] or Silver carbonate [Ag_2CO_3]	acanthite (Ag_2S)**	1
Ferroso-ferric oxyhydroxide	mackinawite (approx. FeS) greigite (Fe_3S_4) pyrrhotite ($Fe_{1-x}S$)	2
Goethite [FeO(OH)]	mackinawite pyrite (FeS_2) marcasite (FeS_2)	2
Zn^{2+}	sphalerite	3
Fe^{2+}	greigite, pyrite	3

* Identified by X-ray diffraction; ** originally erroneously identified as argentite (MacDiarmid, 1966). References: 1 = Baas Becking and Moore (1961); 2 = Rickard (1969); 3 = Hallberg (1972).

(in the order of 25 p.p.m.) of metallic ions in solution to be employed. He concluded that greigite could be an intermediate in pyrite formation as did Sweeney and Kaplan in a purely chemical study of pyrite synthesis. Freke and Tate (1961) reported the formation of a magnetic iron sulphide in intermittent continuous cultures of a mixed microbial population presumably containing sulphate-reducing bacteria. The material was assigned the empirical formula Fe_4S_5 by Freke and Tate but was subsequently identified by Rickard (1969a) as a mixture of greigite and Fe_2O_3.

Since the average oxidation state of sulphur in polysulphur-iron minerals is greater than -2, their formation from hydrogen sulphide requires an oxidant to convert some H_2S to the level of elemental sulphur or polysulphide. The role of oxidant can be filled by oxygen or ferric iron (Berner, 1964a; Roberts et al., 1969) and the latter, in the form of goethite or ferroso-ferric hydroxide, was presumably operative in the experiments of Rickard (1969a). In Hallberg's (1972a) experiments, however, the situation is not so clear since ferrous iron was used and "careful precautions. . . were taken against oxidation by air". The possibility should be considered, therefore, that an intermediate of the sulphate-reduction pathway at a higher oxidation state than sulphide might take part in synthesis of polysulphur-iron minerals. One such intermediate is elemental sulphur which a brief, but unconfirmed report (Iverson, 1967) suggests may be formed in small amounts during growth of sulphate-reducing bacteria. A second is thiosulphate which is a product of sulphite reduction by *Desulfovibrio* (Findley and Akagi, 1969; Kobayashi et al., 1969; Suh and Akagi, 1969) and which has been reported to accelerate pyrite formation from FeS possibly by the reaction described in eq. 3 (Volkov and Ostroumov, 1957):

$$S_2O_3^{2-} + FeS \rightarrow FeS_2 + SO_3^{2-} \tag{3}$$

Berner (1967), however, was unable to verify this mechanism at the pH of marine sediments. The present author (unpublished results) has also observed pyrite formation from ferrous iron in oxygen-free static cultures containing *Desulfovibrio* sp. together with a mixed-soil microflora. Under similar conditions pure cultures of *Desulfovibrio* yielded only amorphous acid-soluble iron sulphides. These experiments suggested that anaerobic oxidation of sulphide by the soil microflora coupled to reduction of an organic constituent in the medium may have been responsible for pyrite formation. Anaerobic production of elemental sulphur by bacteria was proposed by Volkov (1961) to account for pyrite formation in the reducing sediments of the Black Sea. These suggestions, however, must be viewed with some caution since anaerobic biological oxidation of sulphide has so far only been observed in the presence of nitrate (see Trudinger, 1967) or photosynthetic bacteria (vide infra). Neither of these cases applies to the Black Sea or to the experiments described above. Berner (1970) suggested that the most likely cause of sulphur production in the Black Sea was the merging of H_2S-containing deep waters with oxygenated surface waters where both biological and chemical oxidation reactions could operate (Sorokin, 1970).

Rickard (1969a,b) detected no difference in the mode of formation of iron sulphides between chemical and biological systems and, in general, we may assume that the formation of metal sulphides in biological systems is primarily a chemical process between metals and biologically-produced hydrogen sulphide. In certain instances, however, there are indications of a more direct role for the organism. Baas Becking and Moore (1961), for example, noted that native copper was an intermediate product in the formation of covellite from malachite and that native silver appeared transiently during the conversion of AgCl or Ag_2CO_3 to acanthite. Kramarenko (1962) reported that the level of hydrogen sulphide required to precipitate molybdenum from solutions containing bacterial cultures was less than 10% of that necessary to effect precipitation in a simple inorganic mixture. This suggested the possibility that sulphate-reducing bacteria may be able to create a microenvironment in or around their cells in which metallic ions and sulphide are concentrated to a sufficient extent to exceed the solubility product of the metal sulphides. A general ability of sulphate-reducing bacteria to concentrate metals has been demonstrated in the Baas Becking Geobiological Laboratory (H.E. Jones, P.A. Trudinger, and L.A. Chambers, unpublished results).

A characteristic feature of many stratiform sulphide deposits is the presence of large numbers of closely spaced metal sulphide bands which are conformable within their host sedimentary strata. Bubela and McDonald (1969) and Lambert and Bubela (1970) demonstrated that metal ions can be separated by diffusion ("Liesegang") processes and precipitated by biogenic hydrogen sulphide as essentially monomineralic bands reminiscent of those found in stratiform deposits. Similar results were obtained when sodium sulphide was substituted for biogenic hydrogen sulphide. Formation of metal sulphide bands involving ion diffusion had earlier been reported for biological (Temple and Le Roux, 1964a; Berner, 1969a) and purely chemical systems (Watanabe, 1924; Weiss and Amstutz, 1966). Lambert and Bubela (1970, p.102) concluded that "the most likely role for the ionic migration process is in the formation of the very thin sulphide bands within homogeneous layers of unconsolidated sediments or slowly settling suspensions". Once again, however, the function of microorganisms appears to be simply that of providing hydrogen sulphide for subsequent chemical reactions.

The synthesis of sulphide minerals in bacterial systems demonstrates that sedimentary environments in which base-metal sulphide minerals form are not of necessity incompatible with biological activity. Davidson (1962, 1962/63) had criticized the concept of biogenic sulphide ores on the grounds that the concentrations of metals required for significant amounts of mineralization would inhibit biological sulphate reduction. *Desulfovibrio* spp. do indeed appear to be inherently sensitive to some metal ions particularly those of copper, lead and zinc (Table VII). Temple and Le Roux (1964b), however, demonstrated that, since copper sulphides are themselves non-toxic, copper toxicity becomes a limiting factor only when the rate of introduction of the metal to the environment of sulphate reduction exceeds that of H_2S production. It is likely that this situation applies also to other metals which form highly insoluble sulphides. Other factors, such as metal-organic complexing and ion interactions, which modify the toxic effects of metals were discussed by Sadler and Trudinger (1967).

TABLE VII

Sensitivity of *Desulfovibrio* spp. to metals (data give concentrations, in p.p.m., required for inhibition)

Cu^{2+}	Zn^{2+}	Pb^{2+}	Fe^{2+}	Reference
20–50				1
0.25–2.5	15–90		1,000–24,000	2
10–15	>0.54	>0.56		3

References: 1 = Booth and Mercer (1963); 2 = Römer and Schwartz (1965); 3 = Wallhäusser and Puchelt (1966).

TABLE VIII

Calculated times for formation of stratiform sulphide ore bodies (from Trudinger et al., 1972)

Ore deposit	Age	Total sulphide sulphur $(g\,m^{-2})$	Range of times of formation* (years)	Average rates of sulphide precipitation based on columns 3 and 4 $(g\,S\,m^{-2}\,yr^{-1})$
McArthur River (total ore body)	Middle Proterozoic	$1.9 \cdot 10^7$	$0.4-1.7 \cdot 10^6$	$46-11.5$
Mount Isa (No.10/25 ore body)	Middle Proterozoic	$6 \cdot 10^5$	$1.5-6 \cdot 10^4$	$40-10$
Copperbelt [1] (Roan Antelope)	Upper Proterozoic	$6 \cdot 10^5$	$1.5-6 \cdot 10^5$	$4-1$
Kupferschiefer [2] (Mansfeld)	Upper Permian	$2.3 \cdot 10^4$	$1.9-7.5 \cdot 10^3$	$12-3$

* Lower and upper values are based respectively on 1 cm of lithified ore per 75 years (fast accumulation) and 300 years (fairly slow accumulation).

Recently, Trudinger et al. (1972) and Rickard (1973) have shown that present-day rates of sulphate reduction (e.g., Table V) are probably comparable with those of sulphide precipitation during the formation of synsedimentary sulphide ores, including a number of important base-metal sulphide deposits (Table VIII). A biogenic origin for these deposits thus becomes a theoretically feasible proposition (Temple, 1964). Suckow and Schwartz (1968) devised an experimental system for laboratory studies on biologically controlled metal sulphide precipitation in sapropelic muds. Simultaneous determinations of Eh, pH and H_2S profiles throughout the system partially defined the physicochemical requirements for metal sulphide deposition. From the results of experiments on iron- and copper-sulphide formation in sapropelic muds from the Baltic Sea, Suckow and Schwartz concluded that biogenic sulphide could adequately account for sulphurization of metals during genesis of the Mansfeld Kupferschiefer.[2]

While experimental approaches of this type show promise of providing an understanding of mineralization processes in sediments, great care must be exercised in extrapolating quantitative data from short-term, idealized laboratory studies to long-term, complex field situations. Mineralization in sediments is the end result of the interplay of a large number of chemical, physical and biological factors and detailed comprehensive kinetic studies on specific euxinic basins would seem to be necessary for the construction of a realistic quantitative model of sulphide mineral formation. Hallberg (1972b) has attempted to describe the cycle of heavy metals in sea water and the formation of

[1,2] Editor's notes: [1] The origin of the Copperbelt has been discussed by Fleischer et al. in Chapter 6, Vol. 6. [2] For details on the origin of the Kupferschiefer, see Chapter 7 by Jung and Knitzschke in Vol. 6.

sedimentary heavy-metal deposits in terms of energy-circuit systems which incorporate many of the chemical, physical and biological parameters involved. His systems help to define the nature of the problem but they also emphasize the formidable amount of information required for its solution.

It should be pointed out that while the formation of iron minerals, principally pyrite, from biogenic sulphide in modern sediments is now well documented (e.g., Sugawara et al., 1954; Barghoorn and Nichols, 1961; Kaplan et al., 1963; Berner, 1964b, 1970) there appear to be no recorded instances of significant sulphide mineralization of other metals in the present-day biosphere. This presumably reflects the generally poor availability of metals other than iron in most sedimentary situations but whether environments favouring the accumulation of non-ferrous minerals are, for unknown reasons, incompatible with the development of active sulphate reduction remains to be determined. In this respect it may be relevant that the modern sediments of the Red Sea Deeps, which contain economic amounts of Zn, Cu, Pb, Ag and Au (Bischoff and Manheim, 1969) appear to be sterile (Watson and Waterbury, 1969; cf. also Chapter 4 by Degens and Ross in Vol. 4).

Pyritic framboids

While the hypothesis of large-scale sulphide mineralization by sulphate-reducing bacteria is reasonable on theoretical grounds, its application to the genesis of mineral deposits in past geological periods must depend on uncovering evidence for the presence of these bacteria at the time of mineral deposition. The existence of framboidal pyrite has been considered by some to be significant in this regard. Pyritic framboids are spherules ranging in diameter from 1–100 microns and composed of many euhedral pyrite crystals which impart a raspberry-like, or "framboidal" appearance. They are common constituents of both modern and ancient sediments and have been recognized in economic sulphide deposits such as those of Rammelsberg, Mt. Isa and the Kupferschiefer (Love and Amstutz, 1966).

Schneiderhöhn (1923) believed that framboids are the fossilized remains of aggregated cells of sulphur bacteria within which iron and other metals had precipitated as sulphides. This idea was supported by Ramdohr (1953) and subsequent findings by Love and his colleagues (Love, 1957, 1962a,b,c, 1963; Love and Zimmerman, 1961; Love and Murray, 1963) of the presence of pyrite spherules in argillaceous rocks and sediments were in line with the view that they represented fossil organisms. In some instances organic "structures" were revealed on removal of mineral matter (Love, 1962c; Love and Amstutz, 1966). Pyrite crystals in globules in pelagic ooze from the Pacific Ocean have been attributed to metabolic products of globular bacteria (Skripchenko, 1970).

Schouten (1946), on the other hand, while accepting that sulphate-reducing bacteria played a dominant role in the formation of syngenetic pyrite, questioned the existence of fossil sulphur bacteria. He believed framboidal pyrite to be of inorganic origin. The discoveries of pyritic framboids in andesites (Love and Amstutz, 1969), Kuroko ores

(Matsukuma and Horikoshi, 1970) and possibly also in hydrothermal veins (Schouten, 1946) clearly indicate that organic matter is not essential for their formation. Vallentyne (1962, 1963) also reported that framboids from surface sediments of Little Round Lake, Ontario, contained insufficient organic matter for them to be regarded as organic micro-fossils.

Purely inorganic formation of framboids has been confirmed by laboratory experiments. Berner (1969b) prepared framboidal microspheres at 65° and neutral pH by the reaction of FeS with elemental sulphur in saturated H_2S solutions. Framboidal sulphides of iron and a number of other metals were synthesized by Farrand (1970) by bubbling H_2S through dilute solutions of metal sulphates or chlorides buffered with $CaCO_3$. The hydrothermal formation of framboidal pyrite was described by Sunagawa et al. (1971). Sweeney and Kaplan (1973) studied the sulphidization by elemental sulphur of freshly precipitated iron sulphide (FeS) under a variety of conditions. They concluded that spherical textures, which develop when the initial iron sulphide precipitate is transformed to greigite, are prerequisites for framboid formation. During conversion of greigite to pyrite the spherical structure is either retained or is changed by internal nucleation of pyrite crystals to form framboids.

These results do not exclude the possibility that bacteria may be involved in the genesis of framboids, particularly those containing organic material, and Berner (1969b) considers that most framboidal pyrite is a sedimentary mineral formed during early diagenesis of biogenic hydrogen sulphide. Rickard (1970) in a theoretical discussion suggested that framboids may form by replacement or infilling of globules of organic matter (see also Papunen, 1966). That these globules could be bacteria is suggested by the report of Issatchenko (1929) of "pyrite granules" within bacterial cells and by the intra-cellular deposition within sulphate-reducing bacteria of particles of, apparently, iron sulphide (H.E. Jones, personal communication).

Nevertheless, while biogenic framboidal pyrite may still remain a distinct possibility, the chemical synthesis of framboids must inevitably cast doubts on the diagnostic value of these structures with respect to the biological versus chemical origins of sulphide deposits.

Isotopic evidence[1]

The stable sulphur isotope compositions of natural sulphur-containing minerals are now widely used in speculations on their modes of origin. Inherent in these speculations is the assumption that primordial terrestrial sulphur had a constant composition which is thought to be close to, if not identical with, that of meteoritic sulphur (Macnamara and Thode, 1950; Vinogradov, 1958; Ault and Kulp, 1959). Wide variations in the isotopic compositions of minerals, however, are encountered at the present time and, in particular, sedimentary sulphides exhibit $\delta^{34}S$ values[1] ranging from nearly $-50‰$ to $+50‰$ (e.g.,

---

[1] For general contributions on isotopes, see Chapters 7, 8, 9 of this Volume.

TABLE IX

Preferred sulphur-isotope utilization in biochemical reactions (after Kaplan and Rittenberg, 1962)

$^{32}S > {}^{34}S$	$^{34}S > {}^{32}S$	$^{34}S = {}^{32}S$
$SO_4^{2-} \to HS$		
$SO_4^{2-} \to$ organic S	$H_2S \to S_nO_6^{2-}$	$S^0 \to SO_4^{2-}$
$SO_3^{2-} \to H_2S$	$H_2S \to SO_4$	$S^0 \to H_2S$
$H_2S \to S^0$	(photosynthetic)	
$H_2S \to SO_4$		
(chemosynthetic)		

Ault, 1959; Thode, 1963). These variations are now generally considered to be due to isotopic discrimination during transformation of sulphur from its "primordial" state.

Fractionation of stable sulphur isotopes takes place to varying extents during both chemical and biological transformations of sulphur. Table IX, based on experiments of Kaplan and Rittenberg (1962), shows the general trends of biological isotope discrimination. In the majority of cases ^{32}S is preferentially utilized by the organisms but polythionates ($S_nO_6{}^{2-}$) arising during the oxidation of sulphide may become enriched in ^{34}S.

In the geochemical cycle of sulphur all biological reactions will presumably have some effect on the isotopic distribution but in the present context the main interest relates to fractionation associated with sulphate reduction. Theoretically the latter could be due to both kinetic and equilibrium isotope effects. Tudge and Thode (1950) calculated that the following exchange reaction (eq. 4):

$$H_2{}^{34}S + {}^{32}SO_4^{2-} \rightleftarrows H_2{}^{32}S + {}^{34}SO_4^{2-} \tag{4}$$

would favour the formation of $H_2{}^{32}S$ with a fractionation factor of about 1.074 at equilibrium and 25°C. Equilibrium, however, appears to be unattainable in purely chemical systems at 25° (Thode, 1964) or at higher temperatures. Using ^{35}S, Voge (1939) failed to demonstrate exchange between sulphate and sulphide at 100° and similar negative results were reported by Vinogradov et al. (1962) for the system $BaSO_4 + H_2S$ at temperatures between 200 and 400°.

---

[1] Note:

$$\delta^{34}S\%_0 = \left[\frac{{}^{34}S/{}^{32}S \text{ (sample)}}{{}^{34}S/{}^{32}S \text{ (standard)}} - 1 \right] \times 1000$$

For naturally-occurring minerals the standard is the troilite from the Canon Diablo meteorite ($^{32}S/^{34}S = 22.21$). The fractionation factor (f) for isotopic variations between pairs of sulphur compounds (a and b), is given by the expression:

$$f = \frac{{}^{32}S/{}^{34}S \ (a)}{{}^{32}S/{}^{34}S \ (b)}$$

It has been suggested that, in some natural environments, the equilibrium has been established by recycling of sulphur through the cooperative activities of sulphate-reducing and sulphide-oxidizing organisms (Thode et al., 1953). The possibility that equilibrium might be approached through reversibility of the sulphate reduction pathway in bacteria was considered by Tudge and Thode (1950) and Kaplan and Rittenberg (1964). Although Trudinger and Chambers (1973) demonstrated anaerobic oxidation of $H_2{}^{32}S$ to sulphate by the dissimilatory sulphate reducer *D. desulfuricans*, there is, at present, no direct evidence that the isotopic exchange equilibrium can be achieved at low temperatures in either chemical or biological systems.

Kinetic isotope effects are generally considered to be important controlling factors in low-temperature fractionation of stable isotopes during sulphate reduction. Kinetic effects are due to differences in bond strengths which lead to unequal rates of reduction of light and heavy sulphates:

$$\left.\begin{array}{l} {}^{32}SO_4^{2-} \xrightarrow{k_1} H_2{}^{32}S \\[6pt] {}^{34}SO_4^{2-} \xrightarrow{k_2} H_2{}^{34}S \end{array}\right\} \text{ where } k_1 \neq k_2$$

Harrison and Thode (1957) demonstrated experimentally that chemical reduction of sulphate favoured ${}^{32}S$ and obtained an average factor (*f*) of 1.022 over the temperature range of 18–50°C. They concluded that the rate-controlling step was the reduction of $SO_4{}^{2-}$ to $SO_3{}^{2-}$ and that fractionation was associated with the initial S–O bond scission. The observed fractionation factor was in agreement with theoretical calculations based on statistical mechanical methods if it were assumed that the structure of the activated complex was about midway between the extremes of no bond stretching and complete bond rupture.

Reduction of sulphate by dissimilatory sulphate-reducing bacteria also results in enrichment of ${}^{32}S$ in H_2S but the reported results vary considerably and the degree of fractionation appears to depend on temperature, electron donor, substrate concentration and other factors (Thode et al., 1951; Jones and Starkey, 1957, 1962; Harrison and Thode, 1958; Kaplan et al., 1960; Nakai and Jensen, 1964; Kaplan and Rittenberg, 1964; Kemp and Thode, 1968; Smejkal et al., 1971). In general, there appears to be an inverse correlation between rates of sulphate reduction per unit cell mass and the degree of fractionation. Harrison and Thode (1958) working with washed suspensions of *Desulfovibrio desulfuricans* reported that little or no fractionation took place at high reduction rates and that fractionation increased as the rate of reduction was lowered: a maximum fractionation factor of about 1.025 was observed. Kaplan and Rittenberg (1964), however, in similar experiments found significant fractionation at the highest rates of reduction studied. A general increase in fractionation with decreasing reduction rates was, however, confirmed when organic electron donors were used (Fig. 1): there was an apparent reversal of this trend with H_2 as the reductant. Generally similar results were reported by Kemp and Thode (1968).

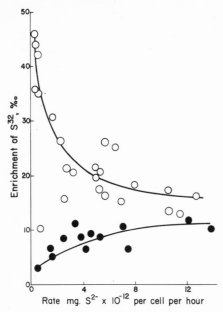

Fig. 1. Enrichment of ^{32}S in H_2S as a function of rate of sulphate reduction by *Desulfovibrio desulfuricans*. (Redrawn after Kaplan and Rittenberg, 1964.) Rate controlled primarily by temperature (range 10–45°C). Open circles: organic electron donors, lactate or ethanol; closed circles: hydrogen as electron donor.

Harrison and Thode (1958) attempted to explain variations in bacterial isotope fractionation in terms of two rate-controlling steps: (*1*) the formation of the sulphate-enzyme complex; and (*2*) the bond breaking reaction:

$$SO_4^{2-} \text{ (solution)} + \text{enzyme} \underset{k_{-1}}{\overset{k_1}{\longrightarrow}} \text{enzyme} - SO_4^{2-}$$
$$k_2 \downarrow$$
$$SO_3^{2-}$$
$$k_3 \downarrow$$
$$S^{2-}$$

The constant, k_3, was assumed to be large with respect to k_1 and k_2. It was postulated that at high reduction rates or low sulphate concentrations k_1 would be rate-limiting and that no discrimination between $^{32}SO_4^{2-}$ and $^{34}SO_4^{2-}$, or even a slight bias towards $^{34}SO_4^{2-}$, would apply. At low rates of reduction, k_2 would become rate-limiting and a kinetic isotope effect due to the initial S–O bond scission ($f = 1.022$) would be observed. Intermediate isotope effects would be obtained when conditions were such that the overall rate of sulphate reduction was influenced by both k_1 and k_2.

This model fails to account for the observations (e.g., Kaplan and Rittenberg, 1962, 1964) that bacterial sulphate reduction may, under some conditions, cause isotope effects which are considerably higher than those associated with S–O bond rupture during chemical reduction (see Fig. 1).

Kemp and Thode (1968) recognized that, in bacterial systems, additive isotope effects associated with two or more enzymes of the sulphate reduction pathway might account for high fractionation factors. This idea has been amplified by Rees (1973) who developed a mathematical treatment for the multistage, steady-state reaction shown in the following sequence:

$$\text{stage 1} \qquad \text{stage 2} \qquad \text{stage 3} \qquad \text{stage 4}$$

$$\text{external } SO_4^{2-} \underset{(O)}{\overset{(-3\%_0)}{\rightleftharpoons}} \text{internal } SO_4^{2-} \underset{(O)}{\overset{(O)}{\rightleftharpoons}} APS \underset{(O)}{\overset{(25\%_0)}{\rightleftharpoons}} SO_3^{2-} \overset{(25\%_0)}{\longrightarrow} H_2S$$

The numbers in brackets are the presumed kinetic isotope effects (expressed as $f - 1$) for each stage and are based on theoretical considerations and available information from studies on the reduction of sulphate and sulphite (e.g., Harrison and Thode, 1958; Kemp and Thode, 1968). Rees (1973) derived the expression:

$$\alpha = \alpha_1 + (\alpha_2 - \alpha_{-1})X_1 + (\alpha_3 - \alpha_{-2})X_1 X_2 + (\alpha_4 - \alpha_{-3})X_1 X_2 X_3$$

where α is the overall isotope effect α_1, α_{-1}, etc. are the kinetic isotope effects for the forward (+ 've) and back (− 've) reactions of stages 1–4, and X_1, X_2, and X_3 are the ratios of backward and forward flows in stages 1–3 respectively. It was also assumed that $0 \leqslant X_3 \leqslant X_2 \leqslant X_1 \leqslant 1$.

The main consequences of this model are:

(1) When the rate of reduction of sulphite (stage 4) *is high relative to all other rates,* the backward flow from sulphite is not established and X_3 is zero. Fractionation is then dependent upon X_1 and X_2. If sulphate uptake is completely rate-limiting (e.g., at very low sulphate concentrations) *no* backward flows are established, X_1 and X_2 are zero, and the overall isotope effect is given by α_1 ($f = 1.003$ in favour of $^{34}SO_4$). As the sulphate concentration is increased backward flows in stages 1 and 2 become significant, X_1 and X_2 are no longer zero, and the overall isotope effect rises to a maximum of 1.022 (in favour of ^{32}S) when X_1 and X_2 are each close to unity, that is when the overall reduction is no longer dependent on sulphate uptake.

(2) When the rate of reduction of sulphite is comparable to those of other steps in the sequence, the backward flow between sulphite and APS will assume importance (i.e., X_3 will be no longer zero) and isotopic effects associated with sulphite reduction will be superimposed on those resulting in stages 1–3. In the extreme case, when X_1, X_2 and X_3 approach unity, the overall isotope effect will be close to the sum of α_1, α_3 and α_4 (since all other isotope effects are assumed to be zero), that is 1.047 favouring $H_2{}^{32}S$.

Rees' model can account for most of the isotopic effects reported for sulphate reduction by dissimilatory sulphate-reducing bacteria which range from a little below 1 to 1.046 (Kaplan et al., 1960; Kaplan and Rittenberg, 1964). The model has also been applied (Rees, 1973) to account for isotopic distributions in the Black Sea where sulphide is enriched in ^{32}S over sulphate by a factor of about 1.050 (Vinogradov et al., 1962).

Rees considers that this isotope effect is largely "due to the internal mechanisms of *D. desulfuricans*". Schwarcz and Burnie (1973) have discussed the depositional environments of a number of sedimentary sulphide deposits in terms of Rees' model. Considerably greater enrichments of ^{32}S in sulphide over coexisting sulphate, however, have occasionally been reported. Examples are the sulphides in recent marine sediments off southern California (f = 1.062; Kaplan et al., 1963) and in waters of Green Lake, New York (f = 1.056–1.059; Nakai and Jensen, 1964). Fractionation factors of this magnitude cannot be accommodated by Rees' (1973) model if his presumed isotope effects for the various stages are close to reality. On the other hand Kaplan et al. (1963) considered that sulphur recycling, a possible cause of high fractionation (vide infra), was unlikely in the southern Californian sediments.

Smejkal et al. (1971) drew attention to the fact that sulphite reduction by assimilatory organisms such as yeast and *Salmonella* may result in ^{32}S enrichment in sulphide with fractionation factors as high as 1.044 (Kaplan and Rittenberg, 1964; Krouse and Sasaki, 1968; Krouse et al., 1967, 1968). They also presented evidence that in certain spring waters of western Canada complete reduction of sulphate to sulphide was catalysed by the "symbiotic" activities of two *Clostridium* sp., one reducing sulphate to sulphite, the other converting the latter to hydrogen sulphide. In such symbiotic situations therefore, the additive effects of isotope discrimination between sulphate and sulphite (1.022) and between sulphite and sulphide (up to 1.044) could result in overall fractionation factors between sulphate and sulphide as high as 1.066. Nevertheless, until a significant contribution of assimilatory sulphate reduction to the formation of sulphides in sediments has been established this explanation for high enrichment values must be treated with some reserve.

The foregoing discussion has dealt with kinetic isotope effects as they apply in so-called "open systems" where there is essentially an infinite supply of sulphate, the isotopic composition of which changes little during the course of reduction. Further variations in isotopic differentiation between sulphate and sulphide are to be expected in "closed systems" where appreciable amounts of sulphate are utilized and, as a consequence, the latter becomes enriched in ^{34}S (Nakai and Jensen, 1964; Jensen and Nakai, 1964; Kemp and Thode, 1968).

The situation is analogous to fractional distillation and the change in isotopic composition of the residual sulphate is given by the expression:

$$R_t = R_0 \times F^{1-k_2/k_1}$$

where R_t and R_0 are the ^{32}S/^{34}S ratios of sulphate at times t and zero respectively, F is the fraction of sulphate remaining at time t, and k_2 and k_1 are the unidirectional rate constants for reduction of ^{34}SO$_4$$^{2-}$ and ^{32}SO$_4$$^{2-}$ respectively (Nakai and Jensen, 1964). Similarly the isotopic composition of the accumulated sulphide will change according to the expression:

$$r = \frac{F^{(k_2/k_1)^{-1}} - F}{1 - F}$$

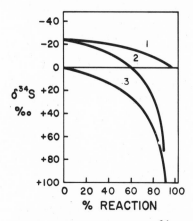

Fig. 2. Theoretical changes in δ^{34}S (relative to starting sulphate) during sulphate reduction in a closed system with a constant fractionation factor (f) of 1.025. Curves: 1 = accumulated H_2S; 2 = H_2S generated during the course of reaction; 3 = residual sulphate.

where r is the ^{32}S/^{34}S ratio of sulphide at time t relative to that of the remaining sulphate.

Fig. 2 shows graphically the theoretical relationships between the isotopic compositions of residual sulphate (curve 3) and accumulated sulphide (curve 1) and the extent of sulphate reduction with a kinetic isotope effect of 1.025 (Kemp and Thode, 1968). Also shown (curve 2) is the expected pattern of isotopic composition of sulphide produced at any instant in the overall reduction sequence. It may be noted that:

(1) the isotopic differentiation between accumulated sulphide and remaining sulphate increases with the extent of reaction: when 95% of the sulphate is reduced, sulphide would be enriched about 82‰ with respect to remaining sulphate;

(2) if sulphide were continually or intermittently removed from the system sulphides produced after about 60% reaction would be enriched in ^{34}S with respect to the starting sulphate.

These predictions have been confirmed experimentally by Nakai and Jensen (1960, 1964), Jensen and Nakai (1964) and by Kemp and Thode (1968).

In sedimentary environments where rates of sulphate reduction are likely to fluctuate (e.g., Sorokin, 1970) and interstitial waters in the sediments equilibrate to varying degrees with the overlying waters, considerable variations in the isotopic composition of biogenic sulphides are to be expected. This situation is realized, for example, in the recent marine sediments off southern California where reported δ^{34}S values (with respect to meteoritic) in pyrite range from +11.2‰ to −42‰ with a mean of about −21‰ (Kaplan et al., 1963). All pyrite samples were also considerably enriched in ^{32}S with respect to sea water sulphate (δ^{34}S w.r.m. = +20.2 to +20.6) which is consistent with biological sulphate reduction in an open or partially closed system (see also Thode et al., 1960).

Enrichment in ^{32}S and/or wide variations in isotopic composition of sulphide minerals

TABLE X

Sulphur isotope compositions of deposits for which a biogenic origin has been proposed

Deposit	Type	δ^{34}S sulphide ‰ relative to meteoritic	δ^{34}S sulphate ‰ relative to meteoritic	Reference
Dzhezkasgau (U.S.S.R.)	Cu–Fe–Pb–Zn	−4.7 to −52.2 (Av −17.6; S.D. 10.3)		1
Udokan (U.S.S.R.)	Cu–Fe–Pb	−7.7 to −22.6 (Av −12.9; S.D. 5.4)		1
Olekma-Vitim (U.S.S.R.)	stratified Cu–Fe	−21.3 to +12.3 (Av −8.0; S.D. 7.8)		2
Rammelsberg (Germany)	Fe–Zn–Pb–Cu	pyrite −15 to +23.2 (Av +8.4; S.D. 12.9) other sulphides +5.1 to +22.1 (Av +14.0; S.D. 4.2)	+13.2 to +36.7 (Av +24.2; S.D.4.4)	3
Nairne (South Australia)	sedimentary pyrite	−11.7 to −20.6 (Av −17.2; S.D. 2.8)		4
Amjhore (India)	stratabound pyrite	+4.5 to +20.0 (Av +9.3; S.D. 3.15)		5
S. Giovanni (Sardinia)	Fe–Zn–Pb	−14.4 to 37.2 (Av +21.9; S.D. 6.3)		6
Mt. Isa (Queensland)	sedimentary Fe–Cu–Pb–Zn–Ag	+0.2 to +30.8 (Av +16.0; S.D. 6.6)		7
Various	sandstone uranium	Jurassic −41.0 to +17.6 (Av −16.9; S.D. 16.7) Triassic −50.0 to +10.8 (Av −35.8; S.D. 11.5) Tertiary −28.8 to −48.2 (Av −36.3)		8
Southeast* (Missouri)	stratabound Fe–Pb–Zn	−9.8 to +34.9 (Av +12)	**	9
Kupferschiefer* (Germany)	stratiform	−3 to −45	−30 to +26	10
White Pine (Michigan)	cupriferous shale	−5.8 to +32.6 (Av +3.5; S.D. 11.7)		11
Wieslock* (Germany)	Fe–Pb–Zn	−22 to +23	+28 to +90	12
Mount Gunson (South Australia)	Cu	−15.6 to +26	+13.3 to +19.1 (ground waters) +14.6 to +17.8 (gypsum)	13

* Approximate data from graphical report; ** average δ^{34}S of galena with respect to coexisting barite = −24‰ (Ault and Kulp, 1960).

References: 1 = Chukrov (1971); 2 = Bogdanov and Golubchina (1971); 3 = Anger et al. (1966); 4 = Jensen and Whittles (1969); 5 = Guha (1971); 6 = Jensen and Dessau (1966); 7 = Solomon (1965); 8 = Jensen (1958); 9 = data from Ault (1959), biogenic interpretation from Jensen and Dessau (1967) and Jensen (1971); 10 = Marowsky (1969); 11 = Burnie et al. (1972); 12 = Gehlen (1966); 13 = Donnelly et al. (1972).

are now often taken as being indicative of biological association with large-scale mineralization. Some examples of deposits to which this reasoning has been applied are listed in Table X. In some instances sulphate from the same deposits has been shown to be enriched in ^{34}S relative to sulphides which is an additional factor in favour of a biogenic origin for the sulphide. The examples cited are primarily of pyrite and base metal sulphide mineralization but include sandstone-type uranium deposits where the presence of associated light iron and copper sulphides has been taken as evidence of biological activity which, indirectly, might have promoted uranium mineralization (see Introduction).

The isotopic ranges for the Permian Kupferschiefer deposit in Central Europe shown in Table X are rather misleading since the major portion of sulphide samples examined had δ^{34}S values between about −15‰ and −40‰ while roughly 50% of sulphate samples were enriched in ^{34}S with respect to meteoritic (Marowsky, 1969). Many of the sulphate samples had δ^{34}S values above +11‰ which is considered to be the value for Permian marine sulphate (Holser and Kaplan, 1966) while oxygen isotopic evidence suggested that light sulphate may have arisen from oxidation of sulphides. The general isotopic distribution in the Kupferschiefer deposit is compatible with bacterial sulphate reduction of marine sulphate in a closed basin (Marowsky, 1969; Wedepohl, 1971), a view which is reinforced by examination of the δ^{34}S profiles of sulphides and sulphate of the Kupferschiefer and overlying Zechstein-Kalk. In the example illustrated in Fig. 3, except for two anomalous peaks in the sulphate profile, there is a remarkable parallelism in the profiles with the sulphides displaced by about −30 to −40‰ relative to sulphate. There is also a general trend from light to heavy sulphur from the base of the Kupferschiefer to the top of the Zechstein as would be predicted for a closed system (see Fig. 2). The parallel

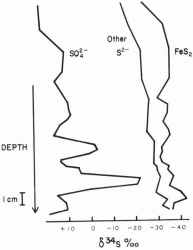

Fig. 3. Vertical profiles of δ^{34}S values (with respect to meteoritic) for sulphides and sulphate of the Kupferschiefer, Ibbenbüren, Westfalen, Germany. (After Marowsky, 1969.)

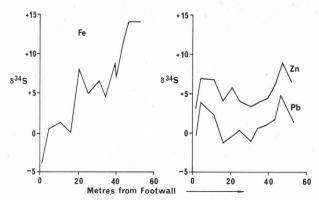

Fig. 4. Vertical profiles of $\delta^{34}S$ values (with respect to meteoritic) of FeS_2, PbS and ZnS in a core from the mineralized zone of the McArthur River deposit, Northern Territory, Australia.

trends in $\delta^{34}S$ for pyrite and the Cu + Pb + Zn sulphides (Fig. 3) strongly suggest syngenetic deposition from a common solution. The preferential enrichment of ^{32}S in pyrite has been attributed to selective precipitation of metals due to their different solubilities with increasing sulphide production (Marowsky, 1969; Wedepohl, 1971).

Isotopic profiles have also provided important information on the mineralization in the Pb–Zn–Ag deposit at the McArthur River in northern Australia (Smith and Croxford, 1973).[1] There is again a general trend towards ^{34}S enrichment in pyrite as one proceeds upwards from the footwall (Fig. 4) which is consistent with bacterial sulphate reduction in a closed basin. The McArthur River deposit differs from the Kupferschiefer, however, in that the isotopic patterns for galena and sphalerite are quite distinct from that of pyrite indicating different modes of mineral deposition. The distinction has been ascribed to differing sources of sulphur for syngenetic Pb–Zn and Fe mineralization (Smith and Croxford, 1973) or to epigenetic replacement of biogenic pyrite by Pb and Zn accompanied by homogenization of sulphur (Williams and Rye, 1974). Pyrite is also distinguished isotopically from galena, sphalerite and chalcopyrite in the Rammelsberg deposits of Germany (Anger et al., 1966; see Table X). In this instance it was concluded that PbS, ZnS and possibly $CuFeS_2$ were of syngenetic magmatic hydrothermal origin and that FeS was subsequently transformed to pyrite by reaction with sulphur derived biogenically from sulphate.

While in the foregoing examples the isotopic data can be interpreted in terms of biogenic or partly-biogenic models, it should be emphasized that alternative explanations can generally be advanced. A recent theoretical study by Ohmoto (1972), for example, on the effects of temperature, oxygen fugacity and pH on isotope partitioning in hydro-thermal solutions demonstrated convincingly that bacterial sulphate reduction need not be the only cause of wide sulphur isotope variations in nature. A striking illustration of

---

[1] Editor's note: For additional discussions, see Chapter 12 by Lambert on the McArthur deposits in Vol. 6.

differing interpretations placed on isotopic data is that of the relationship between sulphides and ocean sulphate of different geological ages. Sangster (1968, 1971) noted that,
while the isotopic composition of ocean sulphate has fluctuated widely since the Precambrian (Nielsen, 1966; Holser and Kaplan, 1966), the average isotopic compositions of
stratabound volcanic and sedimentary sulphides are generally displaced by about $-17\%_0$
and $-14\%_0$ respectively relative to coeval ocean sulphate. On this evidence Sangster
(1971, p. 298) concluded that "sulphur in stratabound sulphide deposits is largely or
entirely of biogenic origin and had as its source sulphate dissolved in coeval sea water".
On the other hand Sasaki (1970) and Sasaki and Kajiwara (1971) have demonstrated that,
at least in the case of volcanic stratabound sulphides, isotopic exchange between sulphide
and ocean sulphate under hydrothermal conditions could be an acceptable alternative
explanation for Sangster's observations.

Clearly then isotopic evidence per se cannot provide a basis for definitive conclusions
regarding the origins of sulphide deposits but must be viewed in the light of all attendant
geological, geochemical and mineralogical information.

SULPHUR DEPOSITS

Bacterial sulphate reduction has also been implicated in the formation of certain
economic deposits of elemental sulphur. Hunt (1915) and Dessau et al. (1962), for
example, suggested that the Sicilian deposits may have arisen from oxidation of bacterially produced sulphide under conditions similar to those existing in the Black Sea. Later
Schneegans (1935) and ZoBell (1946) proposed a microbiological origin for some French
deposits and the Texas and Louisiana deposits, respectively (see also Starkey and Wight,
1945; Jones et al., 1956).

Sulphur deposition, almost certainly involving biological factors has been observed in
several present-day environments. Lake Ain-ez-Zauia, a saline lake in the northern part of
the Lybian Desert in Africa, was extensively studied by Butlin and Postgate (1954). The
sediments of this lake contain about 50% sulphur which is used by the local Arabs. The
lake is fed by warm springs and its waters are essentially saturated with respect to calcium
sulphate. Hydrogen sulphide concentrations range from 15 to 20 ppm at the surface to
100 ppm in the bottom waters. Butlin and Postgate noted the presence of sulphate-
reducing bacteria in Lake Ain-ez-Zauia and the prolific development of the photosynthetic sulphide-oxidizing bacteria, *Chlorobium* and *Chromatium*, in its warm shallow
waters. In the presence of excess sulphide the latter organisms catalyse the formation of
elemental sulphur which is precipitated either within the bacterial cell (*Chromatium*) or
in the external milieu (*Chlorobium*). Butlin and Postgate proposed the model shown in
Fig. 5 whereby the photosynthetic organisms supply organic matter for reduction of
sulphate to sulphide and at the same time oxidize the latter to elemental sulphur. They
were able to reproduce this model in laboratory experiments. A similar situation to Lake
Ain-ez-Zauia appears to apply to the sulphur-bearing Sernoye Lake in the Kuybyhsev

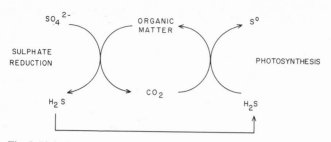

Fig. 5. Biological production of elemental sulphur from sulphate. (Butlin and Postgate, 1954.)

region of the Volga River. This lake is fed by sulphidic spring-waters in which sulphate-reducing bacteria, algae, and photosynthetic sulphide-oxidizing bacteria (quoted by Pankina and Mekhtiyeva, 1969) and non-photosynthetic sulphide-oxidizing bacteria (Sokolova, 1962) have been detected. From the results of studies on the oxidation of ^{35}S-sulphide by the lake waters, Ivanov (1957) concluded that the latter organisms were primarily responsible for sulphur production, only about one third being due to chemical oxidation of sulphide.

Ivanov (1968) questioned the Butlin-Postgate model for sulphur production on the grounds that the known stoichiometry between CO_2 fixation and sulphide oxidation by photosynthetic bacteria is about an order of magnitude less than would be required to sustain the cycle outlined in Fig. 5. On the other hand isotopic analysis of the sulphur compounds in Lake Ain-ez-Zauia (Macnamara and Thode, 1951) and Sernoye Lake (Pankina and Mekhtiyeva, 1969) show δ^{34}S values for sulphur and/or sulphide in the order of -30% relative to sulphate strongly suggesting that bacterial sulphate reduction played an integral part in sulphur production. This would imply the existence of additional unknown sources of organic matter, a possibility which was recognized by Butlin and Postgate (1954).

SubbaRao et al. (1947) suggested that bacterial sulphate reduction followed by iron-catalysed oxidation of hydrogen sulphide may account for the formation of sulphur in clays of coastal areas in India containing up to 35% of the element. They demonstrated the formation of sulphur in field trials using sulphate-reducing bacteria. Extensive micro-biological and chemical studies on native sulphur deposits throughout the Soviet Union have been summarized by Ivanov (1968). In general these deposits are associated with the presence of sulphate-reducing bacteria and non-photosynthetic sulphide-oxidizing bacteria of the genus *Thiobacillus*. The latter organisms proliferate when H_2S-rich anaerobic ground waters merge with aerobic surface waters and they are suggested to play an important part in sulphur deposition.

Feely and Kulp (1957) and Krauskopf (1967), amongst others, have drawn attention to the fact that native sulphur in sedimentary rocks generally occurs in regions where gypsum or anhydrite is, or has recently been, in close contact with petroleum and natural gas. Deposits such as those of Texas and Louisiana are frequently associated with salt domes and the sulphur is found within and under "cap rock", which consists largely of gypsum or anhydrite, together with calcite, and which forms a mantle overlying the salt core. In view of the association of native sulphur with organic matter and sulphate minerals and the demonstration of sulphate-reducing bacteria in sulphur-bearing salt domes (ZoBell, 1947; Miller, 1950a), the idea of a biogenic origin for the sulphur is attractive although not universally accepted. Petroleum was suggested by Schneegans (1935) to be a possible source of organic matter for the biological formation of French sulphur deposits.

There have been a number of reports of sulphate-reducing bacteria metabolizing hydrocarbons and crude oils (Tausson and Alishima, 1932; Tausson and Veselov, 1934; Novelli and ZoBell, 1944; Rosenfeld, 1947; ZoBell, 1950; Davis and Yarbrough, 1966). As mentioned elsewhere many of these reports are possibly dubious since authentic pure cultures were not used (Postgate, 1959). This criticism, however, is probably not serious when applied to a natural situation in which mixed microbial flora are involved. A more serious criticism raised by Postgate was that the purity of reagents was rarely specified. Some doubts must be held, therefore, whether the hydrocarbons themselves rather than impurities were metabolized. Such doubts appear not to apply to the experiments of Davis and Yarbrough (1966), who demonstrated the formation of radioactive carbon dioxide from ^{14}C-labelled methane, ethane and n-octadecane in cultures of *D. desulfuricans*. Nevertheless, even in this instance, metabolism of hydrocarbons proceeded to a very small extent and only in the presence of additional organic matter (lactate). The question as to whether hydrocarbons per se may be primary organic sources for bacterial sulphate reduction remains, therefore, to be resolved.

A number of isotopic studies have provided support for the concept of biogenic sulphur deposits. Sulphur in many Texan and Louisianan (Thode et al., 1954; Jones et al., 1956; Feely and Kulp, 1957; Davis and Kirkland, 1970), Sicilian (Dessau et al., 1962) and Polish (Jensen, 1962) deposits are highly enriched in ^{32}S with respect to the associated anhydrite. Thode et al. (1954), for example, reported an average δ^{34}S (with respect to coexisting sulphates) of $-39‰$ for sulphur in ten salt dome deposits. Calcite associated with the sulphur zones is also generally enriched in ^{12}C and has a carbon isotopic composition similar to that of coexisting petroleum (Thode et al., 1954; Feely and Kulp, 1957; Davis and Kirkland, 1970). These observations strongly suggest that organic matter rather than the atmosphere was the source of carbon dioxide and the overall process of sulphur formation may thus be described by eq. 5 and 6:

$$CaSO_4 + (C + 4H) \xrightarrow{\text{bacteria}} H_2S + CaCO_3 + H_2O \tag{5}$$
$$\text{petroleum}$$

$$H_2S + \tfrac{1}{2}O_2 \xrightarrow[\text{or chemical}]{\text{bacteria}} S^0 + H_2O \tag{6}$$

PHOSPHORITE DEPOSITS

Aside from the guano deposits and bone beds, which have obvious associations with the animal kingdom, major sources of the world's phosphate minerals are the bedded pelletal or nodular phosphorites of marine origin (McKelvey, 1967). The latter are discussed in detail by Cook (Chapter 11, Vol. 7) who draws attention to the alternative chemical (e.g., Kazakov, 1937; Dietz et al., 1942; McKelvey et al., 1953; Sheldon, 1964) and biochemical mechanisms which have been proposed for their formation (see discussions by Charles, 1953; Wilcox, 1953; Bushinskii, 1966; Gulbrandsen, 1969). Oblique, and far from convincing, evidence for the existence of microbiogenic phosphorites is the presence in some deposits of "fossilized bacteria" (Cayeux, 1936; Oppenheimer, 1958; S.R. Riggs, quoted by Pevear, 1967). Somewhat more compelling evidence is the presence in certain phosphorite deposits of stromatolites—structures which are thought to be formed by the secretion and accretion of mineral matter by benthonic algae in intertidal and shallow sublittoral zones (Logan et al., 1964; Monty, 1965). Examples of stromatolitic phosphorites are the Precambrian Aravallian Succession in India (Banerjee, 1971a,b), the Late Precambrian Calc Zone of Pithoragarh Kumaon Himalaya (Valdiya, 1969) and the phosphorite-bearing Sinian rocks of China and in some parts of U.S.S.R. (Bushinskii, 1969).

Generally, however, both calcareous and phosphoritic structures are found and the significance of the latter is obscure in view of claims (Ames, 1959; Youssef, 1965; Pevear, 1966, 1967) that some phosphorites may be formed by secondary phosphatization of carbonate minerals rather than by direct precipitation.

As discussed by Cook (Chapter 11, Vol. 7) present-day accumulations of phosphate minerals often occur in regions of ocean upwelling where cold deep waters, rich in phosphate and other elements, merge with warmer shallow waters. Deep ocean waters may contain up to 0.3 ppm PO_4^{3-} compared with about 0.01 ppm for surface waters (McKelvey, 1967). Because of the high concentrations of phosphorus and other nutrients upwelling regions support the growth of massive amounts of plankton (Ryther, 1963), with an average phosphorus content in the order of 3,500 ppm (Table I). Proponents of "biochemical accumulation" conclude that precipitation of dead plankton followed by release of cellular phosphorus may be the mechanism by which phosphate becomes concentrated in sediments (e.g., Charles, 1953; Wilcox, 1953; Youssef, 1965; Bushinskii, 1966; Gulbrandsen, 1969). A parallel trend between loss of organic matter and phosphatization during lithification of phosphatic pellets in sediments from the continental shelf of southwest Africa was noted by Romankevich and Baturin (1972; see Fig. 6). The stoichiometry, however, does not allow firm conclusions to be drawn regarding the role of organic matter in the phosphatization process.

On the basis of estimates of total phosphorus and the area of deposition for the Permian Phosphoria Formation (McKelvey et al., 1953), and a phytoplankton productivity of about 1.3 g C m^{-2} day^{-1}, Gulbrandsen (1969) calculated that all the phosphorus

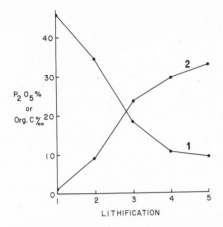

Fig. 6. Phosphate (curve *1*) and organic matter (curve *2*) composition of phosphorite nodules at different stages of lithification. (Romankevich and Baturin, 1972.) Lithology: *1* = diatomaceous ooze enclosing concretions; *2* = phosphatized diatomaceous ooze; *3* = friable phosphorite concretion; *4* = globular phosphorite concretion; *5* = dense phosphorite concretion.

in this formation could have been derived from planktonic sources in about $2 \cdot 10^5$ years, a relatively short period within Permian time. A similar estimate was derived by Cressman and Swanson (1964) for formation of the Retort Member of the Phosphoria Formation. Since planktonic production in regions of ocean upwelling may reach as high as 4 g C m^{-2} day^{-1} (Steeman-Nielsen and Jensen, 1957), the above estimates do not appear to be unreasonable. They were based, however, on the assumption that essentially all cellular phosphorus became mineralized. The phosphorus of living microorganisms is contained largely in nucleic acids, phospholipids, organic phosphate esters and, occasionally, storage material in the form of inorganic polyphosphates (Luria, 1960; Zajic, 1969). All these compounds are rapidly degraded on death of the cell and large-scale incorporation of phosphorus into sediments would seem to require:

(*1*) shallow-water conditions allowing for rapid deposition of dead organic remains with the minimum of degradation during their descent to the bottom (Bushinskii, 1969); and

(*2*) efficient mineralization of released phosphate which may be facilitated by the coexistence of relatively large amounts of calcium in the cellular structures (Table I).

The biochemical-accumulation hypothesis also demands the efficient removal of organic matter which amounts to some 100-fold excess of carbon over phosphorus (Bowen, 1966, pp. 173–210). The conditions leading to degradation of organic matter are not clear. Gulbrandsen (1969) considered that an abundant supply of oxygen is necessary for the rapid decomposition of the bulk of organic matter (see also Berge, 1972). This view is consistent with the observations of Dietz et al. (1942) that phosphatic nodules off California appear to have formed in sediments which are relatively low in organic matter and which are situated in some of the most oxidizing environments of the sea floor. On the other hand, Krumbein and Garrels (1952) concluded that marine phosphorites were probably formed in restricted anaerobic basins with low redox potential and mildly alkaline pH. Support for this conclusion is the fact that uranium in sea-floor phosphorites off California exists mainly in the tetravalent state (Altschuler et al., 1958) which is not stable in marine environments in which the Eh is positive (Garrels, 1960). Blackwelder (1916) and Mansfield (1927) had earlier suggested a relationship between low oxygen tension and phosphorite deposition while Youssef (1965), in a discussion on the genesis of bedded phosphates, noted the frequent association of phosphates with organic matter, pyrite and black shales all of which point to an essentially reducing environment (Pettijohn, 1957).

While planktonic accumulation and deposition of phosphorus has been most widely discussed in relation to phosphorite formation other hypotheses have been advanced. Gulbrandsen (1969) proposed a novel biochemical mechanism for phosphate accumulation based on the fact that calcium salts of reduced phosphorus compounds, particularly hypophosphite (PO_2^{3+}) are appreciably more soluble than those of phosphate (Gulick, 1965). He suggested that bacterial reduction of phosphate under strongly reducing conditions could lead to high concentrations of soluble phosphorus which, on subsequent oxidation, would lead to apatite precipitation.

Biological oxidation of phosphite has been reasonably well documented. Adams and Conrad (1953) reported that orthophosphite added to soils disappears with a corresponding increase in phosphate: the transformation was prevented by the addition of a biological inhibitor, toluene. Casida (1960) demonstrated that orthophosphite could provide the phosphorus requirements for the growth of fourteen strains of bacteria and yeasts. In only one instance did orthophosphate accumulate: in others phosphorus was presumably utilized for the formation of cellular constituents. Oxidation of orthophosphite by inducible enzyme systems was demonstrated in a number of bacteria by Malacinski (1968; see also Malacinski and Konetska, 1966) who suggests that organisms may be normally exposed to reduced phosphorus compounds in the natural environment. The enzyme from *Pseudomonas fluorescens* was studied in some detail (Malacinski and Konetzka, 1967): it appeared to be specific for orthophosphite being inactive towards arsenite, hypophosphite, nitrite, selenite, sulphite and tellurite. A hypophosphite oxidase which shows no activity towards orthophosphite has recently been demonstrated in *Bacillus caldolyticus* (Heinen and Lauwers, 1974).

Rudakov (1927, 1929) reported that soil organisms, including pure cultures of bacilli, reduced orthophosphate and presented evidence for a reduction sequence through orthophosphite and hypophosphite to phosphine:

$$PO_4^{2-} \rightarrow PO_3^{2-} \rightarrow PO_2^{3-} \rightarrow PH_3$$

Phosphine has been found in polluted springs (Lüning and Brohm, 1933) and the production of phosphite and orthophosphite in model paddy fields has been described by

Tsubota (1959). In the latter instance organisms such as *Clostridium butyricum* and *Eschericia coli,* capable of generating an extremely low oxidation-reduction potential appeared to be responsible.

Thus there appears to be at least circumstantial evidence for a biological oxidation-reduction cycle of phosphorus in nature but in view of the limited information available, the quantitative significance of this cycle from a geochemical standpoint must be viewed with considerable caution. In particular, reduced forms of phosphate have not been recognized in the modern marine environment (Gulbrandsen, 1969), although Gulick (1965) suggests that they may have been more prevalent in Precambrian seas.

Another role for microorganisms in the formation of phosphate minerals is suggested by studies on calcification by oral bacteria. In their natural peridental environment, as well as in vivo, oral bacteria provide a matrix for hydroxy-apatite formation which may be intracellular or related to the cell surface (Rizzo et al., 1963). Ennever (1963) reported that *Bacterionema matrichotii,* a prominent organism in dental plaque, converts intracellular calcium salts to calcium hydroxyapatite: a factor responsible for the mineralization was demonstrated in extracts of the organism.

McConnell et al. (1961, 1962) had earlier reported that the enzyme, carbonic anhydrase, catalysed the formation of carbonate-hydroxyapatite in the presence of sodium phosphate, calcium chloride and carbon dioxide over a period of several days: mineral formation was prevented by the enzyme inhibitor sulfanilamide. Carbonic anhydrase catalyses the reversible dissociation of carbonic acid (eq. 7)

$$H_2CO_3 \rightleftharpoons H_2O + CO_2 \tag{7}$$

and so has the potential to modify reactions which depend upon the concentration of carbon dioxide. The enzyme is found in highest concentration in tissues, such as erythrocytes, gastric mucosa and photosynthetic tissues, where rapid H_2CO_3/CO_2 equilibration is physiologically important (Davis, 1961) but it has also been detected in bacteria (Veitch and Blakenship, 1963) and a number of fresh water and marine algae (Litchfield and Hood, 1962). Berger and Libby (1969a) have shown that small quantities of carbonic anhydrase increase up to 20-fold the exchange of atmospheric carbon dioxide with sea water and suggest that variation in enzyme content may account for observed differences in rates of equilibration with atmospheric CO_2 for ocean waters from different localities (Berger and Libby, 1968, 1969b). Direct evidence for the presence of significant quantities of carbonic anhydrase in sea water is, however, lacking.

McConnell et al. (1961) proposed that carbonic anhydrase is a controlling factor in bone and tooth mineralization and later (McConnell, 1965) extended this idea to account for the formation of carbonate fluoroapatite in marine phosphorite deposits. McConnell's hypothesis was challenged by Bachra and Trautz (1962) who were unable to demonstrate carbonic anhydrase-catalysed apatite formation. They also pointed out that chemical hydration of CO_2 occurs within seconds and suggested that stimulation of the hydration rate by carbonic anhydrase would have little significance in experiments lasting several days. This criticism is equally pertinent with respect to natural, phosphate-depositing environments, and the proposed role for carbonic anhydrase in phosphorite formation must, therefore, be viewed with scepticism. At the present time the biological phosphate accumulation hypothesis appears to be the most acceptable alternative to purely chemical mechanisms.

BIOLOGICAL PRECIPITATION OF MANGANESE

The importance of biological agencies in the oxidation of manganous ions has long been recognized (Beijerinck, 1913; Söhngen, 1914; Gerretsen, 1937; Leeper and Swaby, 1940). Mann and Quastel (1946), who studied the effects of biological inhibitors on manganous sulphate oxidation by soils, concluded that the reaction was almost entirely accomplished by biological means. Beijerinck (1913) isolated a bacterium, *Bacillus manganicus,* which oxidized $MnCO_3$ to manganic oxides in agar media and since then the ability to promote manganese oxidation has been shown to be widespread amongst bacteria, fungi and algae (see Silverman and Ehrlich, 1964; Alexander, 1965; Schweisfurth, 1970). In some instances symbiotic association of two bacterial species proved necessary for manganese oxidation. Bromfield and Skerman (1950) and Bromfield (1956), for example, reported the oxidation of manganous sulphate by the combined action of a *Corynebacterium* sp. and *Chromobacterium* sp. neither of which by itself was able to catalyse the reaction. A similar situation involving two species of *Pseudomonas* was described by Zavarzin (1962).

Bacteria belonging to the orders Chlamydobacteriales and Hypomicrobiales have particurlarly been associated with deposition of manganic oxides from fresh waters and have been the subject of a number of reviews (Pringsheim, 1949a; Wolfe, 1964; Zavarzin, 1967). Chlamydobacteria (e.g. *Sphaerotilus, Leptochrix, Clonothrix*) are colourless, filamentous bacteria consisting of cells enclosed within a sheath. The latter is composed of an organic matrix that is frequently impregnated with ferric or manganic oxides and hydroxides (Pringsheim, 1949a). The genus *Hyphomicrobium* or *Pedomicrobium* (Tyler and Marshall, 1967a) consists of single-celled organisms that are attached to long stalks at the distal ends of which daughter cells, or buds, form during reproduction (Hirsch and Conti, 1964). Manganese precipitation appears to take place on the outer surfaces of the stalks. *Metallogenium* spp. (Perfil'ev and Gabe, 1965) are also considered by Zavarzin (1967) to belong to Hypomicrobiales. They consist of thin filaments which radiate from a common centre and which are enclosed in an envelope of manganese oxide. In nature they appear to be closely associated with fungi (Zavarzin, 1961, 1967; Mirchink et al., 1970). It is noteworthy that presence of stalks, sheaths or filaments is characteristic of the above bacteria. This may be of considerable significance in manganese deposition since the structures provide a basis on which an intricate network of cellular material and metal precipitate may be built up.

The biological oxidation of manganese appears to require molecular oxygen (Ehrlich, 1966, 1968) and a possible expression for the reaction is given by eq. 8:

$$Mn^{2+} + \tfrac{1}{2}O_2 + 2OH^- \to MnO_2 + H_2O \qquad (8)$$

There is uncertainty, however, about the exact nature of the end product. According to Zavarzin (1967) and Kossaya (1967), manganese oxidation by *Metallogenium* gives rise to Mn^{IV} but Alexander (1965) suggests that oxides produced microbiologically seem to be intermediate between the trivalent and tetravalent forms, Mn_2O_3 and MnO_2. Kossaya (1967) concluded that during the growth of *Leptothrix* mixtures of manganese oxides are formed which ultimately approximate MnO_2 in composition. Schweisfurth and Gatlow (1966) reported that the composition of manganese oxides in natural black sludges of biological origin ranged from $MnO_{1.324}$ to $MnO_{1.800}$.

As a group manganese-oxidizing bacteria are active between 5° and 37°C (Rouf and Stokes, 1964) and about pH 6–10 (Cataldi, 1939; Ganter and Schwartz, 1956; Rouf and Stokes, 1964). Biological manganese oxidation, however, is generally most rapid at pH 6.0–7.5 (Leeper and Swaby, 1940) where chemical autooxidation of manganous ions is minimal (Alexander, 1965). Iron oxides are also frequently produced by manganese-oxidizing organisms but as discussed below deposits in which these organisms have been implicated are generally enriched in manganese. Tyler (1970) briefly reported that pure cultures of some strains of *Hyphomicrobium* preferentially oxidized manganese in culture media in which ferrous ions outnumbered manganous ions by several orders of magnitude.

There is considerable uncertainty as to whether manganese oxidation is physiologically significant and whether manganous ions are, in fact, metabolized by the bacteria. Beijerinck (1913) considered that manganese-oxidizing bacteria were autotrophic, that is capable of utilizing the energy of the oxidative reaction for growth in completely inorganic media. Similarly, Sartory and Meyer (1947) reported that *Leptothrix* and the iron-oxidizing bacterium *Gallionella ferruginea* could utilize iron and manganese as energy-yielding substrates in the absence of organic matter. The autotrophic nature of manganese-oxidizing organisms, however, has not been confirmed. Cataldi (1937, 1939), for example, showed organic matter was required for the growth of *Sphaerotilus* and she was unable to demonstrate growth stimulation by Mn^{2+} with this organism (see also Pringsheim, 1949b).

According to Johnson and Stokes (1966) oxidation of manganous ions by *Sphaerotilus discophora* is prevented by boiling the organisms and is carried out only by bacteria which were grown in the presence of manganese. These results suggested that manganese oxidation is catalysed by a specific enzyme whose production is induced by the presence of its substrate. Ehrlich (1968) studied manganese oxidation by extracts of an *Arthrobacter* species isolated from an Atlantic manganese nodule. The oxidation had an optimum temperature of about 18°C and was inhibited by mercurials and by heating the extract at 100°C. Catalytic activity was associated with a particular protein fraction which was partly separated from the extracts by gel filtration. Again these results are indicative of a specific enzyme-catalysed reaction. With extracts of *Arthrobacter* and with intact bacteria (Ehrlich, 1963, 1966), the presence of preformed manganese oxide, or certain other solid materials, was necessary for manganese oxidation. Ehrlich (1968, 1972) concluded that the initial reaction was an adsorption of Mn^{2+} to preexisting manganese oxide followed by bacterial oxidation of adsorbed Mn^{2+} creating new adsorption sites (eq. 9, 10).

$$H_2MnO_3 + Mn^{2+} \xrightarrow{\text{inorganic}} Mn\,MnO_3 + 2H^+ \tag{9}$$

$$Mn\,MnO_3 + 2H_2O + \tfrac{1}{2}O_2 \xrightarrow{\text{bacterial}} (H_2MnO_3)_2 \tag{10}$$

While the foregoing observations are indicative of enzymic catalysis of manganese oxidation, others have concluded that non-specific chemical catalysts are involved. Mulder (1964) described manganese oxidation by "toluene-killed" cells of *Leptothrix* and also reported that deposition of manganic oxides occurred at some distance from colonies of *Leptothrix* growing on agar plates suggesting the production of some diffusible oxidation catalyst. Söhngen (1914) noted that malate, citrate and other hydroxyacids catalyse the oxidation of manganous ions and suggested that microbial oxidation of manganese in the presence of citrate is due to the production of an alkaline pH which favours the hydroxyacid-catalysed reaction. Bromfield and Skerman (1950) demonstrated that many bacteria which promoted the oxidation of manganous salts in citrate-containing media failed to do so in soils where the concentration of hydroxyacids was probably low. Möse and Brantner (1966) and Brantner (1970) also believe that organic compounds are important in manganese oxidation and suggested that the principal function of microorganisms is to metabolize the organic moieties of soluble mangano-organic complexes leading to the release and precipitation of the metal. Suggestions to this effect have also been made by Baylis (1924), Aristovskaya (1961) and Sorokin (1972). It is possible that all of the proposed mechanisms can, under certain conditions, play a role in manganese precipitation in the natural environment.

Many studies on biological manganese oxidation have related to manganese deposition in pipe lines (see Tyler and Marshall, 1967a, for references). Zapffe (1931) attributed the formation of manganese-rich deposits in water supply pipes of Brainerd, Minnesota, to the activities of the chlamydobacteria, *Leptothrix* and *Crenothrix*. Tyler and Marshall (1967a–d) and Tyler (1970) concluded that *Hypomicrobium* spp. were intimately associated with ferro-manganese deposition in pipelines in various parts of the world, while Schweisfurth and Mertes (1962) and Möse and Brantner (1966) isolated various manganese-oxidizing cocci and rod-shaped bacteria from similar deposits. Tyler and Marshall (1967a) developed a laboratory system that simulated manganese deposition in pipelines. Sterilization of the circulating waters by autoclaving or by the addition of sodium azide prevented manganese precipitation clearly indicating the importance of biological factors at least for initiation of the depositional process.

TABLE XI

Iron and manganese compositions of pipeline deposits*

Source	Fe(%)	Mn(%)	Mn/Fe	Reference
Brainerd, Minnesota, U.S.A.**	15.3	29.3	1.92	1
Trier, Germany***	7.3	36.3	4.97	2
Newton, Mass., U.S.A.	12.6	21.5	1.71	2
Unstated, U.K.	24.9	25.8	1.04	2
Tarraleah, Tasm., Australia ***⁻	4.4	24.3	5.52	3
Clarence, Tasm., Australia⁻	3.7	15.2	4.11	3
Kareeya, Qld., Australia***	9.0	23.0	2.65	3
Queens' Park, W.A., Australia⁻.	7.1	15.3	1.90	3
Tumut No.1, N.S.W. Australia⁻	5.2	12.9	2.48	3
Nelson, New Zealand	4.6	22.4	4.87	3
Australian Capital Territory**	4.3	14.9	3.48	4
Gloucester, U.K.⁻	8.9	5.9	0.66	3
Sheffield, U.K.⁻	11.8	8.7	0.74	3
Rowallan, Tasm., Australia⁻	10.7	2.1	0.20	3
Tarago, Vic., Australia⁻	2.7	0.6	0.22	3

* Other constituents include SiO_2, Al, Ca, Mg, and organic matter; ** average of three values; *** average of two values; ⁻calculated from values given for sesquioxides.
References: 1 = Zapffe (1931); 2 = Tyler and Marshall (1967a); 3 = Tyler (1970); 4 = Trudinger (1971).

Some examples of the chemical composition of pipeline deposits are shown in Table XI. Two characteristics of these deposits are of particular significance:

(1) The deposits are generally enriched in manganese with respect to iron and in some cases the Mn/Fe ratio may be the reverse of that in the waters from which they were derived. For example, the deposits from the Australian Capital Territory with a Mn/Fe ratio of about 3.5 were formed from waters with an average ratio of 0.5 (Trudinger, 1971). Similar results were reported by Zapffe (1931) for waters and deposits in the Brainerd water supply system. The selectivity may be more marked than appears from Table XI since Tyler and Marshall (1967a) noted that the iron content in deposits is quite variable possibly because of variable contamination from rust in steel pipelines. Tyler (1970) suggests that the selective action of bacteria towards manganese may be the explanation for the existence of manganese ore bodies otherwise unaccountably low in iron. Krauskopf (1957) has also considered selective precipitation by bacteria to account for separation of iron and manganese in small deposits in some Swedish lakes (Ljunggren, 1955).

(2) The deposits represent extraordinary enrichments of metals from the associated waters. Wolfe (1960) cited an example in which bacteria, probably Clonothrix sp., obtained from filters of the Richmond Water Works Corporation, Indiana, contained 7.5% Mn and 6.8% Fe on a dry wt. basis compared with less than 0.01 p.p.m. of Fe and undetectable Mn in the original waters.

Manganese-oxidizing organisms have been detected in mineral springs (Hariya and Kikuchi, 1964) and a number of soils and manganese-rich fresh water sediments (Gabe et al., 1965; Shapiro, 1965; Ten, 1967, 1968; Sokolova-Dubinina and Deryugina, 1967a,b, and earlier references).[1] *Metallogenium* and other manganese bacteria have been recorded in the sediments of numerous lakes and, in some instances, were directly colonizing the ferro-manganese nodules extracted from lake bottoms (Sapozhnikov, 1970). Sapozhnikov also reported laboratory experiments which simulated the biological precipitation of manganese in the upper-oxidizing zones of lake sediments. These and other observations have led to suggestions that bacteria play a major role in the diagenesis of manganese and in the formation of large-scale manganese mineralization. Of particular interest in this respect are the ferro-manganese nodules which are commonly found in deep-ocean sediments, and which are potentially of economic importance since they contain significant quantities of elements such as cobalt, nickel, copper and lead (Mero, 1965, 1972). Purely physico-chemical mechanisms have been advanced for the formation of nodules (e.g., Goldberg and Arrhenius, 1958; Mero, 1965) but biological factors have also been implicated (Butkewitsch, 1928; Kalinenko, 1946; Kalinenko et al., 1962; Schweisfurth, 1971; Sorokin, 1972). Ehrlich (1963, 1966) isolated manganese-oxidizing bacteria from manganese nodules from the Atlantic Ocean. He suggested (Ehrlich, 1971) that the importance of these organisms in manganese deposition becomes significant when the mass of the nodule increases to the extent where the oxidation step becomes rate-limiting. Biological degradation of soluble organo-metallic complexes was considered by Sorokin (1972) to be an important factor in nodule formation. A similar suggestion was made by Graham and Cooper (1959) who reported the presence of manganese-rich coatings on Foraminifera isolated from deep-ocean muds during the Chain Cruise No. 2 and of organic matter in manganese nodules from the Blake Plateau (Graham, 1959). However, Goldberg (1965) suggested that the presence of organic matter in manganese nodules might be more readily explained by sorption from sea water or by the activities of benthic organisms and bacteria on the surfaces of the nodules. Recently, Monty (1973) proposed a biogenic origin for manganese nodules from the Blake Plateau. This proposal was based on the presence in the nodules of laminated structures reminiscent of stromatolites (vide infra) and of networks of "bacterial" filaments which are characteristic of biogenic deposits (Shapiro, 1965; Tyler and Marshall, 1967a).

The iron and manganese contents of ferro-manganese deposits from marine, lake and pipeline sources are compared in Table XII. In each case there are wide variations but there are general similarities in the average compositions of the various deposits and all, on average, show a significant preferential enrichment of manganese. These chemical similarities may reflect similarities in the mechanisms of formation of the various deposits.

Thiele (1925) and Zapffe (1931) recognized the possible implications of biological

---

[1] For details on Recent marine and fresh-water Mn-deposits, see in Vol. 7 Chapters 7 and 8 by Glasby/Read, and Callendar/Bowser, respectively.

Table XII

Comparison of manganese deposits from various sources

Source	Fe(wt. %)		Mn(wt. %)	Mn/Fe	Reference
Pacific Ocean	range	1.1 – 39.5	7.6 – 57.1	0.2 – 61.8	1
	av.	18.5	32.4	4.11	
Atlantic Ocean	range	1.54 – 25.9	2.4 – 18.9	0.2 – 12.3	1
	av.	15.7	13.1	1.5	
Indian Ocean	range	4.5 – 32.3	9.4 – 28.3	0.5 – 6.5	2
	av.	12.4	17.8	1.6	
Pipelines	range	2.7 – 24.9	0.6 – 50.0	0.2 – 8.5	3
	av.	8.1	18.8	2.9	
Lake deposits*	range	9.1 – 31.5	38.3 – 59.9	1.22 – 8.14	4
	av.	21.5	48.4	2.8	

* Developed in situ in peloscope.
References: *1* = Mero (1965); *2* = Bezrukov and Andruschenko (1973); *3* = Trudinger (1971); *4* = Shapiro (1965).

manganese precipitation with regard to the formation of sedimentary manganese ores. Specific ore bodies for which a biological origin has been proposed are the Nikopol deposits of the Ukraine (Bateman, 1959), those in the Tannengebirge district in Austria (Cornelius and Plöchinger, 1952) and the Chiatura deposits in Georgia (A.G. Betekhtin cited by Sokolova-Dubinina and Deryugina, 1966). In the latter instance *Metallogenium* was found to be associated with the mineralized zone (Sokolova-Dubinina and Deryugina, 1966) but, for the most part, evidence for participation of organisms in manganese ore genesis is at best circumstantial. (For particulars on ancient Mn-deposits, refer to Chapter 9 by Roy, Vol. 7.)

MICROBIAL PRECIPITATION OF IRON

There are two general classes of bacteria which are active in the precipitation of ferric iron. The first is exemplified by *Thiobacillus ferrooxidans* (or *Ferrobacillus ferrooxidans*), an autotrophic bacterium which utilizes the energy from oxidation of Fe^{2+} for growth. It is adapted to an acid environment (optimum pH for growth around 2) and was originally isolated from acidic drainage waters in a bituminous coal mine (Colmer and Hinkle, 1947; Colmer et al., 1950). The activities of *T. ferrooxidans* and related sulphide-oxidizing bacteria are now exploited in the extraction of metals from low-grade base-metal sulphides (Beck, 1967; Trudinger, 1971) and the organisms may play a significant role in weathering of sulphide minerals in nature (Kuznetsov et al., 1963). Because of their requirement for highly acidic conditions, however, these bacteria are unlikely to be important in the formation of iron ore.

The second class of iron bacteria consists of a group of varied organisms which grow at near neutral pH. Some are related to those catalysing manganese oxidation (vide infra) and include species of chlamydobacteria, *Hyphomicrobium* and *Metallogenium*. In some instances both iron and manganese can be precipitated by a single organism but for most species of metal-precipitating bacteria this bifunctional property has not been demonstrated. Additional organisms which are specifically associated with ferric precipitation are species of *Gallionella* which are kidney-shaped bacteria that form long, slender, twisted ribbons or stalks (Choldony, 1924). The latter appear to be composed largely of oxidized iron compounds which are presumably bound to some sort of organic matrix. The exact nature of the iron is uncertain. Microdiffraction, X-ray diffraction and Mössbauer analyses indicate that it is not simply $Fe(OH)_3$ (Mardanyan and Balashova, 1971).

The ecology and some of the general characteristics of the neutral iron bacteria have been discussed by Pringsheim (1949b).

Aside from these organisms, which appear to play a direct role in iron precipitation, other microorganisms, particularly the oxygen-evolving algae, participate indirectly by creating an oxidizing environment which facilitates Fe^{2+} oxidation (Choldony, 1926). In some instances the precipitated iron oxide (and manganese oxides) may form incrustations on the cell surface (Pringsheim, 1946).

Even less is known of the mechanism of iron precipitation by the iron bacteria than in the case of manganese. The organisms should perhaps be classified as iron-accumulating rather than iron-oxidizing since unequivocal evidence that they catalyse iron oxidation is lacking. It has, in fact, been suggested that their unique physiological characteristic may be a specific binding capacity for preformed ferric hydroxides (Ellis, 1907; Pringsheim, 1949b). Others (e.g. Clark et al., 1967) claim that ferric-organic complexes are absorbed into the cell and that iron is fixed in capsular material after metabolism of the organic moiety. The problem of the physiological significance of iron oxidation is difficult to resolve since Fe^{2+} undergoes rapid chemical autooxidation over the pH range of 5–8 which supports the growth of the iron bacteria (Baas Becking et al., 1956). It has been pointed out that if iron oxidation is a physiologically important process then the neutral iron bacteria must be regarded as gradient organisms, that is restricted to zones where the redox potential is sufficiently high to allow biological oxidation of iron but also low enough to minimize competition with rapid chemical iron oxidation (Pringsheim, 1949a; Wolfe, 1964). Baas Becking et al. (1956) reported that the natural Eh limits of the iron bacteria are between +200 and +500 mV and Mulder (1964) also alludes to the ability of some iron bacteria to grow and "oxidize" ferrous iron at low oxygen tensions.

The iron bacteria occur almost universally in ferruginous waters and they can grow and precipitate ferric hydroxide from streams containing less than 1 ppm of iron (Starkey, 1945a; Alexander, 1965). Like the manganese-oxidizing bacteria they are associated with metal oxide precipitation in pipelines (Starkey, 1945b; Wilson, 1945; Dondero, 1961). Species of chlamydobacteria appear to be frequent inhabitants of environments in which bog-iron ore is forming (Harder, 1919; Kuznetsov et al., 1963). The latter authors considered that microorganisms may play a dual role in the formation of bog-iron ore. Firstly the general microflora may create a reducing environment in the subsoil causing a reduction of fixed ferric iron to the mobile ferrous form which migrates upwards towards the oxidizing zone. Here iron is reoxidized to ferric hydroxide, a process in which the iron bacteria take part (also Starkey and Halvorson, 1927). The association of *Metallogenium* with lake sediments, mentioned earlier, raises the possibility that these and other iron bacteria are involved in the formation of lake iron ore. In a study of the dynamics of iron in Lake Krasnoe of the Isthmus of Karelia, Drabkova and Stavinskaya (1969) reported that the periodic mass development of iron bacteria (*Metallogenium, Siderocapsa* and *Gallionella*) was accompanied by a decrease in both Fe^{2+} and total soluble iron indicating a role for these organisms in the oxidation and deposition of iron during these periods.

Harder (1919), in an excellent and comprehensive review of the iron bacteria and their geological implications, concluded on theoretical grounds that iron bacteria could be important in the deposition of massive iron-formations. Zapffe (1933) discussed the possible analogies between present-day microbial precipitation of iron and manganese from natural waters and the origin of these metals in the Lake Superior iron-formation. Cloud (1968a,b, 1973) equated the deposition of the banded iron-formations in the period prior to $2 \cdot 10^9$ years B.P. with the development of blue-green algae which generated oxygen for the oxidation of ferrous iron (see also L.Y. Khodyush, 1969, cited by Alexandrov, 1973). Lepp and Goldich (1964) agreed that algae and bacteria may have been important in the genesis of Precambrian iron-formations possibly by modifying pH, Eh and CO_2 concentration. The controlling influences of pH, oxygen and CO_2 tensions on iron precipitation had been discussed earlier by Halvorson and Starkey (1927) and Halvorson (1931) and are treated in detail by Garrels and Christ (1965).

Direct evidence for the participation of microorganisms in the genesis of iron-formations is, however, equivocal and relies heavily on the presence of "biogenic" structures within the depositional environments. Lougheed and Mancuso (1973) believed that hematite framboids found in the Negaunee iron-formation in Michigan were derived from biogenic framboidal pyrite and cited this as evidence that biogenic processes were operative during deposition of the formation. Gruner (1922) observed structures in cherts of the Biwabik formation in the Lake Superior region which he claimed to be fossilized algae and iron bacteria. He suggested that these organisms had been largely responsible for the precipitation of iron and silica. LaBerge (1967, 1973) detected spheroidal "microfossils" in the Biwabik deposit and reported the presence of similar structures in cherts from the following Precambrian iron-formations: Kipalu, Belcher Islands; Temiscamie, Lake Albanel, Quebec; Gunflint, Ontario; Mt. Goldsworthy, Western Australia; Brockman, Western Australia; Marra Manuba, Western Australia and Ironwood, Michigan. The cherts of Gunflint iron-formation also contain structures which resemble present-day iron bacteria such as *Sphaerotilus, Gallionella* and *Metallogenium* (Tyler and Barghoorn, 1954; Barghoorn and Tyler, 1965; Cloud and Hagen, 1965; Cloud, 1965; Cloud and Licari, 1968).

On the basis of fossil evidence both LaBerge (1973) and Moorehouse and Beales (1962) proposed that microorganisms were possibly instrumental in precipitating the original constituents of iron-formations. Walter (1972a) noted similarities between siliceous stromatolites of the Lake Superior region iron-formations and those presently forming by algal and bacterial action in the hot spring and geyser effluents in Yellowstone National Park, Wyoming (Walter et al., 1972). He suggested that the ancient and modern stromatolites were built by similar, though not necessarily identical microbiota. The occurrence of stromatolites in the Hamersley Basin on the northwest coast of Australia, which contains extensive iron-formations, has been summarized by Walter (1972b).

Studies such as those just described are providing at least presumptive evidence for the existence of an active microflora in environments in which some of the major iron-formations arose. However, suggestions that the microflora was intimately related to the depositional process must be regarded as highly speculative. In particular great caution needs to be exercised in interpreting the presence of "microfossils" as evidence of biogenic mineralization. There will always be some doubt whether many of these structures are truly biological in origin and even where there is reasonable assurance that they represent fossilized microorganisms the specific biochemical properties which are assigned to them will have to be inferred from the nature of the environment in which they are found. Bacterial forms especially are, in general, not sufficiently distinctive to allow conclusive identification of "fossilized" organisms on the basis of structure alone. Finally, given that some of the reported structures are indeed fossilized-iron bacteria the importance of their contribution to the genesis of iron-formations will remain problematical at least until the role of bacteria in present-day iron deposition is more clearly defined.

CONCLUDING REMARKS

In this review an attempt has been made to illustrate the roles which microorganisms may have played in the formation of some of the main types of economic mineral deposits. The reader may be excused for concluding that, for the most part, these roles are equivocal and that speculation far outweighs the available evidence. Such a conclusion unfortunately reflects the present state of the art with respect to the mineral deposits in question. To the author's mind there is no single instance in which a biogenic origin can be claimed with confidence although in the case of some sulphide and sulphur deposits the possibility is quite high. On the available evidence cogent arguments, which are beyond the scope of this discussion, can be raised in favour of purely chemical mechanisms. Indeed in many instances proof of biological participation may be impossible to achieve due to lack of distinctive features of biologically-produced minerals over those of purely chemical origin. In these cases the final decision will rest on an assessment of the overall geological, mineralogical and chemical features of the deposit in question.

Nevertheless it is worthwhile pointing out that, while the concept of biogenic mineral deposits is by no means modern (e.g., Breger, 1911), it is only in recent years that serious research in this area has been undertaken and that more sophisticated techniques and instrumentation needed for this research has become available. One may look forward with some confidence to more persuasive information becoming available in the near future.

NOTE ADDED IN PROOF

A number of publications relevant to the present discussion have appeared since this Chapter was prepared.

Sulphides

The range of organic compounds oxidized by sulphate-reducing bacteria (Table IV) has been extended by the discovery of a *Desulfovibrio* sp. from sheep rumen, which utilizes glucose, fructose, galactose, ribose and mannose in addition to lactate and pyruvate (Huisingh et al., 1974).

Ramm and Bella (1974) described anaerobic microbial utilization of soluble organic matter derived from an algal mat from the surface of an intertidal mud flat. About two thirds of the organic matter was in the form of mannose and glucose and there were high levels of propionate, butyrate and acetate. In many instances the stoichiometry of soluble organic carbon consumed versus sulphide produced indicated that bacterial sulphate reduction was the predominant mechanism whereby organic matter was utilized. In this instance, however, non-sulphate-reducing microorganisms may well have played a role in converting complex organic matter to simpler forms prior to its oxidation by sulphate respiration.

Goldhaber and Kaplan (1974) reviewed the sulphur cycle in sediments and discussed the importance of stable sulphur isotope data in the interpretation of the diagenetic history of sulphur. On the subject of stratiform ore deposits they concluded that rates of sedimentation and organic contents of sediments in near-shore environments are such that bacterial sulphate reduction need not be a limiting factor in the formation of deposits of the same magnitude as many stratiform sulphide ores.

Further studies on fractionation of sulphur isotopes by yeast have been undertaken by McCready et al. (1974). They imply (p. 1252) that sulphate reduction by heterotrophic assimilatory organisms may contribute significantly to the distribution of sulphur isotopes in natural environments.

Renfro (1974) suggested that the depositional environments of Kupferschiefer, Roan and certain other evaporite-associated stratiform metalliferous deposits may be analogous to coastal sabkhas. He proposed that oxidized terrestrial formation waters at low pH mobilized metals from underlying rocks and then, during periods of regression, passed upward through H_2S-charged layers of decomposing algal mats where metals are immobilized as sulphides.

Algae were also considered by Roberts (1973), to be involved in the genesis of the Woodcutters lead–zinc prospect in the Northern Territory of Australia. According to Roberts, dense algal growth and evaporitic conditions in a shallow restricted basin caused coprecipitation of lead and zinc with precursors of the dolomitic host rocks of the Woodcutters Deposit. During the transformation of these precursors to dolomite, the metals formed complexes with products of algal deposition and were released to pore solutions which then migrated to sites of metal sulphide deposition.

The recent discovery of microfossils of bacterial dimensions in the mineralized zones of the McArthur River lead–zinc–silver deposit, Northern Territory, Australia (Hamilton and Muir, 1974), provides some evidence that microorganisms were present at the time of formation of this deposit if not actively involved in the mineralization process.

Iron

Puchelt et al. (1973) reported the presence of *Gallionella ferruginea* in the iron-oxide sediments which are forming at the present time in bays of the Kameni Islands of the Aegean Sea. The organisms occurred in such masses that Puchelt et al. concluded that they must play a part in the process of iron deposition.

References

Goldhaber, M.B. and Kaplan, I.R., 1974. The sulphur cycle. In: E.D. Goldberg (Editor), *The Sea. 5.* Wiley, New York, N.Y., pp. 569–655.

Hamilton, L.H. and Muir, M.D., 1974. Precambrian microfossils from the McArthur River lead–zinc–silver deposit, Northern Territory, Australia. *Miner. Deposita,* 9: 83–86.

Huisingh, J., McNeill, J.J. and Matrone, G., 1974. Sulfate reduction by a *Desulfovibrio* species isolated from sheep rumen. *Appl. Microbiol.,* 28: 489–497.

McCready, R.G.L., Kaplan, I.R. and Din, G.A., 1974. Fractionation of sulfur isotopes by the yeast *Saccharomyces cerevisiae. Geochim. Cosmochim. Acta,* 38: 1239–1253.

Puchelt, von H., Schock, H.H., Schroll, E. and Hanert, H., 1973. Rezente marine Eisenerze auf Santorin, Griechenland, I. Geochemie, Entstehung, Mineralogie, II. Bakteriogenese von Eisenhydroxidsedimenten. *Geol. Rundsch.,* 62: 786–812.

Ramm, A.E. and Bella, D.A., 1974. Sulfide production in anaerobic microcosms. *Limnol. Oceanogr.,* 19: 110–118.

Renfro, A.R., 1974. Genesis of evaporite-associated stratiform metalliferous deposits–a sabkha process. *Econ. Geol.,* 69: 35–45.

Roberts, W.M.B., 1973. Dolomitization and the genesis of the Woodcutters lead–zinc prospect, Northern Territory, Australia. *Miner. Deposita,* 8: 35–56.

ACKNOWLEDGEMENTS

The Baas Becking Geobiological Laboratory is supported by the Bureau of Mineral Resources, the Commonwealth Scientific and Industrial Research Organization and the Australian Mineral Industries Research Association Limited.

REFERENCES

Abd-el-Malek, Y. and Rizk, S.G., 1963. Bacterial sulphate reduction and the development of alkalinity, 3. Experiments under natural conditions in the Wadi Natrûn. *J. Appl. Microbiol.*, 26:
20–26.
Adams, F. and Conrad, J.P., 1953. Transition of phosphite to phosphate in soils. *Soil Sci.*, 75:
361–371.
Alexander, M., 1965. *Introduction to Soil Microbiology.* Wiley, New York, N.Y., 472 pp.
Alexandrov, E.A., 1973. The Precambrian banded iron-formations of the Soviet Union. *Econ. Geol.*,
68: 1035–1062.
Altschuler, Z.S., Clarke, R.S. and Young, E.J., 1958. Geochemistry of uranium in apatite and phos-
phorite. *U.S. Geol. Surv., Prof. Pap.*, 314D: 90 pp.
Ames, L.L., 1959. Genesis of carbonate apatites. *Econ. Geol.*, 54: 829–841.
Anger, G., Nielsen, H., Puchelt, H. and Ricke, W., 1966. Sulfur isotopes in the Rammelsberg ore
deposit (Germany). *Econ. Geol.*, 61: 511–536.
Aristovskaya, T.V., 1961. Accumulation of iron in the breakdown of organomineral humus complexes
by microorganisms. *Dokl. (Proc.) Acad. Sci. U.S.S.R.*, 136: 111–114.
Ashida, J., 1965. Adaptation of fungi to metal toxicants. *Annu. Rev. Phytopathol.*, 3: 153–174.
Ault, W.U., 1959. Isotopic fractionation of sulfur in geochemical processes. In: P. Abelson (Editor),
Researches in Geochemistry. Wiley, New York, N.Y., pp. 241–259.
Ault, W.U. and Kulp, J.L., 1959. Isotopic geochemistry of sulphur. *Geochim. Cosmochim. Acta*, 16:
201–235.
Ault, W.U. and Kulp, J.L., 1960. Sulfur isotopes and ore deposits. *Econ. Geol.*, 55: 73–100.
Baas Becking, L.G.M., 1959. Geology and microbiology. *N.Z. Dept. Sci. Ind. Res. Inf. Ser.*, 22:
48–64.
Baas Becking, L.G.M. and Moore, D., 1961. Biogenic sulfides. *Econ. Geol.*, 56: 259–272.
Baas Becking, L.G.M. and Wood, E.J.F., 1955. Biological processes in the estuarine environment, 1.
Ecology of the sulphur cycle. *Proc. Kon. Ned. Akad. Wetensch.*, 58: 160–180.
Baas Becking, L.G.M., Wood, E.J.F. and Kaplan, I.R., 1956. Biological processes in the estuarine
environment, VIII. Iron bacteria as gradient organisms. *Proc. Kon. Ned. Akad. Wetensch.*, 59:
398–407.
Baas Becking, L.G.M., Kaplan, I.R. and Moore, D., 1960. Limits of the natural environment in terms
of pH and oxidation-reduction potentials. *J. Geol.*, 68: 243–284.
Bachra, B.M. and Trautz, O.R., 1962. Carbonic anhydrase and the precipitation of apatite. *Science
N.Y.*, 137: 337–338.
Banerjee, D.M., 1971a. Aravallian stromatolites from Udaipur, Rajasthan. *J. Geol. Soc. India*, 12:
349–355.
Banerjee, D.M., 1971b. Precambrian stromatolitic phosphorites of Udaipur, Rajasthan, India. *Bull.
Geol. Soc. Am.*, 82: 2319–2330.
Barghoorn, E.S. and Nichols, R.L., 1961. Sulfate-reducing bacteria and pyritic sediments in Antarc-
tica. *Science N.Y.*, 134: 180.
Barghoorn, E.S. and Tyler, S.A., 1965 Microorganisms·from the Gunflint chert. *Science N.Y.*, 147:
563–577.
Bastin, E.S., 1926. A hypothesis of bacterial influence in the genesis of certain sulphide ores. *J. Geol.*,
34: 773–792.
Bateman, A.M., 1959. *Economic Mineral Deposits.* Wiley, New York, N.Y., 2nd ed., 579 pp.
Baylis, J.R., 1924. Manganese in Baltimore water supply. *J. Am. Water Works Assoc.*, 12: 211–233.
Beck, J.V., 1967. The role of bacteria in copper mining operations. *Biotech. Bioeng.*, 9: 487–497.
Beijerinck, M.W., 1913. Oxydation des Mangankarbonates durch Bakterien und Schimmelpilze. *Folia
Microbiol., Delft*, 2: 123–124.
Berge, J.W., 1972. Physical and chemical factors in the formation of marine apatite. *Econ. Geol.*, 67:
824–827.

Berger, R. and Libby, W.F., 1968. U.C.L.A. radiocarbon dates, VII. *Radiocarbon*, 10: 149–160.

Berger, R. and Libby, W.F., 1969a. Equilibration of atmospheric carbon dioxide with sea water: possible enzymatic control of the rate. *Science N.Y.*, 164: 1395–1397.

Berger, R. and Libby, W.F., 1969b. U.C.L.A. radiocarbon dates, IX. *Radiocarbon*, 11: 194–209.

Berger, W., 1950. The geochemical role of organisms. *Mineral. Petrogr. Mitt.*, 2: 136–140.

Berner, R.A., 1964a. Iron sulfides formed from aqueous solution at low temperatures and atmospheric pressure. *J. Geol.*, 72: 293–306.

Berner, R.A., 1964b. Distribution and diagenesis of sulfur in some sediments from the Gulf of California. *Mar. Geol.*, 1: 117–140.

Berner, R.A., 1967. Diagenesis of iron sulfide in recent marine sediments. In: G.H. Lauff (Editor), *Estuaries*. Am. Assoc. Adv. Sci., Washington, D.C., pp. 268–272.

Berner, R.A., 1969a. Migration of iron and sulfur within anaerobic sediments during early diagenesis. *Am. J. Sci.*, 267: 19–42.

Berner, R.A., 1969b. The synthesis of framboidal pyrite. *Econ. Geol.*, 64: 383–384.

Berner, R.A., 1970. Sedimentary pyrite formation. *Am. J. Sci.*, 268: 1–23.

Bezrukov, P.L. and Andruschencho, P.F., 1973. Iron-manganese concretions in the Indian Ocean. *Int. Geol. Rev.*, 15: 342–356.

Bischoff, J.L. and Manheim, F.T., 1969. Economic potential of the Red Sea heavy-metal deposits. In: E.T. Degens and D.A. Ross (Editors), *Hot Brines and Recent Heavy Metal Deposits in the Red Sea.* Springer, New York, N.Y., pp. 535–541.

Blackwelder, E., 1916. The geologic role of phosphorus. *Am. J. Sci.*, 42: 285–298.

Bogdanov, Y.V. and Golubchina, M.N., 1971. Isotopic composition of sulfur in stratified deposits of copper ores in Olekma-Vitim montane area. *Int. Geol. Rev.*, 13: 1405–1417.

Booth, G.H. and Mercer, S.J., 1963. Resistance to copper of some oxidizing and reducing bacteria. *Nature, Lond.*, 199: 622.

Bowen, H.J.M., 1966. *Trace Elements in Biochemistry*. Academic Press, London, 241 pp.

Brantner, H., 1970. Untersuchungen zur biologischen Eisen- und Manganoxydation. *Zbl. Bakteriol. Parasitkd., Abt. 2*, 124: 412–426.

Breger, C.L., 1911. Origin of some mineral deposits by bacteria. *Mining Eng. World*, 35: 289–291.

Bromfield, S.M. 1956. Oxidation of manganese by soil microorganisms. *Aust. J. Biol. Sci.*, 9: 238–254.

Bromfield, S.M. and Skerman, V.B.D., 1950. Biological oxidation of manganese in soils. *Soil Sci.*, 69: 337–348.

Bubela, B. and MacDonald, J.A., 1969. Formation of banded sulphides: metal ion separation and precipitation by inorganic and microbial sulphide sources. *Nature, Lond.*, 221: 465–466.

Burnie, S.W., Schwarcz, H.P. and Crocket, J.H., 1972. A sulfur isotope study of the White Pine mine, Michigan. *Econ. Geol.*, 67: 895–914.

Bushinskii, G.I., 1966. The origin of marine phosphorites. *Lithol. Mineral Resour.*, 3: 292–311.

Bushinskii, G.I., 1969. *Old Phosphorites of Asia and Their Genesis*. Israel Programme for Scientific Translations, Jerusalem, 266 pp.

Butkewitsch, W.A., 1928. Die Bildung der Eisenmangan-Ablagerungen am Meeresboden und die daran beteiligten Mikroorganismen. *Mosk. Nauchni. Inst. Tr.*, 3: 63–88.

Butlin, K.R. and Postgate, J.R., 1954. The microbiological formation of sulfur in Cyrenaican lakes. In: J.L. Cloudsley-Thompson (Editor), *Biology of Deserts*. Institute of Biology, London, pp. 112–122.

Calvert, S.E., 1968. Silica balance in the ocean and diagenesis. *Nature, Lond.*, 219: 919–920.

Casida, L.E., 1960. Microbial oxidation and utilization of orthophosphite during growth. *J. Bacteriol.*, 80: 237–241.

Cataldi, M., 1937. Aislamento de *Leptothrix* ochracea en medios sólidos a partir de cultivos liquidos. *Folia Biol. B. Aires*, 79–82: 337–344.

Cataldi, M., 1939. *Estudio Fisiólogico Sistemático de Algunas Chlamydobacteriales*. Thesis Univ. Buenos Aires.

Cayeux, L., 1936. Existence de nombreuses bactéries dans les phosphates sédimentaires de tout ages. Conséquences. C.R. Acad. Sci. Paris, 203: 1198–1200.

Charles, G., 1953. Sur l'origine des gisements de phosphates de chaux sedimentaires. 19th Int. Geol. Congr., Algiers, 1952, C.R. Sect. 11 (11): 163–184.

Chilingar, G.V., Bissell, H.J. and Wolf, K.H., 1967. Diagenesis in carbonate rocks. In: G. Larsen and G.V. Chilingar (Editors), Diagenesis in Sediments. Elsevier, Amsterdam, pp. 179–322.

Choldony, N., 1924. Zur Morphologie der Eisenbakterien Gallionella und Spirophyllum. Ber. Dtsch. Bot. Ges., 42: 35–44.

Choldony, N., 1926. Die Eisenbakterien, Beiträge zu einer Monographie Pflanzenforschung. G. Fischer, Jena.

Chukhrov, F.V., 1970. Isotopic fractionation of sulfur during lithogenesis. Lithology Mineral Resour., 2: 188–197.

Chukhrov, F.V., 1971. Isotopic composition of sulfur and genesis of copper ores of Zdhezkazgan and Udokan. Int. Geol. Rev., 13: 1429–1434.

Clark, F.M., Scott, R.M. and Bone, E., 1967. Heterotrophic iron-precipitating bacteria. J. Am. Water Works Assoc., 59: 1036–1042.

Cloud, P.E., 1965. Significance of the Gunflint (Precambrian) microflora. Science N.Y., 148: 27–35.

Cloud, P.E., 1968a. Premetazoan evolution and the origins of metazoa. In: E.T. Drake (Editor), Evolution and Environment. Yale University Press, New Haven, Conn., pp. 1–72.

Cloud, P.E., 1968b. Atmospheric and hydrospheric evolution on the primitive earth. Science, N.Y., 160: 729–736.

Cloud, P.E., 1973. Paleoecological significance of the banded-iron formation. Econ. Geol., 68: 1135–1143.

Cloud, P.E. and Hagen, H., 1965. Electron microscopy of the Gunflint microflora: preliminary results. Proc. Natl. Acad. Sci. U.S.A., 54: 1–8.

Cloud, P.E. and Licari, G.R., 1968. Microbiotas of the banded iron formations. Proc. Natl. Acad. Sci. U.S.A., 61: 779–786.

Colmer, A.R. and Hinkle, M.E., 1947. The role of microorganisms in acid mine drainage; a preliminary report. Science N.Y., 106: 253–256.

Colmer, A.R., Temple, K.L. and Hinkle, M.E., 1950. An iron-oxidizing bacterium from the acid drainage of some bituminous coal mines. J. Bacteriol., 59: 317–318.

Cornelius, H.P. and Plöchinger, B., 1952. Der Tennengebirge-N-Rand mit seinen Manganerzen und die Berge im Bereich des Lammertales. Jhber. Geol. Bundesanst. Wien, 95: 145–225.

Couglan, M.P., 1971. The role of iron in microbial metabolism. Sci. Progr., Lond., 59: 1–23.

Cressman, E.R. and Swanson, R.W., 1964. Stratigraphy and petrology of the Permian rocks of South-western Montana. U.S. Geol. Surv., Prof. Pap., 313C: 569 pp.

Davidson, C.F., 1962. The origin of some stratabound sulfide ore deposits. Econ. Geol., 57: 265–274.

Davidson, C.F., 1962/63. Discussion of paper by Brandt, R.T. Relationship of mineralization to sedimentation at Mufulira, northern Rhodesia. Bull. Inst. Mining Metall., 72: 191–208.

Davis, J.B. and Kirkland, D.W., 1970. Native sulfur deposition in the Castile Formation, Culberson County, Texas. Econ. Geol., 65: 107–121.

Davis, J.B. and Yarbrough, H.F., 1966. Anaerobic oxidation of hydrocarbons by Desulfovibrio desul-furicans. Chem. Geol., 1: 137–144.

Davis, R.P., 1961. Carbonic anhydrase. In: P.D. Boyer, H. Lardy and K. Myrbäch (Editors), The Enzymes 5, 2nd ed., pp. 545–562.

Dessau, G., Jensen, M.L. and Nakai, N., 1962. Geology and isotopic studies of Sicilian sulfur deposits. Econ. Geol., 57: 410–438.

Deuser, W.G., 1971. Organic-carbon budget of the Black Sea. Deep-Sea Res., 18: 995–1004.

Dietz, R.S., Emery, K.O. and Shepard, F.P., 1942. Phosphorite deposits on the sea floor off Southern California. Bull. Geol. Soc. Am., 53: 815–848.

Dondero, N.C., 1961. Sphaerotilus, its nature and economic significance. Adv. Appl. Microbiol., 3: 77–107.

Donnelly, T.H., Lambert, I.B. and Dale, D.H., 1972. Sulphur isotope studies of the Mount Gunson copper deposits, Pernatty Lagoon, South Australia. *Mineral. Deposita*, 7: 314–322.

Duursma, E.K., 1965. The dissolved organic constituents of sea water. In: J.P. Riley and G. Skirrow (Editors), *Chemical Oceanography, 1.* London Academic Press, London, pp. 433–475.

Drabkova, V.G. and Stavinskaya, E.A., 1969. Role of bacteria in the dynamics of iron in Lake Krasnoe. *Microbiology*, 38: 304–309.

Ehrlich, H.L., 1963. Bacteriology of manganese nodules, I. Bacterial action of manganese in nodule enrichments. *Appl. Microbiol.*, 11: 15–19.

Ehrlich, H.L., 1966. Reactions with manganese by bacteria from marine ferromanganese nodules. *Develop. Ind. Microbiol.*, 7: 279–286.

Ehrlich, H.L., 1968. Bacteriology of manganese nodules, II. Manganese oxidation by cell-free extract from a manganese nodule bacterium. *Appl. Microbiol.*, 16: 197–202.

Ehrlich, H.L., 1971. Bacteriology of manganese nodules, V. Effect of hydrostatic pressure on bacterial oxidation of Mn^{11} and reduction of MnO_2. *Appl. Microbiol.*, 21: 306–310.

Ehrlich, H.L., 1972. The role of microbes in manganese nodule genesis and degradation. In: D.R. Horn (Editor), *Ferromanganese Deposits on the Ocean Floor.* Office International Decade of Ocean Exploration–Nat. Sci. Found., Washington, D.C., pp. 63–70.

Ellis, D., 1907. A contribution to our knowledge of the thread bacteria. *Zbl. Bakteriol. Parasitkd.*, *Abt. 2*, 19: 502–518.

Ennever, J., 1963. Microbiologic calcification. *Ann. N.Y. Acad. Sci.*, 109: 4–13.

Farrand, M., 1970. Framboidal pyrite precipitated synthetically. *Mineral. Deposita*, 5: 237–247.

Feely, H.W. and Kulp, J.L., 1957. Origin of Gulf Coast salt-dome sulphur deposits. *Bull. Am. Assoc. Petrol. Geol.*, 41: 1802–1853.

Findley, J.E. and Akagi, J.M., 1969. Evidence for thiosulfate formation during sulfite reduction by *Desulfovibrio vulgaris. Biochem. Biophys. Res. Commun.*, 36: 266–271.

Freke, A.M. and Tate, D., 1961. The formation of magnetic iron sulphide by bacterial reduction of iron solutions. *J. Biochem. Microbiol. Tech. Eng.*, 3: 29–39.

Gabe, D.R., Troshanov, E.P. and Sherman, E.E., 1965. The formation of manganese-iron layers in mud as a biogenic process. In: *Applied Capillary Microscopy. The Role of Microorganisms in the Formation of Iron-Manganese Deposits.* Consultants Bureau, New York, N.Y., pp. 88–105.

Ganter, I.R. and Schwartz, W., 1956. Beiträge zur Biologie der Eisenmikroben, II. *Leptothrix crassa* Chol. *Schweiz Z. Hydrobiol.*, 18: 171–192.

Garrels, R.M., 1960. *Mineral Equilibria.* Harper, New York, N.Y., 254 pp.

Garrels, R.M. and Christ, C.L., 1965. *Solutions, Minerals and Equilibria.* Harper and Row, New York, N.Y., 450 pp.

Gehlen, K.V. 1966. Schwefel-isotope und die Genese von Erzlagerstätten. *Geol. Rdsch.*, 55: 178–197.

Gerretsen, F.C., 1937. Effect of manganese deficiency in oats in relation to soil bacteria. *Ann. Bot., N. Ser.*, 1: 207–230.

Goldberg, E.D., 1965. Minor elements in sea water. In: J.P. Riley and G. Skirrow (Editors), *Chemical Oceanography, 1.* Academic Press, New York, N.Y., pp. 163–196.

Goldberg, E.D. and Arrhenius, G.O.S., 1958. Chemistry of Pacific pelagic sediments. *Geochim. Cosmochim. Acta*, 13: 153–212.

Graham, J.W., 1959. Metabolically induced precipitation of trace elements from sea water. *Science N.Y.*, 129: 1428–1429.

Graham, J.W. and Cooper, S.C., 1959. Biological origin of manganese-rich deposits of the sea floor. *Nature, Lond.*, 183: 1050–1051.

Gruner, J.W., 1922. The origin of sedimentary iron formations: the Biwabik formation of the Mesabi range. *Econ. Geol.*, 17: 407–460.

Gubin, W. and Tzechomaskaja, W., 1930. Über die biochemische Sodabildung in den Sodaseen. *Zbl. Bakt. Parasitkd., Abt. 2* 81: 396–401.

Guha, J., 1971. Sulfur isotope study of the pyrite deposit of Amjhore, Shahbad district, Bihar, India. *Econ. Geol.*, 66: 326–330.

Gulbrandsen, R.A., 1969. Physical and chemical factors in the formation of marine apatite. *Econ. Geol.*, 64: 365–382.

Gulick, A., 1965. Phosphorus as a factor in the origin of life. *Am. Sci.*, 43: 479–489.

Hallberg, R.O., 1970. An apparatus for the continuous cultivation of sulfate-reducing bacteria and its application to geomicrobiological purposes. *Antonie van Leeuwenhoek*, 36: 241–254.

Hallberg, R.O., 1972a. Iron and zinc sulfides formed in a continuous culture of sulfate-reducing bacteria. *Neues Jhb. Mineral. Mh.*, 11: 481–500.

Hallberg, R.O., 1972b. Sedimentary sulfide mineral formation – an energy-circuit approach. *Mineral. Deposita*, 7: 189–201.

Hallberg, R.O., Bågander, L.E., Engvall, A.G. and Schippel, F.A., 1972. Method for studying geochemistry of sediment-water interface. *Ambio*, 1: 71–73.

Halvorson, H.O., 1931. Studies on the transformations of iron in nature, III. The effect of CO_2 on the equilibrium of iron solutions. *Soil Sci.*, 32: 141–165.

Halvorson, H.O. and Starkey, R.L., 1927. Studies on the transformation of iron in nature, I. Theoretical considerations. *J. Phys. Chem. Ithaca*, 31: 626–631.

Harder, E.C., 1919. Iron-depositing bacteria and their geologic relations. *U.S. Geol. Surv. Prof. Pap.*, 113: 89 pp.

Hariya, Y. and Kikuchi, T., 1964. Precipitation of manganese by bacteria in mineral springs. *Nature, Lond.*, 202: 416–417.

Harrison, A.G. and Thode, H.G., 1957. The kinetic isotope effect in the chemical reduction of sulphate. *Trans. Faraday Soc.*, 53: 1648–1651.

Harrison, A.G. and Thode, H.G., 1958. Mechanism of bacterial reduction of sulphate from isotope fractionation studies. *Trans. Faraday Soc.*, 54: 84–92.

Harriss, R.C., 1972. Silica-biogeochemical cycle. In: R.W. Fairbridge (Editor), *The Encyclopedia of Geochemistry and Environmental Sciences, IVA.* Van Nostrand Rheinhold, New York, N.Y., pp. 1080–1082.

Heinen, W. and Lauwers, A.M., 1974. Hypophosphite oxidase from *Bacillus caldolyticus. Arch. Mikrobiol.*, 95: 267–274.

Hirsch, P. and Conti, S.F., 1964. Biology of budding bacteria, 1. Enrichment isolation and morphology of *Hyphomicrobium* spp. *Arch. Mikrobiol.*, 48: 339–357.

Holser, W.T. and Kaplan, I.R., 1966. Isotope geochemistry of sedimentary sulfates. *Chem. Geol.*, 1: 93–135.

Hunt, W.F., 1915. The origin of sulphur deposits of Sicily. *Econ. Geol.*, 10: 543–579.

Issatchenko, B.L., 1929. Zur Frage der biogenischen Bildung des Pyrits. *Int. Rev. Hydrobiol. Hydrogr.*, 22: 99–101.

Ivanov, M.V., 1957. Role of microorganisms in depositing sulfur in the hydrogen sulfide springs of Sergiev mineral waters. *Mikrobiologiya*, 26: 338–345.

Ivanov, M.V., 1968. *Microbiological Processes in the Formation of Sulfur Deposits.* Israel Program for Scientific Translations, Jerusalem, 298 pp.

Iverson, W.P., 1967. Disulfur-monoxide: production by *Desulfovibrio. Science N.Y.*, 156: 1112–1113.

Jensen, M.L., 1958. Sulfur isotopes and the origin of sandstone-type uranium deposits. *Econ. Geol.*, 53: 598–616.

Jensen, M.L., 1962. Biogenic sulfur and sulfide deposits. In: M.L. Jensen (Editor), *Biogeochemistry of Sulfur Isotopes.* Yale University Press, New Haven, pp. 1–15.

Jensen, M.L., 1971. Sulfur isotopes of stratabound sulfide deposits. *Soc. Mineral. Geol. Japan*, 3: 300–303.

Jensen, M.L. and Dessau, G., 1966. Ore deposits of southwestern Sardinia and their sulfur deposits. *Econ. Geol.*, 61: 917–932.

Jensen, M.L. and Dessau, G., 1967. The bearing of sulfur isotopes on the origin of Mississippi Valley type deposits. *Econ. Geol. Monogr.*, 3: 400–409.

Jensen, M.L. and Nakai, N., 1964. Large-scale bacteriogenic fractionation of sulphur isotopes. *Pure Appl. Chem.*, 8: 305–315.

Jensen, M.L. and Whittles, A.W.G., 1969. Sulfur isotopes of the Nairne pyrite deposit, South Australia. *Miner. Deposita*, 4: 241–247.

Johnson, A.H. and Stokes, J.L., 1966. Manganese oxidation by *Sphaerotilus discophorus*. *J. Bacteriol.*, 91: 1543–1547.

Jones, G.E. and Starkey, R.L., 1957. Fractionation of stable isotope of sulfur by microorganisms and their role in the deposition of native sulfur. *Appl. Microbiol.*, 5: 111–118.

Jones, G.E. and Starkey, R.L., 1962. Some necessary conditions for fractionation of stable isotopes of sulfur by *Desulfovibrio desulfuricans*. In: M.L. Jensen (Editor), *Biogeochemistry of Sulfur Isotopes*. Yale University Press, New Haven, pp. 61–79.

Jones, G.E., Starkey, R.L., Feely, H.W. and Kulp, J.L., 1956. Biological origin of nature sulfur in salt domes of Texas and Louisiana. *Science N.Y.*, 123: 1124–1125.

Kalinenko, V.O., 1946. Role of bacteria in formation of ferromanganese concretions. *Mikrobiologiya*, 15: 364–369.

Kalinenko, V.O., Belokopytova, O.V. and Nikolaeva, G.C., 1962. Bacteriogenic iron-manganese concretions in the Indian Ocean. *Okeanoliya*, 2: 1050–1059.

Kaplan, I.R. and Rittenberg, S.C., 1962. The microbial fractionation of sulfur isotopes. In: M.L. Jensen (Editor), *Biogeochemistry of Sulfur Isotopes*. Yale University Press, New Haven, pp. 89–93.

Kaplan, I.R. and Rittenberg, S.C., 1964. Microbiological fractionation of sulphur isotopes. *J. Gen. Microbiol.*, 34: 195–212.

Kaplan, I.R., Emery, K.O. and Rittenberg, S.C., 1963. The distribution and isotopic abundance of sulphur in recent marine sediments off California. *Geochim. Cosmochim. Acta*, 27: 297–331.

Kaplan, I.R., Rafter, T.A. and Hulston, J.R., 1960. Sulphur isotopic variations in nature, 8. Application to some biogeochemical problems. *N.Z. J. Sci.*, 3: 338–361.

Kazakov, A.V., 1937. The phosphorite facies and the genesis of phosphorites in geological investigations of agricultural ores. *Leningr. Sci. Inst. Fertil. Insecto-Fungacides Trans. (U.S.S.R.)*, 142: 95–113.

Kemp, A.L.W. and Thode, H.G., 1968. The mechanism of the bacterial reduction of sulphate and sulphite from isotope fractionation studies. *Geochim. Cosmochim. Acta*, 32: 71–91.

Kobayashi, K., Tachibana, S. and Ishimoto, M., 1969. Intermediary formation of trithionate in sulfate reduction by a sulfate-reducing bacterium. *J. Biochem., Tokyo*, 65: 155–157.

Kossaya, T.A., 1967. Composition of manganese oxides in cultures of *Metallogenium* and *Leptothrix*. *Microbiology*, 36: 857–861.

Koyama, T., Nakai, N. and Kamata, E., 1965. Possible discharge rate of hydrogen sulfide from polluted coastal belts in Japan. *J. Earth Sci.*, 13: 1–11.

Kramarenko, L.Ye., 1962. Bacterial biocenoses in underground waters of some mineral fields and their geological importance. *Microbiology*, 31: 564–569.

Krauskopf, K.B., 1955. Sedimentary deposits of rare metals. *Econ. Geol.*, 50th Anniv. Vol., pp. 411–463.

Krauskopf, K.B., 1957. Separation of manganese from iron in sedimentary processes. *Geochim. Cosmochim. Acta*, 12: 61–84.

Krauskopf, K.B., 1967. *Introduction to Geochemistry*. McGraw-Hill, New York, N.Y., 721 pp.

Krouse, H.R. and Sasaki, A., 1968. Sulphur and carbon isotope fractionation by *Salmonella heidelberg* during anaerobic SO_3^{2-} reduction in Trypticase soy broth medium. *Can. J. Microbiol.*, 14: 417–422.

Krouse, H.R., McCready, R.G.L., Husain, S.A. and Campbell, J.N., 1967. Sulfur isotope fractionation by *Salmonella* species. *Can. J. Microbiol.*, 13: 21–25.

Krouse, H.R., McCready, R.G.L., Husain, S.A. and Campbell, J.N., 1968. Sulfur isotope fractionation and kinetic studies of sulfite reduction in growing cells of *Salmonella heidelberg*. *Biophys. J.*, 8: 109–124.

Krumbein, W.E., 1972. Rôle des microorganisms dans la genèse, la diagenèse et al dégradation des roches en place. *Rev. Ecol. Biol. Sol.,* 9: 283–319.

Krumbein, W.E. and Garrels, R.M., 1952. Origin and classification of chemical sediments in terms of pH and oxidation-reduction potentials. *J. Geol.,* 60: 1–33.

Kuznetsov, S.I., Ivanov, M.V. and Lyalikova, N.N., 1963. *Introduction to Geological Microbiology.* McGraw-Hill Book Co., New York, N.Y., 252 pp.

LaBerge, G.L., 1967. Microfossils and Precambrian iron-formations. *Bull. Geol. Soc. Am.,* 78: 331–342.

LaBerge, G.L., 1973. Possible biological origin of Precambrian iron-formations. *Econ. Geol.,* 68: 1098–1109.

Lambert, I.B., 1973. Post-depositional availability of sulphur and metals and formation of secondary textures and structures in stratiform sedimentary sulphide deposits. *J. Geol. Soc. Aust.,* 20: 205–215.

Lambert, I.B. and Bubela, B., 1970. Banded sulphide ores: the experimental production of mono-mineralic sulphide bands in sediments. *Mineral. Deposita,* 5: 97–102.

Leeper, G.W. and Swaby, R.J., 1940. Oxidation of manganous compounds by microorganisms in soil. *Soil Sci.,* 49: 163–169.

Lepp, H. and Goldich, S.S., 1964. Origin of Precambrian iron formations. *Econ. Geol.,* 59: 1025–1060.

Lisitzin, A.P., 1972. Sedimentation in the world ocean. *Soc. Econ. Paleontol. Mineral., Spec. Publ.,* 17: 218 pp.

Litchfield, C.D. and Hood, D.W., 1962. Evidence for carbonic anhydrase in marine and fresh-water algae. *Int. Ver. Theor. Angew. Limnol. Verhandl.,* 15 (2): 817–828.

Ljunggren, P., 1955. Some data concerning the formation of manganiferous and ferriferous bog ores. *Geol. För. Stockh. Förh.,* 75: 277–297.

Logan, B.W., Rezak, R. and Ginsburg, R.N., 1964. Classification and environmental significance of algal stromatolites. *J. Geol.,* 72: 68–83.

Lougheed, M.S. and Mancuso, J.J., 1973. Hematite framboids in the Negaunee iron formation, Michigan: evidence for their biogenic origin. *Econ. Geol.,* 68: 202–209.

Love, L.G., 1957. Microorganisms and the presence of syngenetic pyrite. *Q. J. Geol. Soc. Lond.,* 113: 429–440.

Love, L.G., 1962a. Biogenic primary sulfide of the Permean Kupferschiefer and Marl Slate. *Econ. Geol.,* 57: 350–366.

Love, L.G., 1962b. Further studies on microorganisms and the presence of syngenetic pyrite. *Palaeontology,* 5: 444–459.

Love, L.G., 1962c. Pyrite spheres in sediments. In: M.L. Jensen (Editor), *Biogeochemistry of Sulfur Isotopes.* Yale University Press, New Haven, Conn., pp. 121–143.

Love, L.G., 1963. The composition of *Pyritosphaera barbaria* Love, 1957. *Palaeontology,* 6: 119–120.

Love, L.G. and Amstutz, G.C., 1966. Review of microscopic pyrite. *Fortschr. Mineral.,* 43: 273–309.

Love, L.G. and Amstutz, G.C., 1969. Framboidal pyrite in two andesites. *Neues Jhb. Mineral. Mh.,* 3: 97–108.

Love, L.G. and Murray, J.W., 1963. Biogenic pyrite in recent sediments off Christchurch Harbour, England. *Am. J. Sci.,* 261: 433–438.

Love, L.G. and Zimmerman, D.O., 1961. Bedded pyrite and microorganisms from the Mt. Isa Shale. *Econ. Geol.,* 56: 873–876.

Lüning, O. and Brohm, K., 1933. Über das Vorkommen von Phosphorwasserstoff in Brunnenwässern. *Z. Unters. Lebensmittel,* 66: 460–465.

Luria, S.E., 1960. Bacterial protoplasm: composition and organization. In: I.C. Gunsalus and R.Y. Stanier (Editors), *The Bacteria, 1.* Academic Press, New York, N.Y., pp. 1–34.

MacDiarmid, R.A., 1966. Biogenic Ag_2S: a correction. *Econ. Geol.,* 61: 405–406.

Macnamara, J. and Thode, H.G., 1950. Comparison of isotopic constitution of terrestrial and meteoritic sulphur. *Phys. Rev.,* 78: 307–308.

Macnamara, J. and Thode, H.G., 1951. The distribution of ^{34}S in nature and the origin of native sulphur deposits. *Research, Lond.*, 4: 582–583.

Malacinski, G.M., 1968. *Physiological and Biochemical Aspects of Bacterial Oxidation of Orthophosphite.* Thesis, Indiana University.

Malacinski, G.M. and Konetska, W.A., 1966. Bacterial oxidation of orthophosphite. *J. Bacteriol.*, 91: 578–582.

Malacinski, G.M. and Konetska, W.A., 1967. Orthophosphite-nicotinamide adenine dinucleotide oxido-reductase from *Pseudomonas fluorescens*. *J. Bacteriol.*, 93: 1096–1910.

Mann, P.J.G. and Quastel, J.H., 1946. Manganese metabolism in soils. *Nature, Lond.*, 158: 154–156.

Mansfield, G.R., 1927. Geography, geology and mineral resources of part of south eastern Idaho. *U.S. Geol. Surv. Prof. Pap.*, 152: 409 pp.

Manskaya, S.M. and Drosdova, T.V., 1968. *Geochemistry of Organic Substances.* Pergamon Press, Oxford, 345 pp.

Mardanyan, S.S. and Balashova, V.V., 1971. State or iron in the filaments of *Gallionella*. *Microbiology*, 40: 104–106.

Marowsky, G., 1969. Schwefel-, Kohlenstoff- und Sauerstoff Isotopen Untersuchungen am Kupferschiefer als Beitrag zur genetischen Deutung. *Beitr. Mineral. Petrogr.*, 22: 290–334.

Matsukuma, T. and Horikoshi, E., 1970. Kuroko deposits in Japan, a review. In: T. Tatsumi (Editor), *Volcanism and Ore Genesis.* University of Tokyo Press, pp. 153–179.

McCarty, P.L., 1972. Energetics of organic matter degradation. In: R. Mitchell (Editor), *Water Pollution Microbiology.* Wiley-Interscience, New York, N.Y., pp. 91–118.

McConnell, D., 1965. Precipitation of phosphates in sea water. *Econ. Geol.*, 60: 1059–1062.

McConnell, D., Frajola, W.J. and Deamer, D.W., 1961. Relation between inorganic chemistry and biochemistry of bone mineralization. *Science N.Y.*, 133: 281–282.

McConnell, D., Frajola, W.J. and Deamer, D.W., 1962. A reply to comments by B.M. Bachra and O.R. Trautz. *Science N.Y.*, 137: 337–338.

McKelvey, V.E., 1967. Phosphate deposits. *Bull. U.S. Geol. Surv.*, 1252-D: 21 pp.

McKelvey, V.E., Swanson, R.W. and Sheldon, R.P., 1953. The Permian phosphorite deposits of western United States. *19th Int. Geol. Congr., Algiers, 1952, C.R. Sec. 11* (11): 45–64.

Mero, J.L., 1965. *The Mineral Resources of the Sea.* Elsevier, Amsterdam, 312 pp.

Mero, J.L., 1972. Potential economic value of ocean-floor manganese nodule deposits. In: D.R. Horn (Editor), *Ferromanganese Deposits on the Ocean Floor.* Office International Decade of Ocean Exploration. – Nat. Sci. Found. Washington, D.C., pp. 191–203.

Miller, L.P., 1950a. Rapid formation of high concentrations of hydrogen sulfide by sulfate-reducing bacteria. *Contrib. Boyce Thompson Inst. Plant. Res.*, 15: 437–465.

Miller, L.P., 1950b. Tolerance of sulfate-reducing bacteria to hydrogen sulfide. *Contrib. Boyce Thompson Inst. Plant Res.*, 16: 73–83.

Miller, L.P., 1950c. Formation of metal sulfides through the activities of sulfate-reducing bacteria. *Contrib. Boyce Thompson Inst. Plant Res.*, 16: 85–89.

Mirchink, T.G., Zaprometova, K.M. and Zvyagintsev, D.G., 1970. Satellite fungi of manganese-oxidizing bacteria. *Microbiology*, 39: 327–330.

Monty, C., 1965. Recent algal stromatolites in the Windward Lagoon, Andros Island, Bahamas. *Bull. Soc. Géol. Belg.*, 88: 269–276.

Monty, C., 1973. Les nodules de manganèse sont des stromatolithes océániques. *C.R. Acad. Sci. Paris*, 276: 3285–3288.

Moorehouse, W.W. and Beales, F.W., 1962. Fossils from the Animikie, Port Arthur, Ontario. *Trans. R. Soc. Can., Ser. 3*, 54: 97–110.

Möse, J.R. and Brantner, H., 1966. Mikrobiologische Studien an manganoxydierenden Bakterien. *Zbl. Bakteriol. Parasitkd., Abt.* 2, 120: 470–493.

Mulder, E.G., 1964. Iron bacteria, particularly those of the *Sphaerotilus-Leptothrix* group, and industrial problems. *J. Appl. Bacteriol.*, 27: 151–173.

Nakai, N. and Jensen, M.L., 1960. Biogeochemistry of sulfur isotopes. *J. Earth Sci.*, 8: 181–196.

Nakai, N. and Jensen, M.L., 1964. The kinetic isotope effect in the bacterial reduction and oxidation of sulfur. *Geochim. Cosmochim. Acta*, 28: 1893–1912.

Nicholas, D.J.D., 1967. Biological sulphate reduction. *Mineral. Deposita*, 2: 169–180.

Nielsen, H., 1966. Schwefelisotope im marinen Kreislauf und das δ^{34}S der früheren Meere. *Geol. Rdsch.*, 55: 160–172.

Nissenbaum, A., Presley, B.J. and Kaplan, I.R., 1972. Early diagenesis in a reducing fjord, Saanich Inlet, British Columbia, 1. Chemical and isotopic changes in major components of interstitial water. *Geochim. Cosmochim. Acta*, 36: 1007–1027.

Novelli, G.D. and ZoBell, C.E., 1944. Assimilation of petroleum hydrocarbons by sulfate-reducing bacteria. *J. Bacteriol.*, 47: 447–448.

Ohmoto, H., 1972. Systematics of sulfur and carbon isotopes in hydrothermal ore deposits. *Econ. Geol.*, 67: 551–578.

Oppenheimer, C.H., 1958. Evidence of fossil bacteria in phosphate rock. *Publ. Inst. Mar. Sci., Univ. Texas*, 5: 156–159.

Pankina, R.G. and Mekhtiyeva, V.L., 1969. Isotopic composition of different forms of sulfur in water and sediments of Sernoye Lake (Sergiyevsk mineral waters). *Geochem. Int.*, 6: 511–516.

Papunen, H., 1966. Framboidal texture of the pyritic layer found in a peat bog in southeastern Finland. *C.R. Soc. Géol. Finl.*, 38: 117–125.

Pauli, F.W., 1968. Some recent developments in biochemical research. *Proc. Geol. Assoc. Can.*, 19: 45–49.

Peck, H.D., 1962. Symposium on metabolism of inorganic compounds, 5. Comparative metabolism of inorganic sulphur compounds in microorganisms. *Bact. Rev.*, 26: 67–94.

Perfil'ev, B.V. and Gabe, D.R., 1965. The use of the microbiol-landscape method to investigate bacteria which concentrate manganese and iron in bottom deposits. In: *Applied Capillary Microscopy. The Role of Microorganisms in the Formation of Iron-Manganese Deposits*. Consultants Bureau, New York, N.Y., pp. 9–54.

Pettijohn, F.J., 1957. *Sedimentary Rocks*. Harper and Bros, New York, N.Y., 2nd ed., 475 pp.

Pevear, D.R. 1966. The estuarine formation of United States Atlantic coastal plain phosphorite. *Econ. Geol.*, 61: 251–256.

Pevear, D.R., 1967. Shallow-water phosphorites. *Econ. Geol.*, 62: 562–575.

Postgate, J.R., 1959. Sulphate reduction by bacteria. *Rev. Microbiol.*, 13: 505–520.

Postgate, J.R., 1965. Recent advances in the study of sulphate-reducing bacteria. *Bacteriol. Rev.*, 29: 425–444.

Pringsheim, E.G., 1946. On iron flagellates. *Phil. Trans. R. Soc. Ser. B.*, 232: 311–342.

Pringsheim, E.G., 1949a. The filamentous bacteria *Sphaerotilus, Leptothrix, Cladothrix* and their relation to iron and manganese. *Phil. Trans. R. Soc. Ser. B.*, 233: 453–462.

Pringsheim, E.G., 1949b. Iron bacteria. *Biol. Rev.*, 24: 200–245.

Ramdohr, P., 1953. Mineralbestand Strukturen und Genesis der Rammelsberg-Lagerstätte. *Geol. Jhber.*, 67: 367–495.

Rees, C.E., 1973. A steady-state model for sulphur isotope fractionation in bacterial reduction processes. *Geochim. Cosmochim. Acta*, 37: 1141–1162.

Rickard, D.T., 1969a. The microbiological formation of iron sulphides. *Stockh. Contrib. Geol.*, 20: 49–66.

Rickard, D.T., 1969b. The chemistry of iron sulphide formation at low temperatures. *Stockh. Contrib. Geol.*, 20: 67–95.

Rickard, D.T., 1970. The origin of framboids. *Lithos*, 3: 269–293.

Rickard, D.T., 1973. Limiting conditions for synsedimentary sulfide ore formation. *Econ. Geol.*, 68: 605–617.

Riedel, W.R., 1959. Siliceous organic remains in pelagic sediments. In: H.A. Ireland (Editor), *Silica in Sediments – Soc. Econ. Paleontol. Mineral., Spec. Publ.*, 7: 80–91.

Rizzo, A.A., Scott, D.B. and Blade, H.A., 1963. Calcification of oral bacteria. *Ann. N.Y. Acad. Sci.*, 109: 14–22.

Roberts, W.M.B., Walker, A.L. and Buchanan, A.S., 1969. The chemistry of pyrite formation in aqueous solution and its relation to the depositional environment. *Mineral. Deposita*, 4: 18–29.

Romankevich, Ye.A. and Baturin, G.N., 1972. Composition of the organic matter in phosphorites from the continental shelf of southwest Africa. *Geochem. Int.*, 9: 464–470.

Römer, R. and Schwartz, W., 1965. Geomikrobiologische Untersuchungen, V. Verwertung von Sulfatmineralien und Schwermetall-Toleranz bei Desulfurizieren. *Z. Allgem. Mikrobiol.*, 5: 122–135.

Rosenfeld, W.D., 1947. Anaerobic oxidation of hydrocarbons by sulfate-reducing bacteria. *J. Bacteriol.*, 54: 664–665.

Ross, D.A., Degens, E.T. and McIlvaine, J., 1970a. Black Sea: recent sedimentary history. *Science, N.Y.*, 170: 163–165.

Ross, D.A., Degens, E.T., McIlvaine, J. and Hedberg, R.M., 1970b. Recent sediments of the Black Sea. *Oceanus*, 15: 26–29.

Rouf, M.A. and Stokes, J.L., 1964. Morphology, nutrition and physiology of *Sphaerotilus discophorus. Arch. Mikrobiol.*, 49: 132–149.

Roy, A.B. and Trudinger, P.A., 1970. *The Biochemistry of Inorganic Compounds of Sulphur.* Cambridge University Press, London, 400 pp.

Rudakov, K.J., 1927. Die Reduktion der mineralischen Phosphate auf biologischem Wege. *Zbl. Bakteriol. Parasitkd. Abt. 2*, 70: 202–214.

Rudakov, K.J., 1929. Die Reduktion der mineralischen Phosphate auf biologischem Wege. *Zbl. Bakteriol. Parasitkd. Abt. 2*, 79: 229–245.

Ryther, J.H., 1963. Geographic variations in productivity. In: M.N. Hill (Editor), *The Sea. Ideas and Observations on Progress in the Study of the Seas.* Interscience, New York, N.Y., pp. 347–380.

Sadler, W.R. and Trudinger, P.A., 1967. The inhibition of microorganisms by heavy metals. *Mineral. Deposita*, 2: 158–168.

Sangster, D.F., 1968. Relative sulphur isotope abundances of ancient seas and stratabound sulphide deposits. *Proc. Geol. Assoc. Can.*, 19: 79–91.

Sangster, D.F., 1971. Sulphur isotopes, stratabound sulphide deposits, and ancient seas. *Soc. Mineral. Geol. Japan*, 3: 259–299.

Sapozhnikov, D.G., 1970. Geological conditions for the formation of manganese deposits. In: D.G. Sapozhnikov (Editor), *Manganese Deposits of the Soviet Union.* Israel Program for Scientific Translations, Jerusalem, pp. 9–33.

Sartory, A. and Meyer, J., 1947. Contribution à l'étude des systèmes enzymatiques de deux bactéries ferrugineuses. Rôle du fer et du manganèse comme facteurs d'énergie et de synthèse pour le métabolisme de ces germes. *C.R. Acad. Sci. Paris*, 225: 600–602.

Sasaki, A., 1970. Seawater sulfate as a possible determinant for sulfur isotopic compositions of some strata-bound ores. *Geochem. J.*, 4: 41–51.

Sasaki, A. and Kajiwara, Y., 1971. Evidence of isotopic exchange between seawater sulfate and some syngenetic sulphide ores. *Soc. Mineral. Geol. Japan*, 3: 289–294.

Saxby, J.D., 1969. Metal-organic chemistry of the geochemical cycle. *Rev. Pure Appl. Chem.*, 19: 131–150.

Schneegans, D., 1935. The problem of the bacterial reduction of sulfates in presence of hydrocarbons and carbonaceous material and the origin of sulfur deposits in southern France. *Congr. Int. Mines Météorol. Géol. Appl. 7e Sess., Paris Géol.*, 1: 351–357.

Schneiderhöhn, H., 1923. Chalcographische Untersuchung des Mansfelder Kupferschiefers. *Neues Jhr. Mineral. Geol. Paläontol.*, 47: 1–38.

Schouten, C., 1946. The role of sulphur bacteria in the formation of the so-called sedimentary copper ores and pyritic ore bodies. *Econ. Geol.*, 41: 517–538.

Schwarcz, H.P. and Burnie, S.W., 1973. Influence of sedimentary environments on sulfur isotope ratios in clastic rocks: a review. *Mineral. Deposita*, 8: 264–277.

Schwartz, W., 1972. Uber die Rolle von Mikro-organismen bei der Entstehung und Zerstörung von Lagerstätten. *Mineral. Deposita*, 7: 25–30.

Schweisfurth, R., 1970. Manganoxidierende Pilze. *Zbl. Bakteriol. Parasitkd., Abt. 1*, 212: 486–491.

Schweisfurth, R., 1971. Manganknollen im Meer. *Naturwissenschaften*, 58: 344–347.

Schweisfurth, R. and Gatlow, G., 1966. Röntgenstruktur und Zusammensetzung mikrobiell gebildeter Braunsteine. *Z. Allgem. Mikrobiol.*, 6: 303–308.

Schweisfurth, R. and Mertes, R., 1962. Mikrobiologische und chemische Untersuchungen über Bildung und Bekämpfung von Manganschlamm. Ablagerungen in einer Druckleitung für Talsparrenwasser. *Arch. Hyg. Bakteriol.*, 146: 401–417.

Senez, J., 1962. Some considerations on the energetics of bacterial growth. *Bact. Rev.*, 26: 95–107.

Shapiro, N.I., 1965. The chemical composition of deposits formed by *Metallogenium* and *Siderococcus*. In: *Applied Capillary Microscopy. The Role of Microorganisms in the Formation of Iron-Manganese Deposits*. Consultants Bureau, New York, N.Y., pp. 82–87.

Sheldon, R.P., 1964. Exploration for phosphorite in Turkey – a case history. *Econ. Geol.*, 59: 1159–1175.

Sheridan, R.P. and Castenholz, R.W., 1968. Production of hydrogen sulphide by a thermophilic blue-green alga. *Nature, Lond.*, 217: 1063–1064.

Siebenthal, C.E., 1915. Origin of the lead and zinc deposits of the Joplin region. *Bull. U.S. Geol. Surv.*, 606: 283 pp.

Silverman, M.P. and Ehrlich, H.L., 1964. Microbial formation and degradation of minerals. *Adv. Appl. Microbiol.*, 6: 153–206.

Skripchenko, N.S., 1970. "Mineralized bacteria" in ocean ooze. *Dokl. (Proc.) Acad. Sci. U.S.S.R.*, 192: 131–133.

Smejkal, V., Cook, F.D. and Krouse, H.R., 1971. Studies of sulfur and carbon isotope fractionation with microorganisms isolated from springs of Western Canada. *Geochim. Cosmochim. Acta*, 35: 787–800.

Smith, J.W. and Croxford, N.J.W., 1973. Sulphur isotope ratios in the McArthur lead-zinc-silver deposit. *Nature Phys. Sci.*, 245: 10–12.

Söhngen, N.L., 1914. Umwandlungen von Manganverbindungen unter Einfluss mikrobiologischer Prozesse. *Zbl. Bakteriol. Parasitkd., Abt. 2*, 40: 545–554.

Sokolova, G.A., 1962. Microbiological sulfur formation in Sulfur Lake. *Microbiology*, 31: 264–266.

Sokolova-Dubinina, G.A. and Deryugina, Z.P., 1966. The role of microorganisms in the formation of oxidized ores in the Chiatura manganese deposits. *Microbiology*, 35: 293–298.

Sokolova-Dubinina, G.A. and Deryugina, Z.P., 1967a. On the role of microorganisms in the formation of rhodochrosite in Punnus-Yarvi Lake. *Microbiology*, 36: 445–451.

Sokolova-Dubinina, G.A. and Deryugina, Z.P., 1967b. Process of iron-manganese concretion formation in Lake Punnus-Yarvi, *Microbiology*, 36: 892–899.

Solomon, P.J., 1965. Investigations into sulfide mineralization at Mount Isa, Queensland. *Econ. Geol.*, 60: 737–765.

Sorokin, Yu.I., 1962. Experimental investigation of bacterial sulfate reduction in the Black Sea using ^{35}S. *Microbiology*, 31: 329–335.

Sorokin, Yu.I., 1964. On the primary production and bacterial activities in the Black Sea. *J. Conserv. Perm. Int. Explor. Mer.*, 29: 41–60.

Sorokin, Yu.I., 1970. Interrelations between sulphur and carbon turnover in meromictic lakes. *Arch. Hydrobiol.*, 66: 391–446.

Sorokin, Yu.I., 1972. Role of biological factors in the sedimentation of iron, manganese, and cobalt and in the formation of nodules. *Oceanology*, 12: 1–11.

Starkey, R.L., 1945a. Precipitation of ferric hydrate by iron bacteria. *Science, N.Y.*, 102: 532–533.

Starkey, R.L., 1945b. Transformations of iron by bacteria in water. *J. Am. Water Works Assoc.*, 37: 963–984.

Starkey, R.L. and Halvorson, H.O., 1927. Studies on the transformations of iron in nature, II. Concerning the importance of microorganisms in the solution and precipitation of iron. *Soil Sci.*, 24: 381–402.

Starkey, R.L. and Wight, K.M., 1945. *Anaerobic Corrosion of Iron in Soil*. Am. Gas Assoc., New York., N.Y., 108 pp.

Steeman-Nielsen, E.S. and Jensen, E.A., 1957. Primary oceanic production. *Galathea Rep.*, 1: 49–136.

Strøm, K.M., 1955. Land-locked waters and the deposition of black muds. In: P.D. Trask (Editor), *Recent Marine Sediments*. Am. Assoc. Petrol. Geol., pp. 356–372.

SubbaRao, M.S., Iya, K.K. and Sreenivasaya, M., 1947. Microbiological formation of elemental sulphur in coastal areas. *Abstr. Comm. IVth Int. Congr. Microbiol.*, p. 157.

Suckow, R. and Schwartz, W., 1968. Geomikrobiologische Untersuchungen, X. Zur Frage der Genese des Mansfelder Kupferschiefers. *Z. Allgem. Mikrobiol.*, 8: 47–64.

Sugawara, K., Koyama, T. and Kozawa, A., 1954. Distribution of various forms of sulphur in lake-, river- and sea muds, 2. *J. Earth Sci.*, 2: 1–14.

Suh, B. and Akagi, J.M., 1969. Formation of thiosulfate from sulfite by *Desulfovibrio vulgaris. J. Bacteriol.*, 99: 210–215.

Sunagawa, I., Endo, Y. and Nakai, N., 1971. Hydrothermal synthesis of framboidal pyrite. *Soc. Mineral. Geol. Japan*, 2: 10–14.

Sweeney, R.E. and Kaplan, I.R., 1973. Pyrite framboid formation: laboratory synthesis and marine sediments. *Econ. Geol.*, 68: 618–634.

Tausson, W.O. and Alishima, W.A., 1932. The reduction of sulphates by bacteria in the presence of hydrocarbons. *Mikrobiologiya*, 1: 229–261.

Tausson, W.O. and Veselov, I.J., 1934. The bacteriology of decomposition of cyclical compounds in the reduction of sulphates. *Mikrobiologiya*, 3: 360–369.

Temple, K.L., 1964. Syngenesis of sulfide ores: An evaluation of biochemical aspects. *Econ. Geol.*, 59: 1473–1491.

Temple, K.L. and Le Roux, N.W., 1964a. Syngenesis of sulphide ores: desorption of absorbed metal ions and their precipitation as sulphides. *Econ. Geol.*, 59: 647–655.

Temple, K.L. and Le Roux, N.W., 1964b. Syngenesis of sulfide ores: sulfate-reducing bacteria and copper toxicity. *Econ. Geol.*, 59: 271–278.

Ten Khan-mun, 1967. Iron and manganese-oxidizing organisms in soils of south Sakhalin. *Microbiology*, 36: 276–381.

Ten Khan-mun, 1968. The biological nature of iron-manganese crusts in soil forming rocks in Sakhalin mountain soils. *Microbiology*, 37: 621–624.

Thiele, G.A., 1925. Manganese precipitated by microorganisms. *Econ. Geol.*, 20: 301–310.

Thode, H.G., 1963. Sulphur isotope geochemistry. In: D.M. Shaw (Editor), *Studies in Analytical Chemistry – R. Soc. Can. Spec. Publ.*, 6: 25–41.

Thode, H.G., 1964. Stable isotopes. A key to our understanding of natural processes. *Bull. Can. Petrol. Geol.*, 12: 242–262.

Thode, H.G., Kleerekoper, H. and McElcheran, D., 1951. Isotopic fractionation in the bacterial reduction of sulphate. *Research, Lond.*, 4: 581–582.

Thode, H.G., Macnamara, J. and Fleming, W.H., 1953. Sulphur isotope fractionation in nature and geobiological and biological time scales. *Geochim. Cosmochim. Acta*, 3: 235–243.

Thode, H.G., Wanless, R.K. and Wallouch, R., 1954. The origin of native sulphur deposits from isotope fractionation studies. *Geochim. Cosmochim. Acta*, 54: 286–298.

Thode, H.G., Harrison, A.G. and Monster, J., 1960. Sulphur isotope fractionation in early diagenesis of recent sediments of northeast Venezuela. *Bull. Am. Assoc. Petrol. Geol.*, 44: 1809–1817.

Trudinger, P.A., 1967. The metabolism of inorganic compounds by thiobacilli. *Rev. Pure Appl. Chem.*, 17: 1–24.

Trudinger, P.A., 1969. Assimilatory and dissimilatory metabolism of inorganic sulphur compounds by microorganisms. *Adv. Microbiol. Physiol.*, 3: 111–158.

Trudinger, P.A., 1971. Microbes, metals and minerals. *Mineral. Sci. Eng.*, 3: 13–25.

Trudinger, P.A. and Bubela, B., 1967. Microorganisms and the natural environment. *Mineral. Deposita*, 2: 147–157.

Trudinger, P.A. and Chambers, L.A., 1973. Reversibility of bacterial sulfate reduction and its relevance to isotope fractionation. *Geochim. Cosmochim. Acta*, 37: 1775–1778.

Trudinger, P.A., Lambert, I.B. and Skyring, G.W., 1972. Biogenic sulphide ores: a feasibility study. *Econ. Geol.*, 67: 1114–1127.

Tsubota, G., 1959. Phosphate reduction in the paddy field. *Soil Plant Tokyo*, 5: 10–15.

Tudge, A.P. and Thode, H.G., 1950. Thermodynamic properties of isotopic compounds of sulphur. *Can. J. Res.*, B28: 567–578.

Tyler, P.A., 1970. Hyphomicrobia and the oxidation of manganese in aquatic systems. *Antonie van Leeuwenhoek*, 36: 567–578.

Tyler, P.A. and Marshall, K.C., 1967a. Microbial oxidation of manganese in hydroelectric pipelines. *Antonie van Leeuwenhoek*, 33: 171–183.

Tyler, P.A. and Marshall, K.C., 1967b. Pleomorphy in stalked, budding bacteria. *J. Bacteriol.*, 93: 1132–1136.

Tyler, P.A. and Marshall, K.C., 1967c. Hyphomicrobia – a significant factor in manganese problems. *J. Am. Water Works Assoc.*, 59: 1043–1048.

Tyler, P.A. and Marshall, K.C., 1967d. Form and function in manganese oxidizing bacteria. *Arch. Mikrobiol.*, 56: 344–353.

Tyler, S.A. and Barghoorn, E.S., 1954. Occurrence of structurally preserved plants in Precambrian rocks of the Canadian Shield. *Science N.Y.*, 119: 606–608.

Valdiya, K.S., 1969. A new phosphatic horizon in the Late Precambrian Calc Zone of Pithoragarh Kumaon Himalaya. *Current Sci.*, 38: 415–416.

Vallentyne, J.R., 1962. A chemical study of pyrite spherules isolated from sediments of Little Round Lake, Ontario. In: M.L. Jensen (Editor), *Biogeochemistry of Sulphur Isotopes*. Yale University Press, New Haven, Conn., pp. 144–152.

Vallentyne, J.R., 1963. Isolation of pyrite spherules from recent sediments. *Oceanogr. Limnol.*, 8: 16–30.

Veitch, F.P. and Blakenship, L.C., 1963. Carbonic anhydrase in bacteria. *Nature, Lond.*, 197: 76–77.

Verner, A.R. and Orlovskii, N.V., 1948. The role of sulphate-reducing bacteria in the salt balance of Barbary soils. *Pochvovedenie*, 9: 553.

Vinogradov, A.P., 1958. Isotopic composition of sulphur in meteorites and in the earth. In: R.C. Extermann (Editor), *Radioisotopes in Scientific Research, 2*. Pergamon, New York, N.Y., pp. 581–591.

Vinogradov, A.P., Grinenko, V.A. and Ustinov, V.I. 1962. Isotopic composition of sulfur compounds in the Black Sea. *Geochemistry*, 10: 973–997.

Viragh, K. and Szolnoki, J., 1970. Role of bacteria in the formation and reaccumulation of the uranium ore of Meesek (Hungary). *Földt. Közl.*, 100: 43–54.

Voge, H.H., 1939. Exchange reactions with radiosulfur. *J. Am. Chem. Soc.*, 61: 1032–1035.

Volkov, I.I., 1961. Iron sulfides, their interdependence and transformation in Black Sea bottom sediments. *Akad. Nauk. S.S.S.R., Inst. Okeanol. Tr.*, 50: 68–92.

Volkov, I.I. and Ostroumov, E.A., 1957. The form of sulfur compounds in the interstitial waters of the Black Sea sediments. *Geochemistry*, 4: 397–406.

Wainright, T., 1970. Hydrogen sulfide production by yeast under conditions of methionine, panto-thenate or vitamin B6 deficiency. *J. Gen. Microbiol.*, 61: 107–119.

Wallhäusser, K.H. and Puchelt, H., 1966. Sulfatreduzierende Bakterien in Schwefel- und Gruben-wässern Deutschlands und Österreichs. *Beitr. Mineral. Petrol.*, 13: 12–30.

Walter, M.R., 1972a. A hot spring analogue for the post depositional environment of Precambrian iron formations of the Lake Superior region. *Econ. Geol.*, 67: 965–980.

Walter, M.R., 1972b. Stromatolites and the biostratigraphy of the Australian Precambrian and Cambrian. *Palaeontol. Assoc. Lond., Spec. Pap. Palaeontol.*, 11.

Walter, M.R., Bauld, J. and Brock, T.D., 1972. Siliceous algal and bacterial stromatolites in the hot spring and geyser effluents of Yellowstone National Park. *Science N.Y.*, 178: 402–405.

Watanabe, M., 1924. Zonal precipitation of ores from a mixed solution. *Econ. Geol.*, 19: 497–503.

Watson, S.W. and Waterbury, J.B. 1969. The sterile hot brines of the Red Sea. In: E.T. Degens and D.A. Ross (Editors), *Hot Brines and Recent Heavy Metal Deposits in the Red Sea*. Springer, New York, N.Y., pp. 272–281.

Wedepohl, K.H., 1971. "Kupferschiefer" as a prototype of syngenetic sedimentary ore deposits. *Soc. Mineral. Geol. Japan,* 3: 268–273.

Weil, R., 1955. Réproduction expérimentale des sulfures metalliques des sédiments biogènes. In: *80e Congr. Soc. Savantes.* Gauthier-Villars, Paris, pp. 117–125.

Weil, R., 1958. Recherches expérimentales sur quelques aspects de la géochimie de la biosphère. *Geochim. Cosmochim. Acta,* 14: 166.

Weil, R., Hocart, H. and Monier, J.C., 1954. Synthèses minérales en milieux organiques. *Bull. Soc. Fr. Minéral. Cristallogr.,* 77: 1084–1101.

Weiss, A. and Amstutz, G.C., 1966. Ion-exchange reactions on clay minerals and cation-selective membrane properties as a possible mechanism of economic metal concentration. *Mineral. Deposita,* 1: 60–66.

Wilcox, N.R., 1953. The origin of beds of phosphatic chalk with special reference to those at Taplow, England. *19th Int. Geol. Congr., Algiers, 1952, C.R. Sec., 11* (11): 119–133.

Williams, N. and Rye, D.M., 1974. Alternative interpretation of sulphur isotope ratios in the McArthur lead-zinc-silver deposit. *Nature, Lond.* (in press).

Wilson, C., 1945. Bacteriology of water pipes. *J. Am. Water Works Assoc.,* 37: 52–58.

Wolfe, R.S., 1960. Microbial concentration of iron and manganese in water with low concentrations of these elements. *J. Am. Water Works Assoc.,* 52: 1335–1337.

Wolfe, R.S., 1964. Iron and manganese bacteria. In: H. Heukelekian and N.C. Dondero (Editors), *Principles and Applications in Aquatic Microbiology.* Wiley, New York, N.Y., pp. 82–97.

Woolfolk, C.A. and Whiteley, H.R., 1962. Reduction of inorganic compounds with molecular hydrogen by *Micrococcus lactilyticus,* I. Stoichiometry with compounds of arsenic, selenium, tellurium, transition and other elements. *J. Bacteriol.,* 84: 647–658.

Youssef, M.I., 1965. Genesis of bedded phosphates. *Econ. Geol.,* 60: 590–600.

Zajic, J.E., 1969. *Microbial Biogeochemistry.* Academic Press, New York, N.Y., 345 pp.

Zapffe, C., 1931. Deposition of manganese. *Econ. Geol.,* 24: 799–832.

Zapffe, C., 1933. Catalysis and its bearing on origin of Lake Superior iron-bearing formations. *Econ. Geol.,* 28: 751–772.

Zavarzin, G.A., 1961. Symbiotic culture of a new microorganism oxidizing manganese. *Microbiology,* 30: 343–345.

Zavarzin, G.A., 1962. Symbiotic oxidation of manganese by two species of *Pseudomonas. Microbiology,* 31: 481–482.

Zavarzin, G.A., 1967. Bacteria in relation to manganese metabolism. In: T.R.G. Gray and D. Parkinson (Editors), *Ecology of Soil Bacteria.* Liverpool University Press, pp. 612–623.

ZoBell, C.E., 1946. *Marine Microbiology.* Mass. Chronica Botanica Co., 240 pp.

ZoBell, C.E., 1947. Biennial report for 1945–47 on A.P.I. Research Proj. 43A. In: *Report of Progress-Fundamental Research on Occurrence and Recovery of Petroleum, Am. Petrol. Inst. 1946–47,* pp. 100–106.

ZoBell, C.E., 1950. Assimilation of hydrocarbons by microorganisms. *Adv. Enzymol.,* 10: 443–486.

ZoBell, C.E., 1958. Ecology of sulfate-reducing bacteria. *Producers Mon. Penn. Oil. Prod. Assoc.,* 22: 12–29.

Chapter 7

OXYGEN AND CARBON ISOTOPES IN ORE DEPOSITS IN SEDIMENTARY ROCKS

P. FRITZ

INTRODUCTION

The origin of hydrothermal fluids and their dissolved load has intrigued earth scientists for several centuries as the search for an answer not only presented an academic challenge, but had significant economic implications. In general it was assumed that many thermal waters contained significant amounts of primary magmatic water (e.g. Allen and Day, 1935). However, no analytical technique was available to qualify and quantify such assumptions since, for example, fresh water in contact with igneous rocks can become chemically very similar to magmatic fluids. The breakthrough came with the development of precise mass-spectrometers able to determine isotopic differences as small as 0.1‰ (permil) (Nier 1947, McKinney et al., 1950).

Stable-isotope techniques then permitted the distinction between fluids of different origin and greatly advanced our knowledge about subsurface processes which affect the chemical and isotopic properties of hydrothermal fluids and their origin. Such analyses are now a standard analytical technique in many geological investigations and have been applied in disciplines as different as paleoclimatology, carbonate sedimentology, igneous petrology and hydrology.

In this chapter some aspects of the distribution and processes controlling the abundance of ^{18}O and ^{13}C in hydrothermal systems are discussed. It should be noted that here the term "hydrothermal" is not used in a genetic sense but refers to hot, mineralized fluids which in this specific case deposited economic amounts of ore minerals in sedimentary rocks.

Unfortunately, such ore minerals contain only in exceptional cases significant amounts of oxygen and carbon but both are normally very abundant in the associated gangue minerals: carbonates and silicates. Therefore, most stable isotope investigations which use ^{18}O or ^{13}C are based on the distribution of these isotopes in these gangue minerals and it is evident that the paragenetic relationships between gangue and ore must be fully understood, if such data are to yield information about the formation of ore minerals, and the origin of fluids and metals.

The first comprehensive study of this type was presented by Engel et al. (1958) and since then much has been written on this subject either to elucidate the genesis of ore deposits, to discuss the origin of the mineralizing fluids and their dissolved constituents, or to evaluate this technique as a prospecting tool. Much of the published information on modern and fossil hydrothermal systems in sedimentary rocks is presented here in summarized form and a more detailed discussion concentrates on four genetically different deposits: the Providencia mines in Mexico; the Bluebell Mine in British Columbia (Canada); the Pine Point deposits in the Northwest Territories (Canada); and the Eldorado Mines in the Beaverlodge district in Saskatchewan (Canada).

The deposits in the Providencia mining district in north-central Mexico consist of a series of conformable, pipe-like lead-zinc ore bodies. They occur in Mesozoic limestones adjacent to a granodiorite stock to which they are probably genetically related (Rye, 1966).

The Bluebell Mine is located near the centre of a structural belt known as the Kootenay arc. This arc contains a succession of sedimentary and volcanic rocks in which the Bluebell ore body replaces Paleozoic limestones. It consists of coarsely crystalline and massive bodies of pyrrhotite, sphalerite, galena, associated with quartz and calcite gangue. The hydrothermal solutions most probably had a meteoric origin (Ohmoto and Rye, 1970).

The Pine Point deposits occur as plum-shaped bodies in Middle-Devonian carbonate sediments. Jackson and Beales (1967) postulated that the mineralizing fluids were expelled from an adjacent sedimentary basin and precipitation of ore minerals occurred in an H_2 S-rich environment in porous Middle Devonian "reefs". The mineralogy and paragenesis of these deposits is simple: dolomitization and neoformation of dolomite preceded the deposition of marcasite, pyrite, galena and sphalerite. During the late stages of mineralization calcite was deposited with the ore and continued to form after sulfide mineral deposition had come to an end.

The Beaverlodge district in Saskatchewan is characterized by several large vein-type uranium deposits in which pitchblende is associated mainly with hematite or pyrite, minor amounts of copper sulfides, quartz, dolomite and calcite. The genesis of these deposits is not yet fully understood, but it has been postulated that they formed during metamorphism of a clastic sedimentary sequence (Sassano et al., 1972).

The oxygen and carbon isotopic compositions of carbonates in these deposits are expressed in the conventional δ ‰ notation:

$$\delta = \left(\frac{R \text{ sample}}{R \text{ standard}} - 1 \right) \times 1000$$

where R is the $^{18}O/^{16}O$ or $^{13}C/^{12}C$ ratio in the samples and standards. The $\delta^{18}O$-values refer to the Standard Mean Ocean Water (SMOW), the $\delta^{13}C$-values to the PDB-standard (*Belemnitella americana*, from the Pee Dee Formation in South Carolina, U.S.A.). A $\delta^{18}O = +5$‰ signifies that the sample has 5‰ more ^{18}O than the standard. The analytical error is normally less than $\pm$ 0.2‰ for both ^{18}O and ^{13}O analyses. All dolomite data presented here are corrected for the acid fractionation according to Sharma and Clayton (1965).

ENVIRONMENTAL ABUNDANCES OF ^{13}C AND ^{18}O

To facilitate the understanding of the following discussions it seems appropriate to outline briefly the distribution of oxygen-18 and carbon-13 in materials on the earth's crust. Both isotopes are present only in relatively small amounts and the approximate natural abundance for ^{18}O in the total oxygen reservoir is 0.2%, for ^{13}C in the total carbon reservoir is 1.1%. These values vary by about 7% for both isotopes (Fig. 1,2). The variations are due to isotope fractionation processes which are a function of differences in mass or energy contents (reduced partition functions) of different molecules (Urey, 1947; and others).

The lowest ^{18}O contents are found in the permanent ice sheets of Greenland and Antarctica where values as low as $-60‰$ have been measured (Fig. 1). In temperate climates precipitations in general show δ^{18}O values between -7 and $-20‰$, whereas the ^{18}O contents of rain in tropical climates differ only by a few permil from average ocean water (Dansgaard, 1964).

Groundwater normally will closely reflect the average isotope content of the precipitations in the aquifer recharge areas. However, if fresh water penetrates deeply into the

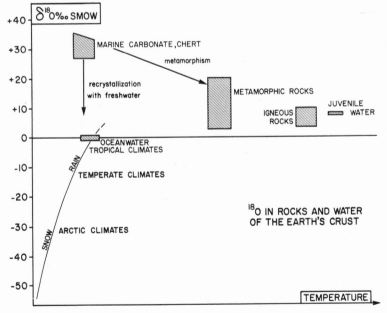

Fig.1. ^{18}O in natural materials on the earth's crust.

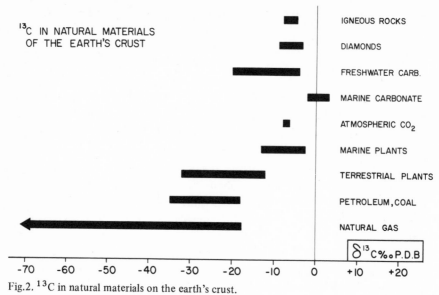

Fig.2. ^{13}C in natural materials on the earth's crust.

subsurface, exchange reactions with carbonates and silicates at elevated temperatures can alter its oxygen isotopic composition. This process is described in more detail in the following sections.

Marine carbonates and cherts have higher ^{18}O contents than most other rocks. Their δ^{18}O values typically lie between +22 and +35‰. The ^{18}O contents of igneous rocks are considerably lower and rarely exceed values of +10‰. Since isotope fractionations between minerals and water decrease with increasing temperature (see below), at high temperatures little or no fractionation between the two phases is possible and both have similar ^{18}O contents. Therefore, it is assumed — though no actual measurements have been made — that juvenile water has a δ^{18}O of close to +7‰, about the average isotopic composition of igneous rocks (Fig. 1).

The ^{18}O contents of metamorphic rocks vary over a wide range because their isotopic compositions are determined by the ^{18}O contents of the parent material, the amount and ^{18}O contents of the water present. The δ^{18}O contents of these rocks in general are, therefore, intermediate between those of marine sediments and igneous rocks.

The ^{18}O contents of silicates, oxides, carbonates and water in the earth's crust are primarily controlled by physicochemical processes. Similarly, the abundance of ^{13}C can be controlled by such processes but biological activities may play an even more important role, because much of the carbon in near surface environments passes through biological cycles.

The large atmospheric and magmatic carbon reservoirs have a fairly uniform carbon isotope composition, both with δ^{13}C-values close to −7‰ (Fig. 2). It is thus not surprising that igneous rocks and inorganic marine carbonates show only small variations in their

^{13}C contents. However, in many fresh-water systems biological processes are of importance, and as such processes tend to decrease the ^{13}C contents in the organic products, for example, a relatively wide range of δ^{13}C values can be expected in fresh-water carbonates. The lowest ^{13}C contents are found in natural hydrocarbons due to the combined effects of biological activity and large isotope fractionation factors between oxidized and reduced carbon in gases (Rosenfeld and Silverman, 1959; Bottinga, 1969).

^{18}O AND ^{13}C IN HYDROTHERMAL SYSTEMS

Theoretical considerations

Oxygen-18. The oxygen-isotopic compositions of oxygen-18 bearing minerals deposited in isotopic equilibrium with the mineralizing fluid depend primarily on the oxygen-18 content of the fluid and temperature dependent ^{18}O-fractionation factors. It is safe to assume that in most hydrothermal systems which lead to ore deposition the atomic ratio of oxygen in water/oxygen in carbonate deposited is very high. Therefore, aqueous carbonate species are of no importance and the physicochemical conditions of the hydrothermal system exert only an indirect influence.

The fractionation factors describe the distribution of ^{18}O in the mineral-water systems and are defined as:

$$\alpha_{\text{min-H}_2\text{O}} = \frac{R \text{ mineral}}{R \text{ water}}$$

where $R = {}^{18}\text{O}/{}^{16}\text{O}$. Their numerical values decrease with increasing temperature, that is, the isotopic differences between mineral and water are higher at low temperatures than at high temperatures. For example, the calcite-water ^{18}O-fractionation factor has a value of about 1.028 at 25°, but becomes close to 1.000 above 600°C. Thus, calcite deposited in isotopic equilibrium with water at 25°C will have 28‰ more ^{18}O than the water and both will have approximately the same ^{18}O contents above 600°C.

O'Neil et al. (1969) determined experimentally the distribution of ^{18}O between calcite and water over the temperature range 0–500°C. The fractionation factor is represented by the equation:

$$1000 \ln \alpha_{\text{calcite-H}_2\text{O}} = 2.78 \, (10^6 T^{-2}) - 3.39 \tag{1}$$

where T is the absolute temperature and:

$$1000 \ln \alpha \simeq (\alpha - 1) \times 1000 \simeq \delta^{18}\text{O}_{\text{calcite}} - \delta^{18}\text{O}_{\text{H}_2\text{O}}$$

(both $\delta^{18}\text{O}_{\text{calcite}}$ and $\delta^{18}\text{O}_{\text{H}_2\text{O}}$ are expressed vs. the standard, SMOW).

A similar relationship has been derived by Northrop and Clayton (1966) for the dolomite/water fractionation at temperatures between 300° and 500°C; whereby:

$$1000 \ln \alpha_{\text{dolomite-H}_2\text{O}} = 3.2 \, (10^6 T^{-2}) - 2.0 \tag{2}$$

The different fractionation factors for calcite/H_2O and dolomite/H_2O should result in about 3–4‰ enrichment in ^{18}O in dolomite with respect to calcite if they were deposited in isotopic equilibrium with one another near 100°C. Inasmuch as this fractionation decreases with increasing temperatures only about 1 ‰ difference in ^{18}O should exist near 400°C.

It should be noted, however, that the dolomite/water fractionation is only poorly defined at temperatures below 300°C and it is well possible that the actual fractionation factors are considerably smaller (Friedman and Hall, 1963; Fritz and Smith, 1970). Unfortunately, the dolomite/calcite fractionations cannot be determined from ^{18}O data obtained from (naturally occurring) carbonate pairs, because it is virtually impossible to demonstrate that pairs of coexisting calcite and dolomite actually formed in isotopic equilibrium with one another or under identical conditions. Furthermore, it is quite common to observe dolomitization preceding the deposition of ore minerals, whereas calcite becomes important during the final phases of mineralization. For this reason, much of the following discussion will be restricted to hydrothermal calcites, and the dolomite data will only serve to strengthen the arguments presented.

Quartz is another oxygen-bearing gangue mineral of possible importance in ore deposits in sedimentary rocks. The equilibrium fractionation factor between quartz and water is described by the following equation (Clayton et al., 1972):

$$1000 \ln \alpha = 3.38 \, (10^6 T^{-2}) - 3.4 \tag{3}$$

These mineral/water fractionation factors permit the calculation of the ^{18}O contents of the mineralizing fluids – which normally cannot be directly measured – provided the temperature of mineral deposition is known and the minerals formed in isotopic equilibrium with the fluids. The temperature of mineralization can be determined by fluid inclusion studies, thermodynamic considerations, and may also be deducted from ^{18}O or ^{34}S data if equilibrium pairs of minerals can be found in a deposit. For a detailed discussion see the article by Sangster in this volume. Under equilibrium conditions the differences in ^{18}O contents between two minerals are a function of temperature alone. Assume, for example, quartz and calcite are found as such a pair, then eq. 1 and eq. 3 can be combined to give the following relationship:

$$\delta^{18}O_{\text{quartz}} - \delta^{18}O_{\text{calcite}} \simeq 0.60 \, (10^6 T^{-2}) \tag{4}$$

In general, the isotope analysis of a single mineral at a locality does not suffice to show that the mineral was deposited in isotopic equilibrium. However, if two or more minerals from the same specimen are analysed, some conclusions can be drawn. This is most easily done by a graphical presentation of the isotope data in which equilibrium pairs will fall in a straight line in a diagram of $\delta^{18}O_{min. A} - \delta^{18}O_{min. B}$ vs. $\delta^{18}O_{min. A}$

or a plot of $\delta^{18}O_{min. A} - \delta^{18}O_{min. B}$ vs. $\delta^{18}O_{min. A} - \delta^{18}O_{min. C}$

In most isotope studies of ore deposits it is impossible to sample and analyse the isotopic composition of the mineralizing fluids. However, as already mentioned their oxygen isotope compositions can be calculated, if the ^{18}O contents of carbonate and the temperature of their deposition are known. These calculations are of great significance since the ^{18}O contents of the hydrothermal fluids reflect the origin of the fluids and/or secondary changes of its ^{18}O contents through exchange reactions with the hostrocks.

As shown above, the total range of ^{18}O contents in natural waters is close to 70 ‰ with precipitation and surface waters showing the largest variations. As the δ^{18}O values of water samples refer to the "Standard Mean Ocean Water" (SMOW) most fresh waters have negative δ^{18}O values and cover the range between 0 and $-60‰$. Juvenile or magmatic water which equilibrated with silicates at elevated temperatures covers the other end of the total range and values between +7 to +9‰ can be expected (Craig, 1963; Taylor, 1967). However, positive δ^{18}O values in hydrothermal fluids do not necessarily indicate that the mineralizing fluids had a deep-seated, magmatic origin as, for example, exchange reactions between water and marine carbonate sediments can result in similarly high δ^{18}O values: if meteoric water with a $\delta^{18}O = -20$ ‰ SMOW equilibrates, within a temperature range from near surface values to about 200°C, with limestone having a $\delta^{18}O = +20‰$ (SMOW), the ^{18}O content of water will increase and δ^{18}O values as high as +10‰ will be reached as evidenced by oil field brines (Hitchon and Friedman, 1969). Such exchange reactions, however, do not affect the deuterium contents of the water, which thus closely reflects the composition of the fluids before they entered the hydrothermal systems. This has been documented by Craig (1963) who observed that the deuterium contents of thermal waters in areas with hot spring activities are close to those of surface waters in the immediate vicinity. Craig (1961) also demonstrated that the ^{18}O and deuterium contents of meteoric waters are closely described, over a wide range of isotopic compositions, by the equation: $\delta D‰ = 8 \delta^{18}O‰ + 10‰$.

It is often possible to obtain from fluid inclusions the minute quantities of water needed for a deuterium analysis. A comparison of calculated or measured ^{18}O values with deuterium values can thus be used to differentiate between the different types of water, and it is then also possible to determine the original ^{18}O contents of the hydrothermal fluid. As an example, data obtained from Salton Sea geothermal brines and the Bluebell and Providencia mines are shown in Fig. 3. Note the constant δD values and the widely spread δ^{18}O values in the Bluebell and Salton Sea system. The intercept between the "meteoric water line" and the "Bluebell-line" gives the isotopic composition of the

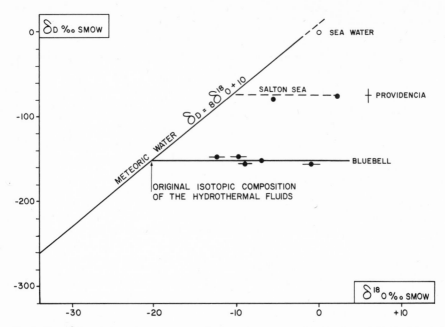

Fig. 3. The ^{18}O and deuterium contents of meteoric waters and hydrothermal fluids. The Salton Sea data are from Craig (1966), the Bluebell data from Ohmoto and Rye (1970) and the Providencia data from Rye (1966). Note that the $\delta^{18}O$ values shown for the Providencia fluids (Rye, 1966) differ slightly from the data shown in Fig. 7. For further explanation see text.

water before it exchanged its ^{18}O with marine carbonates, and the intercept with the Salton Sea line corresponds to the isotope contents of the local meteoric water. A more detailed discussion of Fig. 3 will be given in the following sections.

Carbon-13. In the previous section it has been demonstrated that the oxygen isotope compositions of hydrothermal carbonates are almost exclusively controlled by the ^{18}O contents of the mineralizing fluids and the temperature dependent carbonate-water fractionation factors. Similarly, the ^{13}C contents of such carbonates will be determined by the ^{13}C contents of the carbonates in solution and temperature dependent fractionation factors. However, whereas the ^{18}O contents are not measurably affected by chemical reactions occurring during ore deposition, the ^{13}C contents of their dissolved inorganic carbon species can vary over a wide range, not only because of isotopically variable carbon supplies, but primarily as a function of chemical processes. Aqueous carbon species which may be of importance in hydrothermal systems not exceeding 400°C are HCO_3^-, CO_3^{2-}, CH_4 aq. and H_2CO_3 apparent, the latter being defined as the sum of CO_2 aq. and H_2CO_3. Their abundance is strongly dependent on the physicochemical conditions existing in the system which can exert a significant influence on the ^{13}C contents of carbonates deposited.

The isotopic fractionation factors existing between the different carbon species are only approximately known and are summarized in Fig. 4. As indicated in this figure little or no isotope fractionation exists between CO_2 gas and H_2CO_3 app. (Deuser and Degens, 1967; Wendt, 1968) and between CH_4 gas and CH_4 aq. (Ohmoto, 1972). It is interesting to note that apparently little change in the relative ^{13}C-fractionations between HCO_3^-, CO_3^{2-} and $CaCO_3$ occurs from near surface temperatures to about 400°C. It should be noted, however, that the fractionation between these species at elevated temperatures is poorly understood and, for example, the HCO_3^- line has been derived through extrapolation between high-temperature and low-temperature experimental data (Malinin et al., 1967; Ohmoto, 1972). The $CaCO_3$ curve is based on theoretical calculations by Bottinga (1969). Little is known about the dolomite-CO_2 fractionation which here is assumed to be similar to the calcite-CO_2 fractionation especially at elevated temperatures.

The large isotope fractionation factors existing between the different carbon species and the physiochemical controls, which dominate their relative abundance in solution often make it difficult to discuss the origin of carbon in the hydrothermal carbonates.

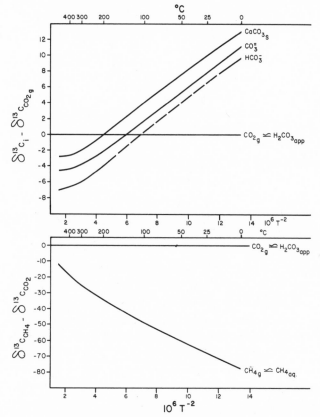

Fig. 4. The carbon isotope fractionation between different carbonate species (top) and between methane and carbon dioxide (bottom) as a function of temperature.

For example, if a large reservoir of CO_2 is available during carbonate deposition and no partial reduction of the CO_2 to CH_4 takes place the ^{13}C in the carbonates will be largely controlled by the CO_2-$CaCO_3$ fractionation. If, however, appreciable amounts of CH_4 are formed, the ^{13}C contents of the CO_2 and the dissolved carbon species will increase, and the ^{13}C in the carbonates can no longer be directly related to the carbon source. A similar argument can be brought forward in the case where the carbon in solution is derived from the dissolution of limestones and where the precipitation, and thus the ^{13}C content, in the carbonates is controlled by the chemistry of the solution.

However, if the controlling parameters, such as salinity, temperature, pH and partial pressure of oxygen are known, or can be estimated, it is possible to calculate the average composition of the carbon in solution, whereby:

$$\delta^{13}C_{\Sigma \text{ carbon}} = (\delta^{13}C \times N)_{H_2CO_3 \text{ app.}} + (\delta^{13}C \times N)_{HCO_3^-} + (\delta^{13}C \times N)_{CO_3^=} +$$

$$(\delta^{13}C \times N)_{CH_4 \text{ aq.}}$$

$(\delta^{13}C \times N)_i$ is the product of the measured or calculated $\delta^{13}C$ value and the mole faction N of the aqueous carbon species i. This has been discussed in detail by Ohmoto (1972).

^{18}O in active geothermal systems

The discovery of active thermal systems as a result of economic interests in geothermal energy was of great significance for the discussions of the origin of mineralizing fluids and the formation of ore deposits in sedimentary rocks. Probably the most exciting find were the Salton Sea brines which White et al. (1963) described as man's first sample of a pure magmatic ore fluid from an active thermal system. Similarly exciting and important for economic geologists and geochemists was the discovery in 1964 of pools of warm saline brine on the floor of the Red Sea. Both systems have been described in great detail as they offered a unique chance to study the geochemistry of ore metal transport and mineral deposition. An extended summary of these investigations was presented by White (1968).

For the arguments discussed here, it is interesting to note that stable isotope analyses were of decisive importance in the analyses and generic description of these systems. Especially noteworthy is that deuterium and ^{18}O analyses provided proof that neither system contained measurable amounts of juvenile or magmatic water (Craig, 1966; 1969). Similar conclusions were possible for almost all known active geothermal systems, and the reader is referred to Craig (1963) for a more detailed discussion.

The assumption at present is that the Salton Sea system is recharged in infiltration of local meteoric water (Craig, 1966) which increased its original ^{18}O contents through high temperature exchange with ^{18}O-rich silicates. As the deuterium content of thin water is not measurably affected by these processes the ^{18}O shift typical for almost all geothermal waters can be observed on a deuterium vs. ^{18}O diagram (Fig. 3).

Fig. 3 also shows almost identical deuterium contents for the Providencia and Salton Sea fluids. At present this agreement must be considered a coincidence. However, it should be mentioned that the Providencia fluids are thought to be of magmatic origin and from stable isotope data alone one could thus conclude that the Salton Sea brines contain as much as 75% of Providencia-type fluids. Other evidence, however, does not support such origin for the Salton Sea brines (White, 1968).

The geochemistry of the cooler Red Sea brines is probably less complicated than the one of the Salton Sea brines and is described in detail in a series of publications edited by Degens and Ross (1969). The ^{18}O–deuterium studies were done by Craig (1966, 1969), who summarizes his findings as follows: (Craig, 1969, pp.239–240) "...seawater circulates downward through evaporite sediments and flows northward, driven by the density difference between columns of brine and sea water, and heated by the local geothermal gradient which is quite sufficient to bring it to the observed temperature. ... The deuterium and oxygen-18 concentrations indicate that the source water originates at least several hundred km south of the depressions, and probably about 800 km south, near or on the southern sill . . . The absence of an oxygen-18 shift in the water shows that temperatures are not much higher than 100°C at most."

^{18}O in gangue minerals and mineralizing fluids

Carbonate and silicate gangue minerals are both of potential interest for this discussion. However, most oxygen-isotope studies of hydrothermal ore deposits in sedimentary rocks concentrate on the distribution of ^{18}O in the carbonate gangue and only very few analyses of ^{18}O in hydrothermal silicates are found in the published literature. This is unfortunate since, for example, quartz is mineralogically and chemically considerably more stable than either dolomite or calcite and would thus yield more reliable ^{18}O data than can be expected from carbonates. Recrystallization of early hydrothermal carbonates under the influence of later, geochemically different solution is thus a serious problem in stable isotope studies of hydrothermal systems.

Furthermore, carbonate deposition commences in many hydrothermal deposits at a very early stage in the "cycle of mineralization" and often continues after the ore minerals have been precipitated. Thus, not only might one expect large temperature variations throughout the mineralization period, but also the composition of the fluids involved may undergo significant isotopic changes. Both will be reflected in the ^{18}O contents of the carbonate gangue minerals as for example documented in the data obtained from the Pine Point deposit (Fig.5).

Therefore, it is a requirement for isotope studies of such hydrothermal systems, that the paragenesis of the minerals deposited be well understood. It is here that a very major problem is encountered, since if is often very difficult to assign a certain generation of gangue minerals to a specific period of mineralization. In deposits where this has been possible, the changes in isotopic composition observed in successive generations of gangue

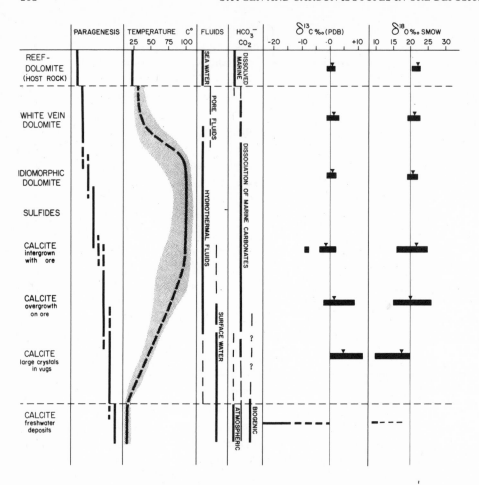

Fig.5. Cycle of mineralization at Pine Point. (After Fritz, 1969.)

reflect the temperature, geochemical and/or isotopic evolution of the system. Ideally, any isotopic studies undertaken to elucidate the genesis of an ore deposit should begin with the analyses of all samples for fluid inclusions, mineralogy, crystal habits and chemistry with the aim of assigning the carbonates to various stages of mineralization. A well documented case for this is found in the very detailed study of the Bluebell Mine in British Columbia, Canada (Ohmoto and Rye, 1970).

In Fig.6 the $\delta^{18}O$ values of major and well studied deposits are presented. The overall range of the $\delta^{18}O$ values is close to 40%. However, the spread within individual deposits is considerably smaller and is only a few permil for carbonate gangue minerals deposited during a particular phase of mineralization (e.g., Providencia). The approximate temperatures of mineralization shown in Fig.6 have mostly been obtained from fluid inclusion

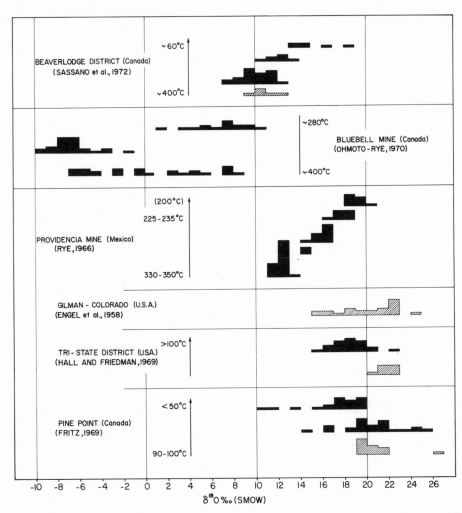

Fig.6. ^{18}O in the carbonate gangue minerals of selected hydrothermal ore deposits. The vertical scale is proportional to the number of samples described. The shaded areas represent dolomite samples, the black ones are calcites. The arrows indicate the successively younger generations.

studies. With this information it is possible to calculate the isotopic composition of the mineralizing fluids, assuming that these carbonates were deposited in isotopic equilibrium and preserved their isotopic composition since deposition. This has been done in graphical form in Fig.7 which shows calculated isodelta lines of the water in equilibrium with calcite at various temperatures.

Also shown in Fig.7 are the isotope and temperature data obtained from calcites in four deposits: the Providencia mines, Pine Point deposits, Bluebell Mine and the Eldorado mines in the Beaverlodge district. These deposits have been selected because they repre-

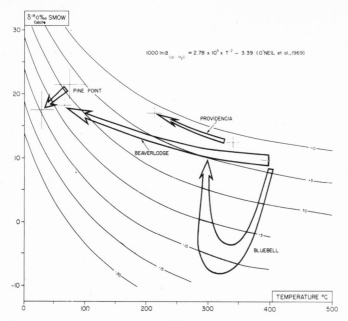

Fig.7. Approximate trends of the ^{18}O concentrations in hydrothermal fluids as deducted from the ^{18}O concentrations in hydrothermal calcites and their temperatures of deposition. The ranges of $\delta^{18}O$ and temperature are given for the earliest and latest calcites only since published data do not permit a more detailed delineation of the changing composition of the fluids.

sent four different systems each of which has its unique characteristics which among others is reflected in the path along which the hydrothermal fluids evolve.

The *Providencia mine* is probably the most simple deposit in this group. As shown in Fig.6, a gradual increase in ^{18}O is observed in successively younger carbonate generations and the temperature of mineralization decreases simultaneously from close to 350°C to approximately 200°C. If temperature and isotope data are plotted on an equilibrium diagram (Fig.7) it can be seen that these carbonates appear to be in equilibrium with a fluid with an approximate $\delta^{18}O$ value of +8‰. This value is closely confirmed by isotope and fluid inclusion data obtained from quartz samples. The changes in isotopic composition, therefore, are a simple function of increasing fractionation factors between the mineralizing fluid and the calcites deposited due to a decrease in temperature. Rye and O'Neil (1968, p.233) suggested that "the oxygen isotopic composition of the hydrothermal fluid was controlled by exchange with a large, high-temperature reservoir of constant isotopic composition at depth and was independent of the temperatures and location of ore deposition". Therefore, the deposition of the Providencia ore occurred with respect to the fluids in an open system. The hydrothermal fluid apparently had a magmatic origin and its composition changed only slightly during ascent. The latter appears to be confirmed by micro-analysis of fluid inclusions for their deuterium contents. These results and the calculated ^{18}O values are shown in Fig. 3: both deuterium and

^{18}O contents in the fluid inclusion are extremely constant in minerals deposited at various stages during the mineralization period.

The $\delta^{18}O$ values of the fluid inclusions determined by Rye (1966) are somewhat lower (-5.8 to $-6.2\%o$) than those assumed for magmatic water ($= -7$ to $-9\%o$) which could indicate that the hydrothermal fluid lost some of its ^{18}O between the source and site of deposition (Rye and O'Neil, 1968). However, because these fluid inclusions contain significant amounts of carbon dioxide, exchange reactions between CO_2 and fluid during cooling could also account for this discrepancy.

A very different system is represented in the uranium deposits in the Beaverlodge district. Fluid inclusion data indicate an almost continuous growth of gangue minerals from about 400°C to less than 100°C. This temperature decrease is paralleled by an increase in salinity in the fluid inclusions (Sassano et al., 1972). The trend to the isotope data shown in Fig. 7 also indicate a continuous decrease in ^{18}O in the mineralizing fluids which is due either to isotopically lighter, but highly saline water, entering the system or the deposition of minerals in a closed system in which only a limited amount of fluid was present. Since it is difficult to explain the origin of such saline water it appears that the latter is the more likely explanation and the ^{18}O data presented would then demonstrate that upon cooling the fluid continuously exchanged its ^{18}O with the host rock and thereby decreased its ^{18}O content due to increasing fractionation factors as the $\delta^{18}O$ of the host rock remains essentially constant. The initial $\delta^{18}O$ value of the mineralizing fluid was between +5 and +10‰ which compares well with values observed or calculated for hydrothermal fluids with temperatures near 400°C. Unfortunately, no deuterium analyses have yet been made, but they would greatly contribute to a deeper understanding of the genesis of this or similar deposits.

The initial fluids which participated in the formation of the *Bluebell ore body* also had $\delta^{18}O$ values close to +5‰ (Fig.7). During the later stages of mineralization, however, a very sharp decline in ^{18}O contents of the fluids is observed accompanied by only little change in temperature (Fig.6). This decline is again reversed during the latest stage of mineralization and the $\delta^{18}O$ of the fluid increases again to about +5‰. Ohmoto and Rye (1970) discussed this drastic change in detail and observed that despite the strong ^{18}O variations the δD values remain very constant at $-152 \pm 5\%o$ (Fig. 3). This suggests that these fluids were meteoric waters which obtained a variety of temperature, salinity and $\delta^{18}O$ values by different degrees of equilibration with the country rocks, or "could be the result of mixing of two meteoric waters—one hot and salty, and high in ^{18}O content, and the other cooler, dilute and low in ^{18}O" (Ohmoto and Rye, 1970, p.433).

A system in which mixing of two different waters occurred is found in the *Pine Point deposit*. As seen in Fig. 7 the initial fluid has a $\delta^{18}O$ somewhere between 0 and +5‰. Jackson and Beales (1967) postulate a stratafugic origin for the mineralizing fluids at Pine Point. This assumption is supported by fluid inclusion studies which indicate a sedimentary origin of the salt in solution (Roedder, 1968, and chapter in this book), and is not contradicted by the isotope data shown in Fig.7 which compare well with oil field brines.

Such brines typically have $\delta^{18}O$ values between 0 and +5‰ (SMOW) though values as high as +10 have been obtained (Hitchon and Friedman, 1969; Kharaka et al., 1973). During the late stages of mineralization the hydrothermal fluids were diluted with surface water which resulted in a significant decrease in salinity and ^{18}O contents (Fig.5). Thus, the ^{18}O contents of the calcites during the last phase of calcite deposition are primarily controlled by the changing isotopic composition of the water which masks any effects temperature changes may have had.

The influx of cool, dilute water into hydrothermal systems during the late stages of mineralization apparently is not an unusual phenomenon. It has been described by Hall and Friedman (1963, 1969) for some Mississippi Valley-type deposits, and by Garlick and Epstein (1966) for the copper deposits in Butte (Montana). Also, the chemical and isotopic similarity of oil field brines and hydrothermal fluids has been noticed not only for the Pine Point deposit but for other Mississippi Valley-type deposits too (Roedder, 1963; Roedder et al., 1963, White, 1968). Thus, if the origin of these hydrothermal fluids and temperature of mineralization is similar for most Mississippi Valley-type deposits, it is not surprising to observe similar ^{18}O contents in their carbonate gangue, as shown in Fig.6.

^{13}C in carbonate gangue minerals

The ^{13}C contents of the carbonates discussed in the previous section are shown in Fig.8. Added to these data are the carbon isotope data of the carbonates from the Biwabik iron-formation in Minnesota (Perry and Tan, 1970). The total spread of the $\delta^{13}C$ values in these deposits is close to 30‰, it is 10—15‰ within individual deposits and normally less than 5‰ for individual generations within a given deposit. Larger variations in individual generations can be attributed either to variable ^{13}C contents in the carbon dioxide or dissolved carbonate species, such as can be caused by redox-reactions in the hydrothermal systems or to difficulties in positioning the carbonates within the paragenetic sequence, a problem already discussed in the section on oxygen isotopes. The "isotopic evolution" of the carbon in the carbonates of the Providencia, Pine Point and Beaverlodge district deposits is shown in Fig. 9, superimposed on the isodelta lines of CO_2 gas in equilibrium with the carbonates at different temperatures.

In hydrothermal systems with isotopically uniform carbon dioxide or dissolved carbonate species variations in the isotopic composition of the carbonates deposited will be primarily a function of temperature. Such is the case in the Providencia deposits. According to Rye (1966), the carbon in these carbonates is controlled by magmatic carbon present in the form of carbon dioxide. The $\delta^{13}C$ values of such CO_2 is believed to lie between -6 and $-9‰$ (Craig, 1963; and others) and a $\delta^{13}C$ value of -7 has been measured in CO_2 from inclusions in sphalerite from the Providence Mine (Rye and O'Neil, 1968). In Fig.9 the $\delta^{13}C$ values of the CO_2 in equilibrium with the carbonates appears to

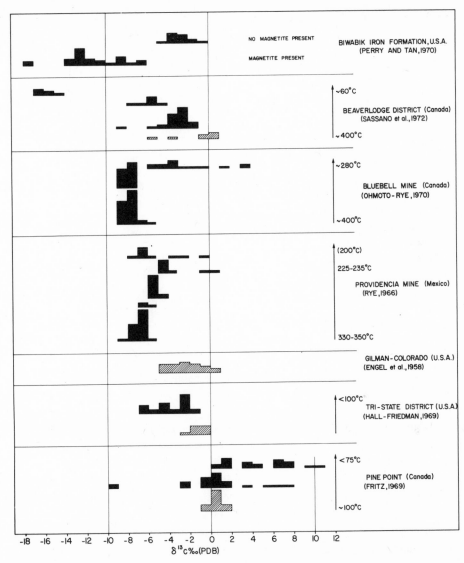

Fig.8. ^{13}C in the carbonate gangue minerals of selected hydrothermal ore deposits. The vertical scale is proportional to the number of samples analysed. The shaded areas represent dolomite samples, dark areas calcites. The arrows indicate the successively younger generations.

have a value close to −5‰, which could indicate that the theoretical fractionation factors calculated by Bottinga (1969) are slightly too low at elevated temperatures and that less than 1‰ fractionation occurs between CO_2 and $CaCO_3$ at temperatures above 200°C. Nevertheless, the data shown in Fig. 9, can confirm the assumption that a significant gas phase was present which controlled the ^{13}C values of the carbonates via increasing frac-

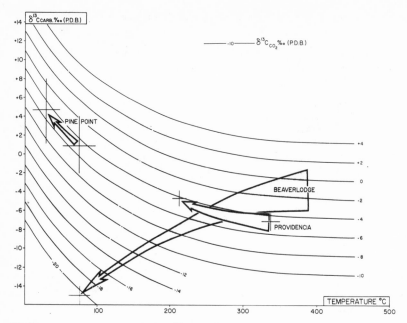

Fig.9. Approximate trends of the $\delta^{13}C$ values of CO_2-gas in equilibrium with hydrothermal calcites as deducted from the ^{13}C concentrations in the calcites and their temperatures of deposition. The isodelta lines have been drawn according to data presented by Bottinga (1969). The ranges of $\delta^{13}C$ and temperatures of deposition are given for the earliest and latest calcites only since published data do not permit a more detailed delineation of the changing composition of the CO_2-gas.

tionation factors at decreasing temperatures. These data confirm the conclusion arrived at from the ^{18}O data, inasmuch as the minerals in the Providencia mines were apparently deposited in an open system.

Similarly the carbon-13 contents of the Pine Point carbonates appear to be controlled by a carbon dioxide of fairly uniform isotopic composition ($\delta^{13}C \simeq -5$). It has to be assumed that this CO_2 is largely limestone-derived (Fritz, 1969) and was generated at depth in sufficient quantities as to be able to control the ^{13}C contents of the hydro-thermal carbonates. The relative large variations of the $\delta^{13}C$ values of these carbonates (Fig.5 and 8) indicate, however, that this hydrothermal system was less uniform than, for example, the Providencia system. Important to remember is, that apparently the ^{13}C fractionation between dissolved and solid carbonate does not change with temperature. Therefore, if no gas phase existed limestone-derived, aqueous carbonate species would act as the only carbon source for the hydrothermal carbonates. The ^{13}C content of the carbonates should then be almost independent of temperature changes. A system with variable amounts of gas at the different loci of mineralization could thus account for the observed temperature dependent trend of the $\delta^{13}C$ values and the large standard deviations.

Ohmoto (1972) proposed that the increasing ^{13}C contents observed in the younger Pine Point calcites (Fig.5, 8) may be explained by increasingly reducing conditions in this system. In view of the ^{13}C trend observed in Fig.9 it appears unlikely, however, that such processes were of major importance throughout the system but they undoubtedly could account for some of the variability observed.

A third system shown in Fig.9 is represented in the Beaverlodge deposits. In the section on oxygen isotopes it was postulated that the deposition of ore minerals in this deposit occurred in a "closed system". This can be confirmed by the ^{13}C contents of the carbonates: in the paragenetic sequence several changes from hematite to pyrite deposition are recognized (Sassano et al., 1972). Such transitions were probably due to continuous change in the geochemical environment with alternating oxidizing reducing conditions. Since the transition from pyrite to hematite can involve changes in the partial pressure of oxygen it may be assumed that CH_4 and CO_2 where present in variable amounts during the cycle of mineralization and that their relative abundance closely controlled the ^{13}C contents of each carbonate generation (Sassano et al., 1972). The ^{13}C content of the carbonate, therefore, does not evolve along an equilibrium line as can be expected for such a closed system in which the ^{13}C of the carbonates are internally rather than externally controlled.

The carbonates of the Bluebell Mine have not been plotted on Fig.9 since, as indicated above this system is rather complex. However, again the ^{13}C does not evolve along equilibrium lines which indicates that internal changes in the geochemistry of the hydrothermal system were of importance.

Juvenile carbon dioxide and limestone-derived biocarbonate are not the only potential sources of carbon present in hydrothermal systems and an inorganic or biological oxidation of organic matter may locally be of importance. Organic matter in general is strongly depleted in ^{13}C and typically δ^{13}C values between -15 and $-30‰$ can be expected (Fig.2). In the deposits discussed above, no evidence for significant amounts of "biogenic" CO_2 is found. However, carbonates associated with Precambrian iron formations typically have low ^{13}C contents and it has been postulated that organic carbon participated in their formation. It is, however, not clear whether this participation is direct inasmuch as the iron is transported or precipitated by some organic process (Becker and Clayton, 1970) or indirect as may occur during the diagenesis of metamorphism of such deposits (Perry and Tan, 1970).

The isotope data obtained by Perry and Tan (1970) on carbonates from the Biwabik iron-formation in Minnesota are shown in Fig. 8. A distinction is made between magnetite-free and magnetite-rich horizons as it is thought that the latter formed during diagenesis or metamorphism of primary ferric-oxide-hydroxide deposits rich in organic matter, thereby producing carbonates depleted in ^{13}C.

Similarly, for the formation of economically important epigenetic deposits, biological activity may be of crucial importance. An example discussed by Cheney and Jensen (1965, 1966) are the uranium deposits in the Gas Hill district in Wyoming. Geology and

TABLE I

Statistical data on the carbon-isotopic compositions of Sicilian, Gulf Coast and Gas Hill calcites (Cheney and Jensen, 1965).

	Mean	Range	Number of samples
Sicilian sulfur limestone	−28.7	− 8.8 to −43.0	27
Gulf Coast salt dome calcite	−37.9	−19.4 to −48.4	35
Gas Hill calcites associated with organic matter	−22.5	−15.2 to −28.5	20

sulfur isotopes suggested a biogenic origin and this finding has been supported through ^{13}C analyses on calcites associated with the ore: the average $\delta^{13}C$ value is −22.5‰.

Anaerobes, which participated in the formation of this deposit, produce both methane and carbon dioxide from organic matter whereby, with an infinite food source, methane will be isotopically lighter and carbon dioxide heavier than the food, in this case wood and asphalt. Cheney and Jensen (1966) show that reducing conditions prevailed throughout the mineralization period and it is, consequently, not surprising to observe in these calcites slightly higher ^{13}C contents than in the organic matter which typically shows $\delta^{13}C$ values between −20 and −35‰.

In deposits where methane is reoxidized lower $\delta^{13}C$ values are observed as is the case for the sulfur deposits in the Gulf Coast salt domes and in Sicily (Feely and Kulp, 1957; Dessau et al., 1962). A statistical comparison of these data is shown in Table I. It demonstrates that in biologically controlled systems the ^{13}C contents of the calcites can vary over a very wide range depending on the composition of the food source and the geochemical conditions existing during the decomposition of the organic matter and the deposition of the calcites.

^{18}O AND ^{13}C AS A GUIDE TO ORE

The passage of hot, carbon-bearing solutions through carbonate rock can initiate a recrystallization of these rocks in the vicinity of loci of ore deposition and/or channelways for the hydrothermal fluids. The degree of recrystallization and thus the size of the halo, which will develop around the ore body or channelway, will depend on the porosity of the wall rock, its mineralogy, temperature, the chemistry of the solutions and other less important factors. For example, in a limestone, solutions already supersaturated with respect to calcite are less likely to affect the wall rock than undersaturated ones. The ^{18}O and ^{13}C halo which may develop during the recrystallization of the wall rock is furthermore governed by the temperatures and isotope contents of the fluids. Solutions which had a long residence time in marine carbonate sediments before ore deposition may already be in isotopic equilibrium with the carbonate and little change in ^{18}O or ^{13}C will occur during recrystallization if no significant temperature variations occur.

Temperature gradients in hydrothermal systems and fluids not in isotopic equilibrium with carbonates can induce the development of isotope halos. Their size will be very variable and will be strongly dependent on the chemistry of the fluids, the local environmental conditions and the permeability and mineralogy of the wall rocks. These halos are structurally and paragenetically related to the ore deposit and may or may not coincide with the halo of recrystallization. In general, a depletion in both ^{18}O and ^{13}C may be observed in the halo with respect to the unaltered wall rock, caused by a temperature gradient, a gradient in the isotopic composition of the fluids in the pores of the wall rock (mixing) or a partial exchange between carbonate and fluid. The significance of these parameters has been discussed in detail by Pinckney and Rye (1972).

The first extensive isotope study of a zone of alteration around an ore deposit or conduits for hydrothermal fluids was carried out in the Gilman mine area in Colorado (Engel et al., 1958). Ore was deposited in the Leadville limestones which were extensively dolomitized during the mineralization period. The unaltered Leadville limestone has an oxygen and carbon isotopic composition similar to other Paleozoic marine carbonates and their δ^{18}O values vary between +21 and +24‰. Similar ^{18}O contents are found in fine grained dolomites even if collected in close proximity to the ore bodies (Fig.10). They do not show signs of recrystallization which indicates that in dolomites isotope exchange occurs only during recrystallization and not in the solid state. Similar observations have been made in the Pine Point area (Fritz, 1972). However, around the Gilman ore body a very significant halo of hydrothermal dolomite has been developed. These dolomites have δ^{18}O values close to 16‰ in the direct vicinity of the ore deposits and show progressively higher ^{18}O contents as the distance from the ore body increases (Fig.10). This gradual change in the isotopic composition of oxygen has been attributed to a temperature gradient in the zone of alteration. Taylor (1967) calculated temperatures of mineralization from quartz-calcite pairs and obtained a temperature of 315°C at the loci of ore deposition and 170°C near the edge of the alteration zone. The δ^{18}O values of the hydrothermal fluids were then calculated to be 8.6 and 6.8‰ in the two cases. This could indicate a magmatic origin for the hydrothermal fluids which possibly lost some of their ^{18}O during their passage through the wall rocks.

Engel et al. (1958, p.393) concluded: "The trends noted in the isotopic composition of oxygen in the hydrothermal dolomite are not reflected in the concentration of constituent elements. For example, detailed studies of the amounts of Cu, Pb, Zn, Mn, Sr, Ti, Al and other elements in the dolomite indicate no systematic trends with respect to ore, inferred ore conduits, or other features. In fact, the recorded differences in δ^{18}O from the margins of hydrothermal dolomite to ore contacts appear to be the one known chemical property of the hydrothermal dolomite that may prove useful as a guide of ore."

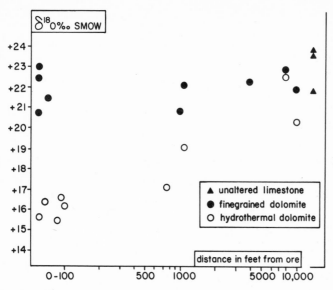

Fig.10. ^{18}O of unaltered and hydrothermally altered carbonates in the vicinity of the Gilman ore body. (After Engel et al., 1958.)

Similar changes in ^{18}O and ^{13}C in the zone of alteration were subsequently observed and discussed by Lovering et al. (1963) for the Drum Mountains, Utah, by Hall and Friedman (1969) for Mississippi Valley deposits and by Pinckney and Rye (1972) for the Cave-in-Rock district, Illinois. A very well defined halo of ^{18}O and ^{13}C observed in the Thomson Temperly Mine, Wisconsin, is shown in Fig.11 (Hall and Friedman, 1969). There is a clear trend from higher $\delta^{18}O$ and $\delta^{13}C$ values in the host rocks outside the zone of alteration to lower isotope values as the ore is approached.

In all deposits discussed here a consistent decrease in ^{18}O by a few permil has been observed in the zone of transition from the unaltered host rock to ore. However, as noted in the previous sections most hydrothermal fluids are enriched in ^{18}O and often have a composition similar to seawater. Therefore, their ability for additional ^{18}O uptake during the recrystallization of the wall rocks, leaving the wall rock depleted in ^{18}O, is limited. They may even lose some of their ^{18}O as shown for the fluids depositing the hydrothermal dolomites in and around the Gilman ore body. Large ^{18}O halos will thus only develop in areas where the physical characteristics of the wall rocks allow the passage of hydrothermal fluids over wide distances and where the mineralogical compositions of the wall rock and chemical composition of the mineralizing fluids is such as to induce a significant recrystallization of the wall rock or deposition of large amounts of hydrothermal minerals. It is, therefore, not surprising to observe that the changes in the wall rock around most ore bodies occur over relatively short distances. For this reason, Hall and Friedman (1969, p.C140) concluded: "The $\delta^{18}O$ and $\delta^{13}C$ of carbonate host rocks

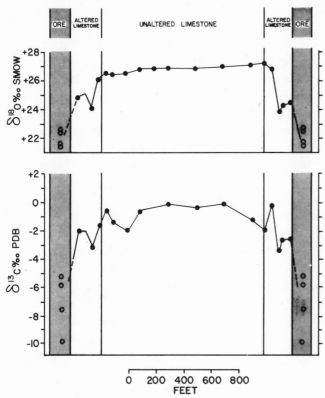

Fig.11. ^{18}O and ^{13}C carbonates from the Thomson Temperly Mine. (After Hall and Friedman, 1969.)

of Mississippi Valley deposits do not seem practical as guides to ore because the area of limestone depleted in those isotopes is small."

The simultaneous changes in ^{18}O and ^{13}C observed in the Thomson Temperly Mine (Fig.11) are not a characteristic feature of alteration zones. Engel et al. (1958) observed only minor differences in the ^{13}C content of the various carbonates which could not be related to a simple model as was proposed for the changes in ^{18}O contents of the hydrothermal dolomites. Lovering et al. (1963, p. B7) found that the " ^{13}C/^{12}C ratios remain relatively constant where limestone is changed to hydrothermal dolomite or manganiferous dolomite but show substantial variations in other types of hydrothermal alteration". Similar observations were made by Pinckney and Rye (1972) which indicate that the factors controlling the carbon isotopic composition of recrystallization carbonates are considerably more variable aerially and temporally than those controlling the ^{18}O contents of the recrystallizing carbonates. This virtually excludes the use of carbon isotopes as guides to ore.

SUMMARY AND CONCLUSIONS

At the time Engel et al. (1958) published their data of the first stable-isotope study of a major ore deposit, it became evident that the abundance of both oxygen and carbon isotopes in host rocks and ore minerals could greatly contribute to the discussion on the origin of hydrothermal fluids and it was hoped that their distribution in future might provide a superior prospecting tool. Unfortunately, subsequent studies on other ore deposits did not support this hope and it appears that only in a very few areas the geochemistry of the solutions and the mineralogy and physical properties of the host rock permitted the development of large isotope halos which might be useful for prospecting purposes.

Not disputable, however, is the significant input which the research on the origin of hydrothermal fluids received from these investigations, particularly if oxygen and hydrogen isotopes were considered together.

The ^{18}O–deuterium relations in thermal systems were first intensively studied by Craig (1963). He observed that the deuterium concentrations in thermal waters were essentially a permanent property of the water, not or only a little affected by physicochemical processes in the subsurface. They, therefore, directly reflect the origin of the water in a thermal system. The oxygen isotope contents, however, are constant only at low temperatures (e.g. in groundwater), but at elevated temperatures exchange reactions with silicate and carbonate rocks can cause a significant shift in the ^{18}O contents in general towards more positive ^{18}O values. A D–^{18}O plot of the fluid composition can thus immediately reveal significant information about the origin and subsurface history of hydrothermal fluids (see Fig.3).

Craig's studies then demonstrated that in none of the active thermal systems were any measurable amounts of magmatic or juvenile water present, and most were recharged by local groundwater systems. Similar conclusions were reached about the participation of magmatic or juvenile fluids in most ore deposits. An apparent exception among the deposits discussed in this presentation are the Providencia deposits for which magmatic fluids were postulated as mineralizing agents. However, no positive proof is possible and it is only through a combined consideration of all geochemical studies in this area that a "magmatic solution" appeared to describe best this system.

A further conclusion from these studies was, that fluids with the geochemical properties of oil field brines apparently participated in the formation of major base metal deposits. This observation provokes immediately the question as to whether ore-minerals are presently being deposited in deep, sedimentary basins. Unfortunately, little work has been done in this direction but it is well known, that for example the limestones and dolomites of the Alberta Basin (Canada) act as host rocks for minor copper–lead–zinc mineralizations.

Hydrothermal fluids in general acquire their oxygen and hydrogen isotope concentrations before ore deposition commences and as a rule little or no change takes place during mineralization. This is not necessarily so if the distribution of carbon isotopes in aqueous carbonate species and carbonate minerals is considered. A strong dependence on chemical reactions is here observed as the fractionation factors can be large between different carbon-bearing molecules in hydrothermal systems. Therefore, carbon isotopes can only yield information about the origin of the dissolved load if the carbon-geochemistry in a specific hydrothermal system is fully understood, for they will always reflect the geochemical conditions existing during the time of carbonate deposition.

One can thus conclude that in geochemically stable systems with a continuous, isotopically constant carbon supply little variation in the ^{13}C contents of different carbonate generations can be expected. Any changes which may occur are usually a simple function of the temperature dependent fractionation factors and the resulting ^{13}C concentrations of the carbonates closely reflect the ^{13}C concentrations of the source.

In systems without significant external carbon supply in which the chemistry of the solutions changes significantly during mineralization or where reservoir effects are of significance, large variations might be expected in the $^{13}C/^{12}C$ ratios in different carbonate generations. As a "rule of thumb" one could say that in most reducing systems in which carbon not only is present in the carbonate species but also, for example, in methane, the ^{13}C concentrations in the carbonate minerals will be higher than under more oxidizing conditions where lower $\delta^{13}C$ values would prevail.

ACKNOWLEDGEMENTS

Drafts of the manuscript have been read by Dr. D.F. Sangster, Ottawa, and Dr. K.H. Wolf, Sudbury. Their criticism and suggestions were greatly appreciated and influenced the final version of this manuscript. This work was supported by the National Research Council of Canada, Grant no. A 7954.

REFERENCES

Allen, E.T. and Day, A.L., 1935. Hot springs of the Yellowstone National Park. *Carnegie Inst. Wash. Publ.*, 466: 525 pp.

Becker, R.H. and Clayton, R.N., 1970. C^{13}/C^{12} ratios in Precambrian banded-iron formation and their implications. *Abstr. AGU Trans.*, 51: 452.

Bottinga, Y., 1969. Calculated fractionation factors for carbon and hydrogen isotope exchange in the system calcite-carbon dioxide-graphite-methane-hydrogen-water vapor. *Geochim. Cosmochim. Acta*, 33: 49–64.

Cheney, E.S. and Jensen, M.L., 1965. Stable carbon-isotopic composition of biogenic carbonates. *Geochim. Cosmochim. Acta*, 29: 1331–1346.

Cheney, E.S. and Jensen, M.L., 1966. Stable isotope geology of the Gas Hills, Wyoming, Uranium district. *Econ. Geol.*, 61: 44–71.

Clayton, R.N., O'Neil, J.R. and Mayeda, T.K., 1972. Oxygen isotope exchange between quartz and water. *J. Geophys. Res.*, 77: 3057–3067.

Craig, H., 1961. Isotopic variations in meteoric water. *Science*, 133: 1702–1703.

Craig, H., 1963. The isotope geochemistry of water and carbon in geothermal areas. In: E. Tongiorgi (Editor), *Nuclear Geology on Geothermal Areas.* Consiglio Nationale delle Ricerche, Spoleto, pp. 17–53.

Craig, H., 1966. Isotopic composition and origin of Red Sea and Salton Sea geothermal brines. *Science*, 154: 1544–1548.

Craig, H., 1969. Geochemistry and origin of the Red Sea brines. In: E.T. Degens and D.A. Ross (Editors), *Hot Brines and Recent Heavy-Metal Deposits in the Red Sea.* Springer, New York, N.Y., pp. 208–242.

Dansgaard, W., 1964. Stable isotopes in precipitation. *Tellus*, 16: 436–468.

Degens, E.T. and Ross, D.A., 1969. *Hot Brines and Recent Heavy-Metal Deposits in the Red Sea.* Springer, New York, N.Y., 600 pp.

Dessau, G., Jensen, M.L. and Nakai, N., 1962. Geology and isotope studies of Sicilian sulfur deposits. *Econ. Geol.*, 57: 410–436.

Deuser, W.G. and Degens, E.T., 1967. Carbon-isotope fractionation in the system CO_2 (gas)–CO_2 (aqueous)–HCO_3 (aqueous). *Nature*, 215: 1033–1035.

Engel, A.E.J., Clayton, R.N. and Epstein, S., 1958. Variations in isotopic composition of oxygen and carbon in Leadville limestone (Mississippi, Colorado) and its hydrothermal and metamorphic phases. *J. Geol.*, 66: 374–393.

Feely, H.W. and Kulp, J.L., 1957. The origin of the Gulf Coast salt-dome sulfur deposits. *Am. Assoc. Petrol. Geol. Bull.*, 41: 1802–1853.

Friedman, I. and Hall, W.E., 1963. Fractionation of O^{18}/O^{16} between coexisting calcite and dolomite. *J. Geol.*, 71: 238–243.

Fritz, P., 1969. The oxygen and carbon-isotopic composition of carbonates from the Pine Point lead-zinc ore deposits. *Econ. Geol.*, 64: 733–742.

Fritz, P., 1972. Geochemical and isotopic characteristics of Middle Devonian dolomites from Pine Point, northern Canada. *24th Int. Geol. Congr.*, Sect. 6: 230–243.

Fritz, P. and Smith, D.G.W., 1970. The isotopic composition of secondary dolomites. *Geochim. Cosmochim. Acta*, 34: 1161–1173.

Garlick, G.D. and Epstein, S., 1966. The isotopic composition of oxygen and carbon in hydrothermal minerals at Butte, Montana. *Econ. Geol.*, 61: 1325–1335.

Hall, W.E. and Friedman, I., 1963. Composition of fluid inclusions, Cave-in-Rock fluorite district, Illinois, and Upper Mississippi Valley zinc-lead district. *Econ. Geol.*, 58: 886–911.

Hall, W.E. and Friedman, I., 1969. Oxygen and carbon isotopic composition of ore and host rock of selected Mississippi Valley deposits. *U.S. Geol. Surv. Prof. Pap.*, 650-C: C140–C148.

Hitchon, B. and Friedman, I., 1969. Geochemistry and origin of formation waters in the western Canada sedimentary basin, 1. Stable isotopes of hydrogen and oxygen. *Geochim. Cosmochim. Acta*, 33: 1321–1349.

Jackson, S.A. and Beales, F.W., 1967. An aspect of sedimentary basin evolution; the concentration of Mississippi Valley-type ores during late stages of diagenesis. *Can. Petrol. Geol. Bull.*, 15: 383–433.

Kharaka, Y.K., Berry, F.A.F. and Friedman, I., 1973. Isotopic composition of oil field brines from Kettleman North Dome, California, and their geologic implications. *Geochim. Cosmochim. Acta*, 37: 1899–1908.

Lovering, T.S., McCarthy, J.H. and Friedman, I., 1963. Significance of O^{18}/O^{16} and C^{13}/C^{12} ratios in hydrothermally dolomitized limestones and manganese carbonate replacement ores of the Drum Mountains, Juab County, Utah, *U.S. Geol. Surv. Prof. Pap*, 475-B: B1-B9.

Malinin, S.D., Kropotova, O.I. and Grinenko, V.A., 1967. Experimental determination of equilibrium constants for carbon-isotope exchange in the system CO_2 (gas)–HCO_3 (sol) under hydrothermal conditions. *Geochem. Int.*, 4: 762–771.

McKinney, C.R., McCrea, J.M., Epstein, S., Allen, H.A. and Urey, H.C., 1950. Improvements in mass-spectrometers for the measurements of small differences in isotopic abundance ratios. *Rev. Sci. Instr.*, 21: 724–730.

Nier, A.O., 1947. A mass spectrometer for isotope and gas analyses. *Rev. Sci. Instr.*, 18: 398–411.

Northrop, D.A. and Clayton, R.N., 1966. Oxygen-isotope fractionation in systems containing dolomite. *J. Geol.*, 74: 174–196.

Ohmoto, H., 1972. Systematics of sulfur and carbon isotopes in hydrothermal ore deposits. *Econ. Geol.*, 67: 551–578.

Ohmoto, H. and Rye, R.O., 1970. The Blubell Mine, British Columbia. 1. Mineralogy, paragenesis, fluid inclusions, and the isotopes of hydrogen, oxygen and carbon. *Econ. Geol.*, 65: 417–437.

O'Neil, J.R. and Epstein, S., 1966. Oxygen isotope fractionation in the system dolomite–calcite–carbon dioxide. *Science*, 152: 198–201.

O'Neil, J.R., Clayton, R.N. and Mayeda, T.K., 1969. Oxygen isotope fractionation in divalent metal carbonates. *J. Chem. Phys.*, 51: 5547–5558.

Perry, E.C. and Tan, F.C., 1970. Significance of carbon-isotope variations in carbonates from the Biwabik iron-formation, Minnesota, and U.N. Internation Symposium on the Geology and Genesis of Precambrian Iron/Manganese Formations and Ore Deposits, SC/CONF. 46/25, Kiev (manuscript).

Pinckney, D.M. and Rye, R.O., 1972. Variation of O^{18}/O^{16}, C^{13}/C^{12}, textures, and mineralogy in altered limestone in the Hill Mine, Cave-in-Rock District, Illinois. *Econ. Geol.* 67: 1–18.

Roedder, E., 1963. Studies of fluid inclusions, 2. Freezing data and their interpretation. *Econ. Geol.*, 58: 167–211.

Roedder, E., 1968. Temperature salinity and origin of the ore-forming fluids at Pine Point, N.W.T., Canada, from fluid inclusion studies. *Econ. Geol.*, 63: 439–450.

Roedder, E., Ingram, B. and Hall, W.E., 1963. Studies on fluid inclusion, 3. Extraction and quantitative analysis of inclusions in the milligram range. *Econ. Geol.*, 58: 353–374.

Rosenfeld, W.D. and Silverman, S.R., 1959. Carbon isotope fractionation in bacterial production of methane. *Science*, 130: 1658–1659.

Rye, R.O., 1966. The carbon, hydrogen, and oxygen isotopic composition of the hydrothermal fluids responsible for the lead-zinc deposits at Providencia, Zacatecas, Mexico. *Econ. Geol.*, 61: 1399–1427.

Rye, R.O. and O'Neil, J.R., 1968. The O^{18} content of water in primary fluid inclusions from Providencia, North-Central Mexico. *Econ. Geol.*, 63: 232–238.

Sassano, G.P., Fritz, P. and Morton, R.D., 1972. Paragenesis and isotopic composition of some gangue minerals from the uranium deposits of Eldorado, Saskatchewan. *Can. J. Earth Sci*, 9: 141–157.

Sharma, T. and Clayton, R.N., 1965. Measurement of O^{18}/O^{16} ratios of total oxygen of carbonates. *Geochim. Cosmochim. Acta*, 29: 1347–1353.

Taylor, H.D., 1967. Oxygen-isotope studies of hydrothermal mineral deposits. In: H.C. Barnes (Editor), *Geochemistry of Hydrothermal Ore Deposits*. Holt, Rinehart and Winston, New York, N.Y., pp. 109–142.

Urey, H.C., 1947. The thermodynamic properties of isotopic substances. *J. Chem. Soc.*, 66: 562–581.

Wendt, I., 1968. Fractionation of carbon isotopes and its temperature dependencies in the systems CO_2-gas-CO_2 in solution and HCO_3^--CO_2 in solution. *Earth Planet. Sci. Lett.*, 4: 64–68.

White, D.E., 1968. Environments of generation of some base metal ore deposits. *Econ. Geol.* 63: 301–335.

White, D.E., Anderson, E.T. and Grubbs, D.K., 1963. Geothermal brim well: mile-deep hole may tap ore-bearing magmatic water and rocks undergoing metamorphism. *Science*, 139: 919–922.

Chapter 8

SULPHUR AND LEAD ISOTOPES IN STRATA-BOUND DEPOSITS

D.F. SANGSTER

INTRODUCTION

Lead and sulphur represent the cationic and anionic portions, respectively, of metallic sulphide deposits. Sulphur, by definition, is common to all such deposits and at least some concentration of lead is found in most, but not all, of them. While the isotopes of other major constituents in these ores, e.g., Cu, Zn, Fe, do not vary significantly in nature, those of sulphur and lead do and hence may be regarded as isotopic "tracers" in the study of ore deposits containing these two elements. Sulphur has the added advantage of occurring in two oxidation states (sulphide and sulphate) and can thereby be used as a "tracer" for the gangue minerals as well.

There is, however, one attribute common to both elements that makes them particularly valuable in the study of mineral deposits and that is their apparent ability to essentially retain their original isotopic composition throughout most, if not all, of the post-depositional processes affecting the deposit in which they are contained. Although some sulphur isotopic exchange takes place between minerals at high metamorphic grades and galena in contact with the host rocks may be contaminated by radiogenic lead from the host, these effects are usually small and, in most cases, can be kept to a minimum by careful sampling. The likelihood of wholesale exchange of lead and/or sulphur with circulating groundwater or other geological agencies such as to completely alter the isotopic composition of the deposit is extremely small, especially when compared with the relative case with which carbon and oxygen isotopic compositions can be altered by post-depositional processes.

SULPHUR ISOTOPES

General theory

Abundance. Sulphur has four stable isotopes with the following average abundances (Macnamara and Thode, 1950):

^{32}S	95.02%
^{33}S	0.75
^{34}S	4.22
^{36}S	0.02

Of these, the two most abundant isotopes, ^{32}S and ^{34}S, differ in mass by about 6% and the ^{32}S/^{34}S ratio can be measured with a precision of about 0.02%.

The theory of sulphur isotopes and their application to the study of mineral deposits has been reviewed by Ault and Kulp (1960), Jensen (1959, 1967), Stanton (1960, 1972), Ohmoto (1972), Hoefs (1973), and Rye and Ohmoto (1974). Normally, sulphur isotope determinations measure the ratio of the two more abundant species, ^{32}S and ^{34}S. Results are reported as permil (‰) difference from a primary standard as follows:

$$\delta^{34}S‰ = \frac{(^{34}S/^{32}S)\ sample - (^{34}S/^{32}S)\ standard}{(^{34}S/^{32}S)\ standard} \times 1000$$

Sulphur from troilite in the Cãnon Diablo meteorite with ^{34}S/^{32}S = 0.0450045 (Jensen and Nakai, 1963), has been widely accepted as the standard and therefore has been assigned a δ^{34}S value of 0.00‰. Positive and negative permil values represent, respectively, enrichment and depletion of ^{34}S relative to this standard.

Fractionation of sulphur isotopes. When metallic sulphides are formed from a sulphur source of fixed isotopic composition, the sulphides will normally have a different composition than that of the source sulphur. That is to say, isotopic fractionation will occur between the sulphur source and the sulphide produced. Two main factors determine the degree and direction of sulphur isotope fractionation: temperature, and biologic action.

(a) *Temperature.* Tudge and Thode (1950) and Sakai (1968) pointed out taht isotopic fractionation between aqueous sulphide species (the sulphur source) and sulphide minerals is temperature-dependent. Fig. 1 shows the degree of isotopic fractionation for several important sulphur species using the example of H_2S as a sulphur source of constant isotopic composition. Note that the fractionation for sphalerite (say) changes in both degree (Δi) and direction (negative to positive) relative to H_2S with decrease in temperature. Galena, in contrast, steadily increases in isotopic fractionation with decrease in temperature. Greatest isotopic fractionation, in general, occurs at lower temperature (for sulphide species) and between sulphide–sulphate species. Fig. 1 also suggests that isotopic fractionation between co-existing sulphur species in ore deposits can be used as a geothermometer. The fractionation between co-existing sulphides was estimated by Sakai (1968) and Bachinski (1969) on theoretical considerations and has since been experimentally determined by Grootenboer and Schwarcz (1969), Rye and Czamanske (1969), Kajiwara and Sasaki (1969), Kajiwara and Krouse (1971), Schiller et al. (1970),

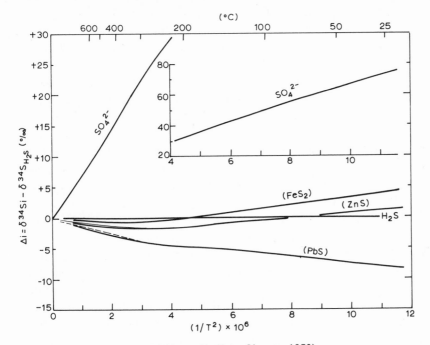

Fig. 1. Isotopic factors for sulphide species (from Ohmoto, 1972).

and Czamanske and Rye (1974). Sulphur isotope fractionation between sulphides is small compared with that between sulphate and sulphide. For example, at 200°C, the fractionation between pyrite and galena is 4.6‰ and between sphalerite and galena is about 3‰. However, at the same temperature, the isotopic fractionation between sulphate in solution (SO_4^{2-}) and pyrite is about 32‰.

The *degree* of fractionation between co-existing sulphide species is temperature-dependent but the *direction* of fractionation is largely a function of the relative strengths of the metal–sulphur bonds. For example, Bachinski (1969) predicted a sequence of minerals from greatest to least enrichment in ^{34}S as follows: pyrite > sphalerite > chalcopyrite > galena, i.e. sulphur in pyrite would be "heavy" with respect to the other minerals in the sequence. Similarly, Harrison and Thode (1957) found, in the reduction of sulphate to sulphide, that $^{32}SO_4^{2-}$ reacts 2.2% faster than $^{34}SO_4^{2-}$, resulting in the reduction product (H_2S) being enriched in the lighter isotope ^{32}S and the remaining sulphate enriched in the heavy isotope ^{34}S. At 25°C under equilibrium conditions a 74‰ enrichment of ^{32}S in hydrogen sulphide relative to sulphate is theoretically possible (Tudge and Thode, 1950).

(b) *Biologic action.* The rate of chemical reduction of sulphate to sulphide, vanishingly low at low temperatures, is catalyzed through the metabolic activity of sulphate-reducing bacteria, notably the anaerobic species *Desulphovibrio desulphuricans.* Thode et al. (1951) were among the first to demonstrate the effectiveness of this process and, since then, other laboratory experiments have demonstrated that during the reduction process, bacteria can produce fractionation up to 25‰ relative to the source composition and that the average fractionation (i.e. under normal biologic conditions) is about 15‰. Factors influencing the degree of bacterial fractionation are mainly those that affect their metabolic activity (i.e. temperature, light, nutrient availability, etc.) and sulphate concentration. The reader is referred to an excellent discussion by Rees (1973) on the rate-controlling mechanisms of bacterial reduction of sulphate. In general, the degree of fractionation is inversely proportional to the rate of sulphate reduction, i.e. the faster that H_2S is generated, the lower the fractionation between source sulphate and hydrogen sulphide produced.

The normal maximum biological fractionation of 25‰ is far less than the theoretical fractionation at 25°C (74‰) and is a measure of the effectiveness of bacteria in reaching equilibrium. The bacterial action also results in considerably more sulphide than would occur by non-biological means, even considering the geologic periods of time that would be involved. In fact, Rees (1973, p. 1141) states "... this bacterium is one of the major agencies by which sulphur is cycled in nature." At low temperatures, then, biologic activity is the catalytic agency by which sulphate is reduced and, in the process, considerable fractionation of the isotopes occurs.

Influence of sulphur species on sulphur isotopic abundances. Rye and Ohmoto (1974) state that composition of source sulphur and the proportions of oxidized and reduced sulphur species in solution are among the major factors controlling the sulphur isotopic composition of minerals. The influence of sulphur source composition was discussed briefly above with regard to fractionation and will be discussed in more detail later. The influence of sulphur species was suggested by Sakai in 1968 and quantitatively evaluated by Ohmoto (1970, 1972). Both workers demonstrated that large isotopic fractionation can be expected among various sulphur species and that largest variations exist between sulphate and sulphide species. At 200°C for example, the sulphur isotopic fractionation between SO_4^{2-} and H_2S is 32‰. In addition to this, however, is the important point that the proportion of (say) sulphate and sulphide species in solution also affects the isotopic compositions of the resultant minerals. To illustrate this effect, Rye and Ohmoto (1974, p. 829) state:

"Suppose the sulfur in an ore fluid is distributed between H_2S and SO_4^{2-}, the sulfur isotopic composition of the fluid system ... is 0‰, $T = 200°C$, and the H_2S/SO_4^{2-} ratio changes from 1/9 to 9/1. As long as isotopic equilibrium is established between the species, the $\delta^{34}S$ value of each aqueous sulfur species varies depending on the H_2S/SO_4^{2-} ratio (e.g. $\delta^{34}S_{H_2S}$ shifts from −28.8 to −16 to −3.2‰, according to the change in the H_2S/SO_4^{2-} ratio from 1/9 to 1/1 to 9/1). If sulfide minerals, such as sphalerite and galena, precipitate from the solutions, and if the amount of sulfur taken out from the solutions is negligible, $\delta^{34}S$ values for the sulfide minerals will also change with the change in H_2S/SO_4^{2-} ratio."

The principles suggested by Sakai (1968) and quantified by Ohmoto (1972) are important in the understanding of sulphur isotopic variations in ore deposits in general; however, as pointed out by Ohmoto (1972, p. 576), many studies have demonstrated that the rates of both chemical and isotopic exchange reactions between reduced and oxidized sulphur species are extremely slow at low temperatures. At temperatures below 200°C, for most stratiform and many strata-bound deposits, other factors, such as bacterial reduction of sulphate, may be the controlling mechanism.

Sulphur isotopes in strata-bound deposits

For the purposes of this paper, and without involving exhaustive definitions, strata-bound deposits are regarded as those which are confined to a single, or a small number of, stratigraphic units. Strata-bound deposits may be either discordant or concordant. Those which are concordant are sometimes referred to as stratiform, i.e., having the form of stratum, and are commonly layered.

Following roughly the lead of Stanton (1960, 1972), strata-bound deposits can be divided into two main categories, those enclosed in predominantly marine sedimentary rocks (carbonates, shales) and those enclosed in predominantly marine volcanic rocks or rocks which are volcanically derived (flows, tuffs, volcanic breccias, etc.) Two other minor types are the sandstone—uranium (Cu, V) type and the conglomerate—gold—uranium type.

Deposits enclosed in marine host rocks. For strata-bound deposits enclosed in marine sedimentary or volcanic rocks, ocean-water sulphate is an obvious source of sulphur. Considerable evidence exists to demonstrate that most strata-bound sulphide deposits have indeed derived their sulphur from sea-water sulphate and that, for many deposits, the reduction to sulphide from sulphate was brought about by bacterial activity. If ocean-water sulphate can be considered a possible source of sulphur, then the isotopic composition of modern and ancient oceans is of paramount importance to an understanding of the sulphur isotopic composition of certain strata-bound sulphide deposits.

Studies have shown that modern sea-water sulphate has a remarkably uniform isotopic composition of about $\delta^{34}S = +20\%_o$ and homogenization by ocean currents on a worldwide scale therefore effectively ensures an infinitely large sulphate reservoir of constant isotopic composition. Evaporite deposits in rocks of all ages from Middle Proterozoic to Recent attest to the long-term presence of sulphate in ocean waters. Furthermore, because chemical precipitation produces no significant isotopic fractionation, the isotopic composition of ancient gypsum deposits will approximate the composition of the ocean at the time of sulphate deposition. This feature has been utilized by several workers to determine the isotopic composition of sulphur in ancient oceans (e.g. Ault and Kulp, 1959; Thode and Monster, 1965; Nielson, 1966; Holser and Kaplan, 1966; Vinogradov and Schanin, 1969; Claypool et al., 1972) and these data show (Table I)

TABLE I

δ^{34}S values for sulphate in ancient oceans (from Thode and Monster, 1965; Claypool et al., 1972; I.R. Kaplan, written communication, 1975)

Period	Epoch	δ^{34}S
Quaternary	Present	+21
Tertiary	Neogene	+22
	Paleogene	+20
Cretaceous	Lower	+15
Jurassic		+17
Triassic		+15
Permian		+10
Carboniferous		+15
Devonian	Upper	+28
	Lower and Middle	+17
Silurian		+25
Ordovician		+27
Cambrian		+31

that the sulphur isotopic composition of sulphate in the oceans varied throughout geologic time. If strata-bound sulphide deposits have, in fact, derived their sulphur from ancient sea-water sulphate, then the isotopic composition of sulphur in such deposits should bear some relation to that of sea-water sulphate source.

Both natural gas and crude petroleum contain large amounts of sulphur (up to 87% H_2S in gas and up to 10% organically bound sulphur in petroleum) and, for purposes of illustration, could be considered a type of strata-bound sulphide deposit. For crude petroleum, Thode et al. (1958) and Thode and Monster (1965) found a correlation

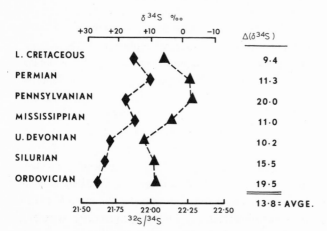

Fig. 2. Average δ^{34}S contents of sulphur in petroleum and coeval sea-water sulphate (from Thode et al., 1958). Diamonds represent compositions of sulphate, triangles represent petroleum. Fractionation factor $\Delta(\delta^{34}$S$) = \delta^{34}$S (sea water) $- \delta^{34}$S (mean for petroleum sulphur).

between the isotopic composition of sulphur in the petroleum and sulphur in coeval seas (Fig. 2). Their conclusions were as follows:

"The pattern of change for petroleum sulphur appears to be parallel to that for the evaporites and ancient seas. However, the petroleum sulphur is, in general, depleted in ^{34}S by about 15‰ with respect to the contemporaneous gypsum and anhydrite deposits. This displacement of 15‰ in the ^{34}S content, which is about the isotope fractionation expected in the bacterial reduction of sulphate, is strong evidence that seawater sulphate is the original source of petroleum sulphur and that it is first reduced by bacterial action in the shallow muds before being incorporated into petroleum."

The possibility of incorporating bacteriogenic sulphur into metallic sulphides is not a new idea; Bastin, in 1926, suggested the bacterial influence in the genesis of sulphide ores and the hypothesis has been kept active ever since by most students of ore genesis (e.g. Temple, 1964; Jensen, 1962). The pattern of isotopic change in sea-water sulphate throughout time, however, has provided the necessary base from which to test the biogenic-source model for strata-bound sulphide ores. If sulphur in these ores is in fact produced by bacteria from sea-water sulphate, then sulphur in the metallic sulphides, like sulphur in petroleum, should show a similar average 15‰ enrichment in the lighter isotope (relative to coeval sea-water sulphate) and a similar parallel pattern throughout geologic time.

The sulphur isotopic composition of strata-bound deposits will correlate with sea-water sulphate only if the latter is the major source of sulphur in the marine environment. Organic sulphur compounds could conceivably constitute an alternate sulphur source, especially in organic-rich muds, a common associate of certain types of strata-bound sulphide deposits. For example, Bowen (1966, p. 47) suggests 0.5% as the mean sulphur content of marine plants and Kaplan et al. (1963) found that the organic sulphur content of marine organisms averages about 1% by dry weight. However, as Berner (1971, p. 131) points out, in many marine sediments the concentration of pyritic sulphur (which forms from H_2S) is far greater than that which could have been originally supplied as organic sulphur. Marine sediments often contain more than 1% pyrite sulphur (dry weight) and, if 1% is accepted as the average organic sulphur content of marine organisms and plants, more than 100% organic matter would be required to produce all the pyrite present in sediments. Also, Berner (1971, p. 202) cites experimental evidence that during bacterial decomposition of biogenic organic matter in sea water, increases in dissolved H_2S are equally matched by decreases in dissolved sulphate. These two lines of evidence support the concept that the major source of sulphur in the marine environment is sea-water sulphate.

The principles used by Thode and Monster (1965) to deduce the origin of petroleum have been applied to strata-bound sulphide deposits by the author (Sangster, 1968, 1971). Sulphur isotope data for 110 strata-bound deposits representing over 2300 individual analyses, have been recorded in computer-processable form at the Geological Survey of Canada. Data were obtained mainly from the literature with the following specifications in mind:

(1) Strata-bound sulphide deposits are considered to be those which have at least one long dimension parallel to nearby bedding planes or extrusive flow contacts, and/or are obviously confined to a particular stratum.

(2) Depending on host rock lithology, the sulphide deposits were assigned to one of two types: (a) sedimentary type in which the host rocks are mainly normal sedimentary rocks (e.g., carbonates, shales, etc.) and (b) volcanic type which are enclosed in volcanic rocks or rocks which are mainly volcanically derived (e.g., flows, tuffs, coarse pyroclastics). This two-fold division was originally proposed by Stanton (1960) to distinguish between non-volcanic and volcanic environments.

(3) Wherever possible, data on individual orebodies or lenses were included. Averages were also computed for co-existing sulphates, where present as major gangue minerals, to serve as independent checks on sea-water values (Table I).

(4) Each deposit is represented by the average of at least five analyzed samples. An orebody having less than five samples was not included.

(5) The age of the host rock had to be reasonably well established.

Data from 110 deposits in rocks of Phanerozoic age were compiled in this manner. Most are metallic sulphide deposits sufficiently large to warrant economic exploitation: a few native sulphur mines in Sicily and large barren pyrite and/or pyrrhotite deposits are also included. Precambrian ores are not considered in this paper because, although data from approximately three dozen deposits are available, there is insufficient evidence concerning the composition of Precambrian sea-water sulphate. Most of the Precambrian ores which have been studied are of Proterozoic age (600–2,560 m.y.); only a very few Archean ores (>2,560 m.y.) have been isotopically studied.

When the data were compiled, it was found that most orebodies showed a near-normal or Gaussian distribution of $\delta^{34}S$ values. However, in an attempt to delete obvious erratic values, for those deposits represented by ten or more sulphur isotope determinations, values falling outside two standard deviations from the mean in either direction (i.e. 95% confidence limits) were rejected. A new mean was then calculated from the somewhat more restricted data. In most cases, only one or two values were rejected by this process so that the new mean differed very little (one or two permil) from the original. Standard deviations ranged from 0.03‰ to 9.9‰ for "sedimentary type" ores and from 0.6‰ to 5.0‰ for "volcanic type" ores. The spread of $\delta^{34}S$ values for individual deposits, after the rejection of values greater than plus or minus two standard deviations, averaged 11.8‰ for "sedimentary type" and 6.5‰ for "volcanic type". In all cases, the mean isotopic composition for individual deposits was depleted in ^{34}S relative to the estimated composition of coeval sea-water sulphate.

In order to assess the magnitude of isotopic fractionation of the orebodies relative to sea-water sulphate, the fractionation factor $[\Delta(\delta^{34}S)]$ for each deposit was calculated as follows:

$$\Delta(\delta^{34}S) = [\delta^{34}S \text{ (sea water)} - \delta^{34}S \text{ (mean for deposit)}]$$

These $\Delta(\delta^{34}S)$ values are summarized in the histograms of Figure 3. The experimentally-determined range and average $\Delta(\delta^{34}S)$ values for sulphur produced by bacterial reduction of sulphate (about 15‰) is also shown for comparison. Sedimentary strata-bound ores showed an average $\Delta(\delta^{34}S)$ value of 13.9; volcanic ores averaged 17.4. Both are in close agreement with the average bacterial fractionation factor of 15‰.

Theoretical considerations and laboratory analyses show that sulphate gangue minerals, such as gypsum, anhydrite, and barite, co-existing with sulphide ores, should be

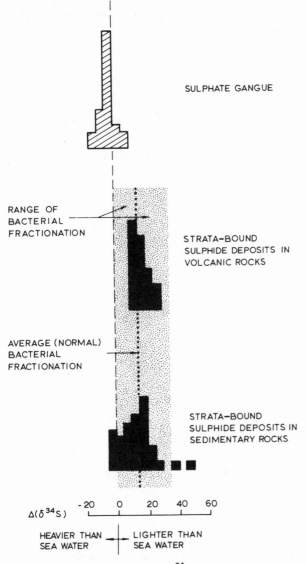

Fig. 3. Frequency diagrams of $\Delta(\delta^{34}S)$ values in strata-bound sulphide deposits enclosed in marine host rocks. Heavy vertical dashed line represents parental sea-water sulphate.

isotopically equal to or slightly heavier than the composition of coeval sea-water sulphate. Enrichment of ^{34}S in gangue sulphate is thought to be produced as a result of precipitation of "light" (^{32}S-enriched) sulphur in the sulphide-rich sediments deposited in sedimentary basins with relatively restricted circulation to the open sea. Enrichment of the lighter isotope in the sulphides results in the sulphate, which remains in solution as SO_4^{2-}, becoming progressively "heavier". If this sulphate were then precipitated as, say, barite, its isotopic composition would indicate enrichment in ^{34}S and δ^{34}S values would be slightly greater than those estimated for open marine sulphate (see discussion in Thode and Monster, 1965; Sangster, 1968). A histogram for gangue sulphate minerals in strata-bound ores is also shown in Fig. 3 and confirms the theoretical considerations. In all cases where data were available on isotopic compositions of gangue minerals the values closely approximated those of previous workers for sea-water sulsulphate, and constitute an independent check of their estimations.

Isotopic compositions of ore sulphides are summarized in Fig. 4 in slightly different form. Here, the mean values for all deposits in each geologic era are compared with coeval sea-water sulphate. Again, the data suggest a relationship between the two similar to that illustrated in Fig. 2 for petroleum sulphur.

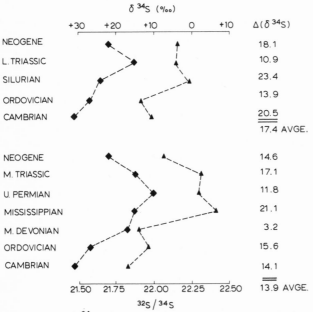

Fig. 4. Average δ^{34}S values of strata-bound marine sulphide deposits in different geologic periods or epochs compared with contemporaneous sea-water sulphate compositions. Upper group is volcanic type ores; lower group is sedimentary type. Diamonds represent sea-water sulphate; triangles represent strata-bound sulphide deposits.

Discussion. The range of δ^{34}S values *within* an individual deposit is likely caused by any or all of the factors discussed previously regarding fractionation, sulphide species, and biologic activity.

The spread in average values *between* deposits for a given geological era, on the other hand, probably reflects slight differences in the δ^{34}S values in the sea-water sulphate in individual sedimentary basins relative to that in the open sea. Basins with only sporadic or restricted circulation with the open ocean would be strongly influenced in their isotopic composition by the amount of sulphate carried in by river waters (contributing sulphate with light sulphur), the rate of sulphide precipitation in the bottom muds (tending to deplete the remaining sulphate in solution in ^{32}S), and the rate of influx of new sulphate from the open sea (tending to restore the isotopic composition to that of the open sea). The effect of "closed" versus "open" sedimentary basins on the composition of source sulphate is illustrated in Fig. 5 in which two contrasting marine basins are schematically illustrated; one with "open" circulation and the other with "restricted" circulation. The former situation results in evaporite deposits of isotopic composition identical to that of the open ocean. Bacterial action in a basin of *restricted* circulation with the open ocean, however, will result in the sulphate left in solution becoming increasingly enriched in the heavier isotope. This is because (1) reduction of sulphate causes isotopically "light" sulphide to be precipitated, thereby rendering the remaining sulphate in solution "heavy", (2) the restricted circulation prevents large influxes of new ocean-water to restore the original isotopic composition. Evaporite deposits in basins of restricted circulation will, therefore, be isotopically "heavy" compared to sulphate in the open ocean.

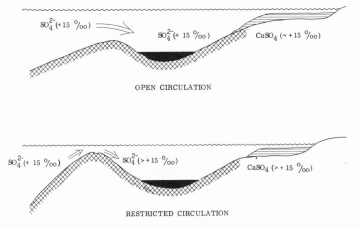

Fig. 5. Schematic representation of two marine basins illustrating the effect of open and restricted circulation on the isotopic composition of evaporite deposits relative to parental sea-water sulphate. Black deposit in the basins represents precipitated metallic sulphides; evaporite deposits are represented by the horizontal lines on the right.

Barium sulphate is extremely insoluble in water and even small amounts of barium introduced into a marine basin would immediately precipitate as barite, thereby preserving in the sediment a record of the isotopic composition of the local marine basin at the time of precipitation. Hence, the ^{34}S content of co-existing barite in the sulphide deposit is probably more representative of local basins than the values suggested for more-or-less open oceans. However, the isotopic composition of even restricted marine basins should not deviate too much from that of the open seas because periodic and frequent influxes of sea water are required to supply the enormous amounts of sulphur in a strata-bound sulphide deposit. These periodic additions of ocean water would tend to restore the average isotopic composition of a basin in which the dissolved sulphate would have been becoming progressively "heavier" as more and more sulphide was produced by bacterial action.

Tending to balance the effect of sulphate in closed basins becoming isotopically heavier as "light" sulphide is precipitated out, is the addition of sulphur from sources other than marine water. Rain water, rivers, or volcanic exhalations could contribute varying amounts of sulphate to the sedimentary basin. Fresh-water sulphate generally gives $\delta^{34}S$ values ranging from about +3‰ to −10‰ (Thode 1961, 1963). Sulphur in volcanic fumaroles, by necessity measured in gases emanating from subaerial volcanoes, generally range between +10‰ and −10‰ (Rafter et al., 1958a,b; Sakai and Nagasawa, 1958; Nakai and Jensen, 1967). Because both fresh water and volcanic sulphur is isotopically light compared to ancient seas, addition of sulphur from these sources to a marine basin could tend to shift the basinal water towards less positive $\delta^{34}S$ values.

Superimposed on the small variations in source sulphur compositions caused by differential access to the open sea, are the probably greater perturbations brought about by variations in the metabolic rate of sulphate-reducing bacteria which, in turn, affects the degree of fractionation of sulphide produced relative to that of the source sulphate. Temperature, light, nutrient and sulphate supply are all factors which affect metabolic rate but, perhaps less obvious, is the fact that the rate of metal supply will also affect the rate of production of H_2S (and hence the degree of fractionation). Although metal ions may be toxic to many organisms (including bacteria), sulphate-reducing bacteria have been shown to envelope themselves in a zone of sulphide of their own making which protect them from all but massive doses of metal. This was demonstrated by Temple and Le Roux (1964) to be the case for copper. They concluded that (p. 276): ". . . a sulfate reducing organism is protected from heavy metal ion toxicity by the sulfide it produces. This principle would apply to any biological sulfate-reducing agent and to any metal that formed an insoluble sulfide." Because it is the metal *ions* which are toxic to bacteria and not the metallic *sulphides,* the bacteria would protect themselves by precipitating the sulphide at a rate sufficient to keep the concentration of metallic ions in solution below the fatal toxic level. Presumably, then, a higher rate of metal supply would result in a higher rate of sulphide production and a lower degree of fractionation.

Returning to sulphur isotopes in petroleum, it is interesting to note that Thode and Monster (1970, p. 628) reflected that "... considering the wide range of sulphur isotope ratios found for crude oils, the same isotope pattern for two oils in a field would be strong evidence that they are derived from the same source rock, regardless of the reservoir in which they are found. Thus sulphur isotope studies should give information about oil migration." Thode and Monster (1970) and Thode and Rees (1970) used the sulphur isotopic composition of total sulphur in crude oil to determine the source-rock for petroleum in Iraq and showed that upward migration into reservoir rocks of a different age had, indeed, taken place, thereby confirming conclusions of earlier workers based on stratigraphic and structural considerations. Because of the possibility of vertical migration of sulphur (as H_2S) between the time of its generation and its fixation as metallic sulphides, ore deposits could also conceivably be found in host rocks of different age than those in which the sulphide was originally formed. This is a very real possibility and may well apply to some strata-bound orebodies which are not stratiform, e.g., the so-called Mississippi Valley type of deposits. If migration of this type took place between host rocks of similar isotopic compositions (e.g., Cambrian and Ordovician), then it probably would not be detected by isotopic means. However, migration of, for example, Cambrian-produced sulphide into Mississippian rocks (Fig. 4) would likely produce a deposit with an average isotopic composition markedly "heavy" with respect to Mississippian sea-water. Situations of this type are not evident, however, and, considering the large number of deposits sampled and the wide range of ages and metamorphic conditions represented therein, one might be justified in concluding that vertical migration of sulphide is not a common phenomenon or, fortuitously, it has occurred between isotopically similar host rocks and has not been detected.

Throughout the collection, compilation, and assessment of the data shown in Figs. 3 and 4, deposits have been assigned to either of two categories: volcanic and sedimentary. This was done in the expectation that the "volcanic ores" would exhibit sulphur isotopic ratios indicative of a volcanic source for sulphur whereas the "sedimentary ores" would likely have ratios indicating a bacteriogenic origin. However, such was not the case; sulphur in both types of ores appears to have as its common source sea-water sulphate reduced to sulphide by bacterial action. The difference in $\delta^{34}S$ for volcanic ores versus sedimentary ores, however, can be shown by a simple statistical method (Student's t test), to be statistically insignificant. That is to say, there is no *real* difference in $\Delta(\delta^{34}S)$ between deposits of sedimentary versus volcanic environments. The two-fold division, therefore, is entirely geological; as far as their isotopic compositions are concerned, strata-bound ores constitute only *one* population relative to sea-water sulphate and one would be statistically justified in combining the two groups shown in Figs. 3 and 4 into one curve. The same statistical test, however, indicate that the difference between sea-water sulphate and the corresponding strata-bound ores is, in fact, statistically real. The geological significance of these observations is discussed below.

In previous papers (Sangster, 1968, 1971), the author concluded (in part): "The sulphur in strata-bound sulphide deposits is largely or entirely of biogenic origin and had as its source sulphate dissolved in coeval sea water" (Sangster, 1971, p. 298). However, Sasaki and Kajiwara (1971) and Sasaki (1970) argued that because deposits of the "volcanic type" may be expected to have formed at higher temperatures than the "sedimentary type", biologic action may not be necessary for sulphate reduction at temperatures of 200°C or higher (the estimated temperature of formation of Kuroko-type volcanic deposits; Sasaki, 1970, p. 46). Sasaki argued that non-equilibrium chemical reduction of sulphate at 200°C could produce the observed average of about 17‰ fractionation between sea-water sulphate and volcanic-type deposits as shown by Sangster (1968). At approximately the same time, Kajiwara (1971) and Kajiwara and Date (1971) quantified Sasaki's model, basing their calculations on Sakai's 1968 data in a manner later popularized by Ohmoto (1972). Kajiwara and his co-workers, like Rye and Ohmoto (1974), postulate a "sea-water mixing with hydrothermal ore-solutions" as a possible mechanism for deposition of Kuroko-type deposits. The observed variation in isotopic composition in the Shakanai No. 1 deposit (Kajiwara, 1971), for example, is explained by changes in the fO_2, pH and total sulphur content of solutions due to the increasing contribution of sea water during ore deposition. Kajiwara's model of high-sulphate ore solutions contrasts with that of Sakai and Matsubaya (1974, p. 988) who postulate ore solutions which are "nearly neutral Na—Ca—Cl-type thermal waters of low sulphate concentration". These contrasting models of ore solutions, the fact that the formation temperature of about 250°C assumed by Kajiwara (1971), Kajiwara and Date (1971), and Rye and Ohmoto (1974) for Kuroko ores may be about 100°C too high for the bulk of the ore contained in the bedded portion of these deposits (Tokunaga and Honma, 1974, p. 387), together with other evidence of possible isotopic non-equilibrium (Sasaki, 1974, p. 391), all suggest that the Kajiwara—Ohmoto—Sakai systematics of sulphur isotopes may not completely apply to Kuroko-type deposits.

However, even if the difficulties outlined above could be accepted or otherwise explained, the concept of ingested and heated sea water as a source of sulphur may apply only to Kuroko-type deposits and not necessarily to all volcanic-type deposits shown in Figs. 3 and 4. Kuroko-type deposits are characterized by a stockwork orebody, underlying the stratiform portion of the ore, which is considered to represent the feeder vent of the ascending ore-forming solution. For Kuroko and other deposits which contain similar feeder pipes, the high-temperature, non-biologic model for the reduction of sea-water sulphate may, indeed, apply. However, a great many deposits included in the volcanic-type of Figs. 3 and 4 do not have associated feeder pipes. Examples would be the Kieslager (Besshi)-type deposits of Japan (Kajiwara and Date, 1971; Kanehira and Tatsumi, 1970) and the Bathurst deposits of Canada (McAllister and Lamarche, 1972). These deposits, although hosted in volcanic rocks (and hence classified as of the volcanic-type) are markedly thinner and more blanket-like in form than typical Kuroko ores and do not seem to have the associated underlying feeder pipes or stockwork ores. These "non-Kuroko" volcanic-hosted deposits, nevertheless, are considered by most workers to have originated as volcanic exhalations in a manner similar to the Kuroko ores, the difference being that the latter were precipitated *on* or *near* their feeder vent whereas the latter were precipitated *away* from the source orifice. The differences in mode of deposition may be caused by variations in the temperature and density of the ore-solution in a manner outlined by Sato (1972; see also paper by Sangster and Scott, Vol. 6, Chapter 5). If these non-Kuroko "Besshi-Bathurst" type of deposits have, in fact, formed at some distance from the source orifice, they likely formed at tempera- tures considerably lower than the 250°C postulated for the Kuroko-type ores. Dense, metal-rich brines may have flowed down a paleoslope from the fumarolic vent and been precipitated in an anoxic, H_2S-rich depression in the sea floor (Sato, 1972, Fig. 5). Under these low-temperature conditions, reduction of sulphate to sulphide may very well have been biologic and Sasaki's original objections to Sangster's 1968 conclusions may apply to only a few of the volcanic-type deposits shown in Figs. 3 and 4. Because most of the geological data used in classification of deposit-type were gathered from the literature and because feeder-zones may not have been recognized by many geolo- gists at the time of writing, it is impossible to estimate how many of the approximately 40 volcanic-type deposits have feeder pipes. However, the Kieslager, Bathurst, and sev- eral Norwegian deposits make up the bulk of the deposits shown and all of these do not have feeder-pipes. Hence, data are weighted in favour of the non-Kuroko deposits. How- ever, it is interesting to note that, although the Heath Steele B-1 deposit in the Bathurst area shows the same copper-to-zinc stratigraphic zoning in the massive ore as do the Kuroko ores (Lusk, 1969), sulphur isotope analyses of about 150 samples (Lusk and Crocket, 1969; Lusk, 1972) do not show the systematic isotopic zoning illustrated by Kajiwara (1971) for the Shakanai No. 1 deposit. The latter was explained by changes in fO_2, pH, and total sulphur content of ore solutions; lack of isotopic zoning may indicate more uniform conditions (syn-sedimentary?) on the ocean floor. It is instruc- tive, perhaps, that $\Delta(\delta^{34}S)$ values for Shakanai, Besshi, and Heath Steele B-1 are, re- spectively, +15, +11, and +13, all of which are close to the average of +15 fractiona- tion factor for bacteriogenic sulphur.

   In summary, then, the weight of sulphur isotopic evidence for strata-bound deposits in volcanic environments supports the concept that the main source of sulphur in these deposits is sea-water sulphate. For deposits such as the Kuroko ores for which feeder pipes can be demonstrated, it is likely that the temperature of ore deposition was suffi- ciently high for chemical reduction of sulphate to be the rate-controlling mechanism in which case the ultimate sulphur isotopic composition of the ore was determined by chemical factors such as fO_2, pH, etc. For other deposits in volcanic terrain, particularly those which do not have associated feeder pipes, which were probably deposited remote from the fumarolic vent and therefore likely formed at temperatures approaching that of contemporaneous sea-water sulphate, depositional conditions of the ore may have been more sedimentary than fumarolic in nature. For these deposits, biologic reduction of sulphur may have been the controlling mechanism and rate of production of hydrogen sulphide could have been enhanced by the influx of metal ions into the depositional site.

Carbonate-hosted, "Mississippi Valley-type" deposits, with their recognized association with petroleum, natural gas, and evaporites (see Sangster, Vol. 6, Chapter 9) are the natural link between Thode's sulphur isotope studies of petroleum (Thode et al., 1958; Thode and Monster, 1965) and native sulphur (Thode et al., 1954) on the one hand and all "sedimentary" sulphide deposits on the other. The association between biogenic sulphur and sedimentary sulphide deposits is not surprising when one considers that 75% of the world's present sulphur supply (elemental sulphur in evaporites, H_2S in sour natural gas, and sulphur compounds in petroleum) is bacteriogenic in origin (Dobbin, 1968; Myers, 1968). Furthermore, even greater tonnages of sulphur compounds of the same derivation are contained in coal, oil shale, and tar sands (Bodenlos, 1973, pp. 606, 610). The bacterial production of H_2S from connate brines or evaporites and its consequent migration to the site of ore deposition where it precipitates base metals from presumed chloride-rich, sulphur-poor brines, is an attractive model of ore genesis for sedimentary deposits of the so-called "Mississippi Valley-type".

Because the correlation between isotopes in sulphide deposits and coeval sea-water is at least as good as that for petroleum it implies that fixation of the sulphur as metallic sulphides took place *before* the H_2S could migrate vertically and fix metals in a rock of different age than that of the sulphur source. It also constitutes independent confirmation that migration of sour gas is probably largely lateral rather than vertical and, presumably, so is the migration of metalliferous brines. Thode et al. (1958) noted that petroleum samples from a single oil pool are remarkably constant in their sulphur-isotopic composition and attributed this homogenization to mixing of the gas during migration and collection. If sour gas (i.e. rich in H_2S) is well mixed prior to fixation as sulphides, one might expect that this would show up as a narrow range in the isotopic composition of the resulting ore deposit. This tendency toward a narrow spread in isotopic compositions of precipitated sulphides due to mixing of the source H_2S is, to some extent, offset by the fractionation effects between the sulphide species. Greatest fractionation, at constant temperature, would occur between galena and pyrite and the theoretical difference in composition between these two minerals, precipitated from hydrogen sulphide of uniform composition, is about 10‰ at 50°C and 7‰ at 100°C (Fig. 1).

The *actual* range would likely be somewhat less than this due to isotopic disequilibrium at these low temperatures. Similarly, syn-sedimentary deposits might be expected to show a greater range of $\delta^{34}S$ values because: (1) the hydrogen sulphide has not had the opportunity to be homogenized in a large reservoir prior to fixation as metallic sulphide, (2) assuming precipitation took place under approximately marine temperatures (25°C), the theoretical range of composition between galena and pyrite is approximately 12‰, (Fig. 1). Factors such as these may explain why Pine Point, for example, shows a narrow isotopic spread (std. dev. 2.4‰; Evans et al., 1968; Sasaki and Krouse, 1969) in contrast to Rammelsberg (std. dev. 5.7‰; Anger et al., 1966) and Mt. Isa (std. dev. 4.5‰; Solomon, 1965; Stanton and Rafter, 1966).

In a recent review of the sulphur isotopes of sulphide deposits in clastic rocks (roughly equivalent to the "syn-sedimentary" type discussed above) Schwarcz and Burnie (1973) attempted to relate the inferred sedimentary environment of a number of deposits to their isotopic distribution. The study was meant to test whether each environment type has a characteristic $\delta^{34}S$ distribution. Two main environments were recognized: (a) a shallow marine or brackish environment considered to be closed with respect to replenishment of sulphate from the open sea, and (b) a euxinic environment containing anoxic water and sediments but which was considered open to normal sea water with respect to sulphate reservoir. The first type of environment gave rise to broad S-isotope distribution ranging from 0 to 25‰ lighter than contemporaneous sea-water sulphate. The second environment produced deposits with narrow peaks averaging about 50‰ lighter than coeval sea-water sulphate. The authors concluded that "The difference in isotope effects can be accounted for by Rees' (1973) model of steady-state sulphate reduction: low nutrient supply and undisturbed, stationary bacterial populations in the open system settings that tend to generate larger fractionations" (Schwarcz and Burnie, 1973, p. 264). The modelling has merit although further testing with more examples is dependent on the correct interpretation of depositional environments for the two types of deposits in order to assign them to either of the two categories. For many deposits this is not yet possible but the model should be kept in mind for future study.

Mention was made earlier that the difference in average $\Delta(\delta^{34}S)$ between volcanic type and sedimentary type ores was found not to be statistically significant. This means that the two groups of deposits (provided they were originally reasonably well classified) *cannot* be distinguished on the basis of their sulphur isotopes and that the original premise, that the volcanic ores would somehow show more "volcanic influence" in their sulphur isotopes compared to the sedimentary ones, is not supported by the data. Therefore, it is fair to conclude, providing the 110 deposits for which isotopic data are available are representative, that, at least for the Phanerozoic, the data show the major source of sulphur in such deposits to be coeval sea-water sulphate. The data do not, however, indicate unequivocally whether sulphate reduction has been bacterogenic or chemical. For sedimentary deposits, geologic deduction may be sufficient to assign a biologic origin to the sulphur. For the volcanic ores the similarity in the *average and range* of fractionation relative to coeval sea-water (Fig. 4) may, as suggested by Sasaki (1970), be entirely fortuitous because of only partial equilibration of isotopes during chemical sulphate reduction at temperatures near 200°. For volcanic type deposits remote from their source orifices, however, (the so-called Besshi–Bathurst type) a bacteriogenic origin may, in fact, be a reasonable possibility. In any case, volcanic sulphur of magmatic origin does not appear to be a major factor in volcanic ore deposits and this may be a consequence of the submarine depositional environment of these deposits implied by other geological indicators such as pillowed lavas and marine fossils in association with these deposits. The correlation in isotopic composition between volcanic ores and sea-water sulphate, incidentally, constitutes independent evidence of the close association between these deposits and submarine (as opposed to subaerial) volcanism. Marine conditions apparently dominate over volcanic conditions for these ores and the overwhelming abundance of marine sulphate is expressed in the isotopic composition of sulphur. Subaerial volcanism *may* contain a higher proportion of magmatic sulphur, but, except for possibly certain native sulphur deposits, few, if any, stratabound deposits are formed under conditions of subaerial volcanism.

Deposits enclosed in rocks of continental or near continental origin. Remarks concerning the characteristics of sulphur isotope abundances in deposits of this type will be confined to a brief assessment of the current literature as the author has not had personal sulphur-isotope experience with this type of deposit.

(a) *Sandstone–uranium deposits.* For purposes of illustration, a few well-known sandstone–uranium type of deposits, so prevalent in the Colorado Plateau area, will be considered. Paraphrasing from Warren (1971), roll-type uranium deposits are found in beds of sandstone at the sides and ends of tongues of altered rock. The ore zone occurs in relatively unaltered sandstone in contact with the end of the altered tongue and invariably contains anomalous amounts of iron sulphides in addition to minerals containing U, Se, V, and Mo. Pyrite and carbonaceous material, normally distributed in a dispersed manner throughout the host sandstone, are virtually missing from the altered zone. A well-developed roll forms a crescent orebody concave in cross-section toward the altered tongue. Ore deposits and associated altered rock are considered to have formed by the action of ore-bearing solutions containing oxygen and uranium which percolated through the permeable host rocks, reacting with and oxidizing the pyrite and organic carbon compounds. The minerals of the ore zone were continually oxidized and dissolved along the upstream edge of the deposit and redeposited a short distance downstream. Because the percolating, oxidizing ore solutions are normally considered to be of meteoric origin for deposits of this type and because the host rock sandstone is of terrestrial origin, comparison of their sulphur isotopes with marine sulphate is not warranted.

Roll-type uranium deposits are characterized by an extremely wide range of S-isotope compositions (Fig. 6) within each deposit, a feature noted originally by Jensen (1958). For four rolls from three Wyoming districts, Austin (1970) showed that "δ^{34}S characteristically increases sharply from barren interior to ore ... The increase is gradual across the arbitrary boundary between ore and protore".

Briefly, two genetic models, the biogenic model and the chemical model, are currently under consideration to explain the observed S-isotope pattern in roll-type uranium deposits. The biogenic model, proposed by Jensen in 1958, postulates that "bacteria in the ore zone of an established deposit chemically reduces the sulfate in the entering ground water to hydrogen sulfide. The organic material which is dispersed throughout

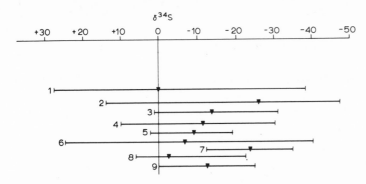

Fig. 6. Sulphur isotopic compositions of nine sandstone–uranium deposits, southwestern United States. From top to bottom, the deposit name, State, and age of host rock are as follows: *1* = Maybell, Colo., Tert.; *2* = Amarillo, Wyom. Tert.; *3* = Kopplin, Wyom., Tert.; *4* = Lucky M^c, Wyom., Tert.; *5* = Rim, Wyom., Tert.; *6* = TSG, Wyom., Tert.; *7* = Day Loma, Wyom., Tert.; *8* = Sunset, Wyom., Tert.; *9* = Western Nuclear, Wyom., Tert. (From Austin, 1970b.)

the ore zone supplies the energy for sulfate reduction. The bacteria feed on the organic material and reduce the sulfate, by their metabolic processes, to hydrogen sulfide which both fixes the uranium and forms the pyrite" (Warren, 1971, p. 920). In contrast to the biogenic model, Granger and Warren (1969) proposed a chemical model in which "... ore-bearing solutions which contain oxygen oxidize the pyrite to unstable or reactive sulfur species such as sulfite and thiosulfate. The compounds decompose spontaneously to form sulfate and sulfide. Chemists call this a disproportionation reaction" (Warren, 1971, p. 921).

Both Jensen (1958) and Austin (1970a) support the biogenic model by which the wide range in S-isotopic composition is due largely to a series of oxidation–reduction reactions involving pyrite and sulphate. Original pyrite in the sandstone is considered to be oxidized to sulphate by the oxidizing ground water. This results in sulphate with essentially the same isotopic composition as the original pyrite because oxidation of sulphide to sulphate results in little or no fractionation. This sulphate is then reduced to hydrogen sulphide by bacterial processes to produce pyrite and the usual bacterial range of fractionation factors applies to this reaction. The original pyrite has a wide range of isotopic compositions and the later bacterial reduction of sulphate derived from it superimposes an even greater diversity in composition, thereby accounting for the wide isotopic range of sulphur in roll-type uranium deposits. Austin (1970a, p. 10), in defence of the biologic model, argues that sulphur of intermediate oxidation states required for the chemical model have not been identified in roll-type uranium deposits and reminds the reader that "No means other than biological is known for reducing a sulfate sulfur at the low temperatures postulated for the Gas Hills deposits". With regard to the observed sharp break in isotopic composition trend shown at the solution front, Austin observes that this is not uncommon for biologic processes and points to the well-known sharp boundary between oxidizing and reducing conditions found in bottom muds of seas and ponds in support of the biologic model.

Warren (1972), on the other hand, while admitting that biogenic reduction of sulphate could produce the isotopic pattern found by Austin (1970a), argues that to produce the observed isotopic pattern, *all* the sulphate sulphur derived from oxidized pyrite would have to be biogenically reduced to sulphide and deposited in the ore zone. This is precluded, according to Warren, by the fact that a maximum of only 2% pyrite is found in the deposit shown as an example for the chemical model. Partial reduction of sulphate (a biogenic steady-state process according to Warren) would not give the observed high $\delta^{34}S$ values found in the subore zone of the deposit studied. For these reasons, Warren (1972) proposes a chemical model.

Points to consider in assessing the two models with respect to sulphur isotopes are:

(1) Intermediate sulphur species have not been detected in ground-water samples from uranium deposits. Granger and Warren (1969) argue that this is due to oxidation of the sample between the time it is collected and submitted to the laboratory for analysis.

(2) Anaerobic, sulphate-reducing bacteria have been reported downstream from the oxidation front of uranium deposits in the USSR but are apparently not common in most uranium deposits. This may merely indicate that a majority of uranium deposits are inactive at the present time.

(3) The S-isotope distribution used by Warren (1972) as a basis for the chemical model was shown by Austin (1970a) to be an almost ideal and unique distribution for the four deposits studied by him. The quantitative modelling proposed by Warren (1972) made no attempt to explain the other, more variable, patterns found by Austin.

(4) Austin's main point, that biologic processes could account for the sharp differences in $\delta^{34}S$ between adjacent or closely spaced samples, has not been shown to be the *only* manner in which these differences could be produced.

In summary, the sulphur isotope abundance in roll-type uranium orebodies have apparently not yet provided the key to understanding the genesis of sulphur in these deposits.

(b) *Sandstone-copper deposits.* These deposits are somewhat similar to sandstone-type uranium deposits in that they occur in terrestrial clastic sequences and contain abundant coalified woody material (Phillips, 1960). The dominantly terrestrial nature of the host rocks precludes comparison of S-isotope abundances (Fig. 7) with contemporaneous sea-water sulphate and the absence of nearby intrusive rocks belies an igneous source for the sulphur. If the data presented in Figs. 6 and 7 are representative of S-isotope abundances of sandstone—uranium and sandstone—copper types of deposits, respectively, then it appears that the latter contain, on the average, sulphur of considerably lighter composition than sandstone—uranium deposits. Furthermore, judging from the data presented, it would appear that the copper deposits possess a considerably narrower spread in composition than the uranium deposits. Remembering

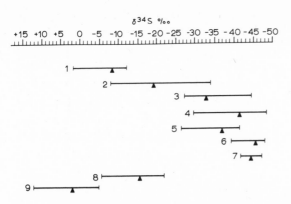

Fig. 7. Sulphur isotope abundances in sandstone-copper type of deposits. Name of deposit, age of host rock and references are as follows: *1* = Warnock, U.S.A., Permian (Phillips, 1960); *2* = Ady Cupb., U.S.A., Permian (Phillips, 1960); *3* = Happy Jack, U.S.A., Triassic (Phillips, 1960); *4* = Eureka, U.S.A., Triassic (Phillips, 1960); *5* = Dorchester, Canada, Pennsylvanian (Jensen, personal communication, 1970); *6* = Chisholm Brook, Canada, Pennsylvanian (Jensen, personal communication, 1970); *7* = Pugwash Bay Prospect, Canada, Pennsylvanian (Jensen, personal communication, 1970); *8* = Dzhez-kazgan, U.S.S.R., Pennsylvanian (Chukhrov, 1971); *9* = Yarrow Creek, U.S.A., Precambrian (Morton et al., 1974). Valves for deposits 3, 4, 5, and 6 are figured in Jensen (1962).

that the sulphide sulphur in the uranium-roll deposits was derived by reduction of sulphate produced by oxidation of previously-deposited protore sulphides (mainly pyrite), it is interesting to note that several workers have suggested that the copper in sandstone-type deposits has replaced early pyrite associated with carbonaceous trash. The fact that the woody material has been replaced by sulphides before compaction (e.g., Phillips, 1960; Woodward et al., 1974) only serves to emphasize the early nature of the sulphides.

Primary copper minerals have commonly been oxidized later to malachite, azurite, chrysocolla, etc. Thus, a model for the copper deposits whereby early sulphides, associated with woody material in permeable sandstones, are later oxidized by ground water is in harmony with that presented for sandstone—uranium deposits in which primary sulphides are oxidized by ground water. Thus the sandstone-type copper deposits may represent cupriferous equivalents of the uranium protore, an hypothesis supported by the many similarities between the two deposit-types as well as the overlap in their compositions (i.e. copper in uranium deposits and vice-versa).

Both Cheney and Jensen (1966) and Austin (1970a) found that, in sandstone—uranium deposits, "... sulfur from sulfides associated with woody or coaly material tends to be light, sometimes extremely light" (Austin, 1970a, p. 19) and distinctly lighter than sulphides in the uranium ore. Austin (op. cit.) found that "for nine samples of sulfides associated with woody or coaly material from sandstone-type uranium deposits, $\delta^{34}S$ ranges from $-27.2‰$ to $-47.2‰$ and averages $-36.7‰$." In both range and average these values are similar to sandstone-type copper deposits (Fig. 7) and are distinctly lighter than sulphur in the uranium ore (Fig. 6). Similarly for wood-associated sulphides in the Wyoming uranium district Cheney and Jensen (1966, Fig. 4) report a range from about $-18‰$ to $-46‰$ with a mean of about $-33‰$.

The similarity in sulphur-isotopic compositions of the wood-associated sulphides of sandstone—uranium deposits to those of the sulphides in sandstone—copper deposits, suggests a common genesis for sulphur in these two deposit-types. The original source of the sulphur is unknown for these terrestrial deposits.

Deposits of the conglomerate—U(—Au) type. The origin of the important Precambrian uranium—gold conglomerate deposits of Witwatersrand, South Africa and similar uraniferous deposits of Elliot Lake, Canada have been the subject of controversy for many years. Variable amounts of pyrite are common in deposits of both areas and two main theories (magmatic—hydrothermal vs. detrital) have evolved to explain its origin. A few isotopic determinations of the pyrite sulphur have been reported in the literature and are shown in Fig. 8. The isotope ratios exhibit an extremely narrow range with averages close to zero permil and well within the range of igneous rocks.

Working apparently independently (neither author refers to the other) both Hoefs and his co-workers (1968) and Roscoe (1968), while acknowledging that the sulphur isotopic composition is indicative of a magmatic origin, reject a magmatic-hydrothermal source for the conglomerate-pyrite on geological evidence found in the deposits as a whole (see references for further discussion). Both authors independently propose a largely detrital origin for the pyrite and would derive the pyrite by erosion of adjacent older Precambrian borderlands. Hoefs et al. (1968) suggest a slight re-adjustment

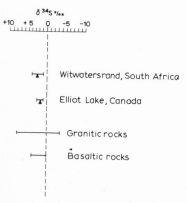

Fig. 8. Sulphur isotope ratios of composite samples from Precambrian conglomerate-uranium deposits of South Africa and Canada. Data from Hoefs et al. (1968) and Roscoe (1968). Range of sulphur isotopes in igneous rocks shown for comparison (Hoefs, 1973, p. 33).

of the sulphur isotopes due to post-depositional recrystallization in the case of Witwatersrand and Roscoe (1968, p. 129) suggests "... that the conglomerates contain a mixture of detrital pyrite and pyrite formed through diagenetic processes".

Both workers (either explicity in the case of Hoefs et al. or implicitly in the case of Roscoe) contrast the narrow range of sulphur isotopes in the conglomerate pyrite and the wide range expected for sedimentary sulphides and compare it with the narrow range for sulphur in igneous bodies. However, sufficient data for sulphur isotope ratios in sediments as old as either the Witwatersrand (>2300 < 2870 m.y.) or Elliot Lake (>2100 < 2500 m.y.) rocks are lacking and comparisons with sulphides in younger Precambrian or Phanerozoic strata may. be misleading, particularly if Roscoe's hypothesis of the necessity of an anoxic environment for formation of these deposits is correct. If the atmosphere was anoxic, perhaps there was not sufficient sulphate available for bacterial reduction (there may not even have been any sulphate reducing bacteria!) and sedimentary sulphides in rocks as ancient as these may, in fact, be characterized by sulphur of narrow isotopic spread.

In summary, while a magmatic hydrothermal origin may be untenable on geologic grounds, the sulphur-isotope data would permit such an interpretation. If the source borderland for the conglomerate (and pyrite) is largely igneous, then the sulphur isotopic composition is consistent with the detrital hypothesis as suggested by Hoefs et al. (1968) and Roscoe (1968). Pending further data on the isotopic composition of sulphides in Lower Proterozoic sediments, a sedimentary theory cannot be precluded.

LEAD ISOTOPES

General theory

The theory of lead isotopes and their application to the study of mineral deposits has been well presented recently by Doe and Stacey (1974) and to more geological problems by Doe (1970). Also Köppel and Saager (Chapter 9, this Vol.) have summarized the physical and mathematical theory of lead isotopes.

Lead isotope abundances in mineral deposits, strata-bound or otherwise, are generally either *single-stage leads* or *anomalous leads*. In the former, the only changes in the value of the parent : daughter ratio are those caused by radioactive decay of the parents (uranium and thorium). Model lead ages determined from single-stage leads approach the time of formation of the mineral deposit. Ores which did not form as a result of such simple conditions give rise to anomalous leads (which are the result of two or more parent : daughter ratios) and model ages calculated from anomalous leads tend to be very much in error. As pointed out by Russell et al. (1961), the distinction between single-stage leads and anomalous leads may or may not be apparent from the isotopic analysis of a single sample.

Single-stage leads. The following description of single-stage leads has been paraphrased from Doe and Stacey (1974, p. 759):

"The single-stage model of lead isotope evolution assumes that the earth began with only one value for primeval lead isotopic composition ($^{206}Pb/^{204}Pb$, $^{207}Pb/^{204}Pb$, $^{208}Pb/^{204}Pb$) and that the only changes in the ratios of the radioactive parent isotopes to the stable daughter index isotope $^{238}U/^{204}Pb$, $^{235}U/^{204}Pb$ (or its equivalent ($^{238}U/^{204}Pb \times 137.88$), and $^{232}Th/^{204}Pb$ are due to radioactive decay. Withdrawal of some lead, uranium, and thorium from the single-stage source by magmas, ore fluids, or other means is not excluded so long as the single stage source is large enough so that the ratios involving the radioactive parent isotopes are not significantly disturbed; that is, the source of the magma or ore metals is an "infinite-reservoir". Single-stage conditions are so stringent that it would be surprising if they could persist unaltered by geologic events throughout the 4.5 b.y. of geologic history, but the conditions are approached for deposits sometimes referred to as the "submarine exhalative" class associated with island arcs. Although single-stage conditions are approached for these deposits, they are not a perfect fit for the model. Kanasewich (1968) has shown that there is a tendency for younger deposits to have generally higher $^{206}Pb/^{204}Pb$ ratios than the expected single-stage systems. The small departures from single stage conditions, shown by even those deposits that most closely approach it, seem best explained by the mixing of several isotopically heterogeneous source materials. This mixing may be done through processes such as multiple reworking of sediments . . . or leaching of lead from very large volumes of rock, in such a way that a single-stage system is approached. Thus, so-called single-stage leads only approximate an evolution under single-stage conditions. The single-stage model of lead isotope evolution, even though imperfect, provides valuable information on the age of formation (model-lead ages), the source of the lead in many ores, and the value of Th/U in the source material".

Anomalous leads (two-stage).

"The simplest process found to produce anomalous leads is a two-stage process. In the first stage, the lead isotopic composition evolved by a single-stage process until some later event occurred that caused the $^{238}U/^{204}Pb$ ratio to be changed into a variety of values. The result is lead isotope data that lie along straight lines (in plots of $^{207}Pb/^{204}Pb$ versus $^{206}Pb/^{204}Pb$ ratios) called secondary isochrons. The general form of the secondary isochron equation is similar to the primary Pb−Pb isochron equation, but an evolved lead is substituted for the primeval lead to begin the second stage and the age of the beginning of the second stage is substituted for the "age of the earth"."

"If the age of mineralization is known approximately, the age (perhaps integrated ages) of the source material for the lead may be calculated, as was done by Stacey et al. (1969). Because the equation for secondary isochrons is transcendental, the solution is one of trial and error. The secondary isochron method may sometimes find application in estimating mineralization ages where a lack of precision is not the key issue, as was done by Kanasewich (1962) in estimating that the age of mineralization of the Thackaringa-type deposits in the Broken Hill area of Australia was about 500 m.y. ago. Secondary isochrons also have application in identifying the source of lead in ores."

(Paraphrased from Doe and Stacey, 1974).

Lead isotopes in strata-bound sulphide deposits

Representatives of both single-stage and anomalous leads are found in strata-bound mineral deposits. In Chapter 9, Köppel and Saager have described lead isotope abundances at Broken Hill, certain carbonate-hosted U.S. lead−zinc deposits, and the Witwatersrand uranium−gold ores. This paper will deal with volcanogenic massive sulphide deposits of various ages (Sangster, 1972b; Sangster and Scott, Vol. 6, Chapter 5) and several examples of carbonate-hosted lead−zinc deposits.

Volcanogenic massive sulphide deposits. Within the meaning of the term "volcanogenic massive sulphide" as used by Sangster (1972b) (i.e. "volcanogenic" stresses the genetic connection between these deposits and volcanism and/or volcanic processes; "massive" refers to mineralization composed almost entirely of sulphides and does not carry any textural connotation), these deposits are equivalent to the "submarine exhalative class associated with island arcs" referred to by Doe and Stacey (1974, p. 759) as approaching single-stage lead conditions.

Lead isotope data for volcanogenic massive sulphide deposits of Archean, Proterozoic, and Miocene age are presented in Tables II–IV and Figs. 9–11. Isochrons in Figs. 9–11 are plotted using primordial values, decay constants, and age of the earth as used by Stacey et al. (1969). For ease of comparison, the three figures have been drawn to the same scale.

The data display the characteristics of single-stage leads in that they tend to cluster together and/or tend to scatter along a trend sub-parallel to the isochrons. The latter effect could be caused either by "204 error" (i.e. laboratory error), by derivation of each deposit (or sample) from several sources of differing $^{238}U/^{204}Pb$ (μ) values, or they could represent, as in the case of the Japanese ores, the normal small spread of compositions expected in any trace-element studies. These patterns of single-stage leads will be contrasted later with anomalous leads but it serves to emphasize that the distinction between single-stage and anomalous leads cannot be achieved on a single determination. In each of these diagrams, the deposits represented can be considered a geological entity. The Japanese ones are from the same age of Miocene volcanics, the (possibly) Proterozoic ones are all from the same greenstone belt, and all of the Archean deposits are from Superior Province and of the 11 only 3 (Geco, Willroy, Willecho) are not from the Abitibi orogen (see Sangster and Scott, Vol. 6, Chapter 5).

The tendency for the Archean values to scatter sub-parallel to the isochrons is, in part, probably due to the fact that the isotopic determinations were done by several laboratories and therefore much of the scatter is caused by "204 error" or interlaboratory scatter. One notes also, that the scatter normal to the isochrons is about the same in all three diagrams. The question arises: does the scatter across the isochrons represent a real age difference or not? In the case of the Kuroko Miocene deposits (Fig. 11), the model lead age difference indicated in the diagram is not real. The host rocks to the ores have been well-dated at about 14 m.y. (Sato, 1974) over the entire strike-length of the Green Tuff volcanic belt, approximately 1500 km. Therefore the scatter in model lead ages represented in Fig. 11 is caused by factors other than age. Some of these might be post-depositional contamination of the ore-leads by radiogenic lead in hydrothermal fluids and/or groundwater, laboratory contamination, syn-ore contamination of ore-forming fluids during their passage through supra-crustal rocks from their source, etc. The average of the nine Kuroko deposits, incidentally, gives a model lead age of 5 m.y. which is in agreement with the age of coeval volcanic rocks. The Archean leads (Fig. 9), all from the Superior Province of the Canadian Shield, represent depos-

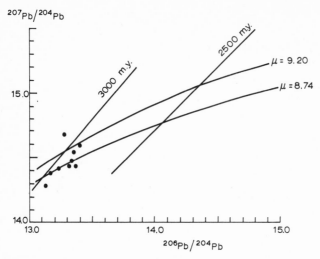

Fig. 9. Lead isotope ratios in Canadian Archean volcanogenic massive sulphide deposits. Data from Table II. In this and all succeeding figures, ratios reported have been corrected to absolute unless otherwise stated.

its over a strike-length of about 800 km and show about the same extent of isotopic scatter across the isochrons. In this case, however, it is not clear whether the model lead age spread is real or not. The associated volcanics *could* have been deposited throughout Superior Province in a short, single episode as were the Japanese ore-containing volcanic rocks; on the other hand, the apparent age spread could represent several volcanic episodes occurring at (say) 10 m.y. intervals. The Pb–Pb data available cannot

TABLE II

Lead isotope ratios in Canadian Archean volcanogenic massive sulphide deposits*

Deposit name	Material				Reference
Horne "H"	galena	13.23	14.41	32.98	Roscoe (1965)
Lake Dufault	galena	13.39	14.60	33.43	Roscoe (1965)
Mattagami Lake	galena	13.16	14.37	33.16	Roscoe (1965)
Orchan	galena	13.16	14.37	33.49	Roscoe (1965)
Coniagas	galena	13.12	14.27	32.85	Roscoe (1965)
Willroy	galena	13.31	14.43	33.16	Stacey et al. (1969)
Barvue	galena	13.27	14.68	33.63	Kanasewich and Farquhar (1965)
West MacDonald	galena	13.35	14.54	33.19	Kanasewich and Farquhar (1965)
Mobrun	galena	13.37	14.43	33.24	Kanasewich and Farquhar (1965)
Lun Echo(Willecho)	galena	13.33	14.46	33.25	Ostic et al. (1967)
Geco	galena	13.33	14.46	33.27	Ostic et al. (1967)

* Ratios in this and in all succeeding tables are reported as corrected to absolute unless otherwise stated.

TABLE III

Lead isotope ratios in Canadian Proterozoic (?) volcanogenic massive sulphide deposits (from Sangster, 1972a)*

Deposit	Material	Isotopic percent			
		204	206	207	208
W. Nuclear	galena	1.510	23.18	22.77	52.54
Flin Flon	mass. ore	1.507	23.10	22.81	52.58
Schist Lk.	mass. ore	1.505	23.13	22.76	52.61
Osborne Lk.	galena	1.508	23.07	22.90	52.52

Deposit	Material	Isotopic ratios				
		206/204	207/204	208/204	207/206	208/206
W. Nuclear	galena	15.35	15.08	34.80	0.9826	2.267
Flin Flon	mass. ore	15.33	15.14	34.90	0.9874	2.276
Schist Lk.	mass. ore	15.37	15.12	34.96	0.9838	2.275
Osborne Lk.	galena	15.30	15.18	34.82	0.9924	2.276

* See also discussion p. 247 concerning age of host rocks.

distinguish such small time-intervals. The Proterozoic ore-leads (Fig. 10), representing a strike-length of about 200 km, show about the same model lead age scatter as do the older and younger examples. However, for neither the Archean nor Proterozoic mineralized belts, are there published zircon or Rb–Sr ages for the host rocks to these massive sulphides.

It is worthwhile to point out that the Japanese deposits have not, of course, been metamorphosed whereas the Archean ones represent a range in metamorphic grade

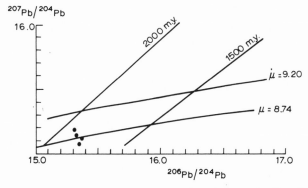

Fig. 10. Lead isotope ratios in Canadian Proterozoic (?) volcanogenic massive sulphide deposits. Data presented in Table III.

TABLE IV

Lead isotope ratios in Japanese Miocene volcanogenic massive sulphide deposits (Kuroko deposits)
(data from Sato and Sasaki, 1973)

Deposit	Material	206/204	207/204	208/204
Uchinotai Western	galena-rich ore	18.35	15.46	38.32
Uchinotai Eastern	galena-rich ore	18.44	15.57	38.78
Uchinotai Uwamyki	galena-rich ore	18.30	15.50	38.19
Aketoshi	galena-rich ore	18.43	15.56	38.58
Osaki	galena-rich ore	18.36	15.46	38.41
Dai-ichi	galena-rich ore	18.55	15.58	38.54
Yoshino	galena-rich ore	18.42	15.58	38.52
Iwami	galena-rich ore	18.23	15.53	38.48
Wanibuchi	galena-rich ore	18.27	15.63	38.84

from greenschist to upper epidote amphibolite due to the effects of the Kenoran orogeny at 2560 m.y. In spite of this, however, the older deposits show the same isotopic
scatter as do the younger ones and, perhaps more importantly, do not appear to have
been isotopically affected by the Kenoran orogeny. They appear to have retained their
isotopic identity in spite of the later metamorphism. The age of the host rocks to Archean
massive sulphide deposits is not known but a U–Pb zircon age by Krogh and Davis
(1971) of 2730 m.y. for two granodiorite bodies that intrude volcanic host rocks in the
Noranda area could be considered minimum ages for the massive sulphide deposits in
that region. Hence, the clustering of Archean leads around 2900 m.y. is not unrealistic
and is compatable with the above-mentioned minimum age. It thus appears that, within
limits, Pb–Pb model ages of volcanogenic massive sulphide deposits can be useful in

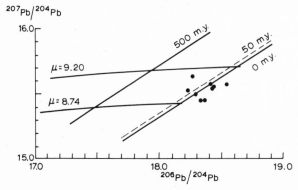

Fig. 11. Lead isotope ratios in Japanese Miocene volcanogenic massive sulphide deposits. Data presented in Table IV.

dating where other, more conventional, techniques are not available. With this principle in mind, Sangster (1972a) was able to demonstrate there is good reason to believe that volcanic rocks of the Snow-Lake—Flin-Flon—Hanson Lake area of the Canadian Shield (see Sangster and Scott, Vol. 6, Chapter 5), previously considered to be Archean in age, are probably Aphebian (Lower Proterozoic). Confirmation of the ore-lead model ages in this area awaits Rb—Sr or zircon studies.

Carbonate-hosted lead—zinc deposits. Although these are examined here as a deposit-type distinct from the volcanogenic massive sulphide deposits discussed previously, carbonate-hosted lead—zinc deposits (excluding skarns) are themselves comprised of two sub-types, the characteristics of which are outlined by Sangster (Vol. 6, Chapter 9). For purposes of lead-isotope discussion, however, carbonate-hosted lead—zinc ores will be examined in the following groups: (a) those within the greater Mississippi Valley, U.S.A.; (b) Appalachian Valley, U.S.A.; (c) Pine Point, Canada; (d) Silesian district, Poland; (e) Central Plain, Ireland.

(a) *Mississippi Valley, U.S.A.* Of all carbonate-hosted lead—zinc deposits, those lying within the drainage of the Mississippi Valley of central U.S.A. have received the most attention with regard to lead isotopes. Within this large region, three major districts contain the bulk of lead—zinc deposits. These are the Tri-State, southeast Missouri, and Upper Mississippi Valley districts. The most recent review of lead isotopes in the entire Mississippi Valley region was·that by Heyl et al. (1974). Previous lead isotope studies on Upper Mississippi Valley deposits by Heyl et al. (1966) were re-examined by Richards et al. (1972) after correcting the old values for systematic bias and using a better regression method. The southeast Missouri lead belt has also been re-studied (Doe and Delevaux, 1972) using modern analytical techniques but earlier studies by Brown (1967) are still the most detailed for this region. A district-wide study of the Tri-State area has apparently never been attempted since the early reconnaissance work of Farquhar (1954) later published by Russell and Farquhar (1960, pp. 149—150).

Lead isotopic analyses for galenas from southeast Missouri, Upper Mississippi Valley, and Tri-State are shown in Figs. 12, 13 and 14, respectively. Data for these areas, corrected to absolute values, are presented in Tables V, VI, and VII, respectively. The reader is referred to Brown (1967) for more data for southeast Missouri which were considered too extensive to reproduce here.

As other workers have noted previously, leads from the Upper Mississippi Valley are very radiogenic, all possess values of $^{206}Pb/^{204}Pb$ greater than 20, and they exhibit a linear trend on the $^{207}Pb/^{204}Pb$ versus $^{206}Pb/^{204}Pb$ diagram which is characteristic of anomalous leads.

The Upper Mississippi Valley district (Fig. 13) exhibits the best linear array and consequently has been examined by both Heyl et al. (1966) and by Richards et al. (1972) in terms of a two-stage model. In such a model, the slope of the regression line through the data of Fig. 13 will be a function of both the age of the source rock for the lead

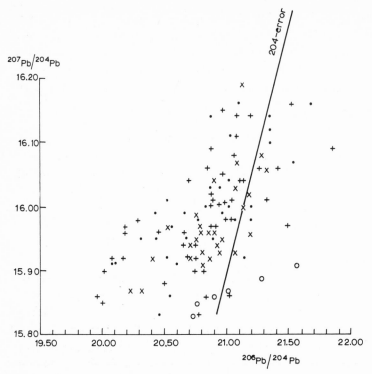

Fig. 12. Lead isotope ratios in galenas from southeast Missouri U.S.A. + = Federal Mine; x = Lead-wood Desloge Mine; • = Bonne Terre Mine (data from Brown, 1967; not corrected to absolute). Open circles are values from Doe and Delevaux (1972) for deposits in the Old and New Lead Belts, corrected to absolute (see Table V).

and the age of mineralization. Neither of these, of course, is known with certainty but, because of at least a *spatial* relationship between basement topography and the lead deposits in overlying Ordovician dolomite (Heyl and King, 1966), an origin of lead in

TABLE V

Lead isotope ratios in galenas from southeastern Missouri (from Doe and Delevaux, 1972)

Mine or area	206/204	207/204	208/204
No. 8, 12/19 contact	20.78	15.85	39.59
No. 8, 15/19 contact	20.91	15.86	39.72
No. 8, 7/10 contact	21.56	15.91	40.56
Bixby area	21.02	15.87	39.81
Bixby area	21.30	15.89	40.20
Bixby area	20.75	15.83	39.60

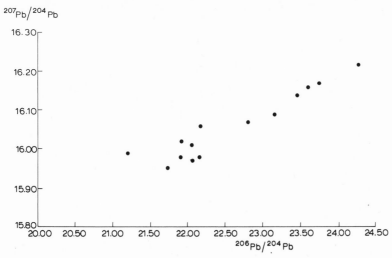

Fig. 13. Lead isotope ratios of galenas from mines in the Upper Mississippi Valley district, U.S.A. (data presented in Table VI).

the basement rocks could be considered a possibility. With this in mind, Heyl et al. (1966) and Richards et al. (1972) considered 1350 m.y. a reasonable approximation of the integrated age of basement to the Upper Mississippi Valley region. If t_1 = 1350 m.y. and using the slope of regression obtained by applying a program developed by

TABLE VI

Lead isotope ratios in galenas from deposits in the Upper Mississippi Valley district, U.S.A. (data from Heyl et al., 1966)

Mine	206/204	207/204	208/204
Capt. Turner	21.20	15.99	40.80
Holmes	21.73	15.95	41.27
Skene	21.92	16.02	41.66
Amelia (av. of 3)	21.90	15.98	41.59
Piquette	22.05	16.01	41.62
Bautsch	22.18	16.06	41.96
Nigger Jim	22.07	15.97	41.77
New Hoskins	22.16	15.98	42.06
Calumet (av. of 2)	22.70	16.07	42.39
Ohlerking	23.16	16.10	42.37
Old Slack	23.46	16.14	42.91
Ivey	23.74	16.17	43.24
North Yellowstone	23.60	16.16	43.39
Demby-Weist	24.27	16.22	43.63

TABLE VII

Lead isotope ratios in galenas from Tri-State district, U.S.A. (from Russel and Farquhar, 1960, pp. 148–150)

Sample No.	Mine	206/204	207/204	208/204
316	Diamond Joe	21.20	15.94	40.65
322	Blue Goose	21.62	15.92	40.94
231	Westside	21.80	15.99	41.34
323	Blue Goose	21.80	16.13	41.45
415	Joplin	21.81	15.92	41.20
314	Webber-Westside	21.89	15.98	41.20
213	Tin-State	21.90	16.01	41.53
318	Webber-Westside	21.93	15.99	41.21
419	Joplin	21.96	15.85	40.86
320	Howe	22.07	16.05	41.38
315	Federal Jarrett	22.45	16.00	41.65
317	Grace B	22.54	16.05	41.79
319	Kitty	22.54	16.08	41.69
321	Otis White	22.55	16.06	41.70

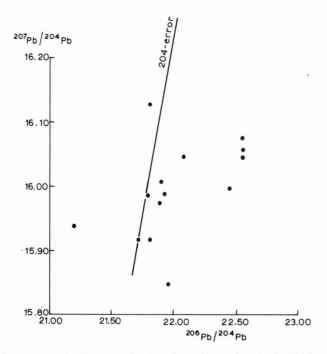

Fig. 14. Lead isotope ratios in galenas from mines in the Tri-State district, U.S.A. (data presented in Table VII).

York (1969), then t_2 = 365 m.y. However, Heyl et al. (1973, p. 18) suggest that mineralization, on structural evidence occurred "sometime between Late Paleozoic and the end of the Mesozoic", i.e., between 250 and 70 m.y.

As Richards et al. (1972) point out, a simple two-stage interpretation of the data requires only that an initial lead be homogenized about 1350 m.y. ago and that this lead be combined with a variable proportion of an added radiogenic component generated in the period between consolidation of the basement (1350 m.y.) and formation of the mineralizing solution. Heyl et al. (1966) also point out that a district-wide lead isotope zoning, with lead showing a progressive increase in "radiogenicity" from southwest to northeast, could be produced by addition of radiogenic lead from country rocks to more normal lead generated from the Forest City and Illinois Basins to the west and south, respectively (see Fig. 17). Such a mixing model would of course, have no time connotations and may explain the poor "fit" between the radiometric mineralization age and that deduced on structural evidence.

The southeast Missouri district (Fig. 12) shows a broad scatter of points probably brought about, in part at least, by 204-error which will cause the points to scatter in a direction indicated by the 204-error line in Fig. 12. The data belong to an earlier instrumental generation and could not be corrected to absolute. They are presented merely to illustrate the range of isotopic compositions present in the district (all data from the Old Lead Belt) and to compare with Doe and Delevaux's (1972) data from six samples from both the Old and New Lead Belts. Brown's data, representing as they do samples from throughout the major mines in the Old Lead Belt, are considerably more representative than are Doe and Delevaux's. It is interesting to note that the southeast Missouri data, falling in the range 20.00 to 21.50 for $^{206}Pb/^{204}Pb$ ratios, would "fit" nicely onto the Upper Mississippi Valley data shown in Fig. 13. From the slope of Doe and Delevaux's data, an integrated "age" of source rock (i.e. basement) lies in the range 1350–1410 m.y. similar to that for Upper Mississippi Valley. These two districts, about 260 miles apart, are floored by Precambrian rocks of the same age range (roughly 1350 m.y.), according to Muehlberger et al. (1967), and it is perhaps significant that, isotopically, they both show a similar relationship to "age" of source rock. Therefore, the lead isotopes in ores from the two districts permit a genetic model such as that described by Richards et al. (1972, p. 289) for Upper Mississippi Valley:

"We follow the general Kanasewich (1968) formalism by postulating thorough mixing of the lead in what is now the basement at some time, before the Cambrian, defined by age measurements on the basement rocks. This lead should be like the "initial ratio" of U—Pb and Th—Pb isochrons (on the same formalism as the well-known Rb—Sr isochron) obtained from the basement. In the succeeding period up to the time of mineralization, further amounts of the 206, 207, and 208 isotopes will have been added in proportion to the amounts of U and Th relative to the Pb. It matters little whether this addition took place locally within the basement, or within the sediments derived from it. As mentioned in the introduction the resulting leads should exhibit a linear pattern, with slope remaining the same so long as the system as a whole neither gained nor lost U or Pb up to the time of mineralization. The resulting lead is finally concentrated, from basement or sediments, by some migrating aqueous solution, and transported to the site of deposition. This solution could bear some

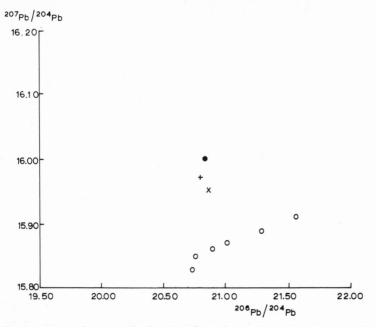

Fig. 15. Average isotope ratios in galena from three mines in the southeast Missouri district, U.S.A. (data presented in Table VIII; not corrected to absolute). For comparison, data from Doe and Delevaux (Table V, and Fig. 12) are also shown. + = average for Federal Mine; x = average for Leadwood-Desloge Mine. ● = average for Bonne Terre Mine.

resemblance to the "oil-field brine (connate water)" of Hall and Friedman (1963). To a first approximation it seems possible that the second meteoric component they envisaged for this district, with its much lower salinity would contribute a smaller proportion of the total lead. In any case this second solution would derive its lead from the same rocks as did the brine, and for present purposes would therefore be virtually indistinguishable."

Note that a short "residence time" in Paleozoic shales and/or carbonates would not affect the isotopic composition (unless these "intermediate hosts" were unusually radiogenic) so that the resulting mineralization would isotopically "appear" to have been derived directly from basement.

TABLE VIII

Average lead isotope ratios of galenas from three mines, Old Lead Belt, S.E. Missouri, U.S.A. (data from Brown, 1967; not corrected to absolute).

Mine	206/204	207/204	208/204	No. of samples
Bonne Terre	20.84	16.00	40.15	33
Leadwood-Desloge	20.87	15.95	40.05	36
Federal	20.79	15.97	40.09	56

The data from southeast Missouri (Fig. 12) may be averaged for each of the three mines sampled by Brown (1967; see Table VIII). When this is done (Fig. 15) the three averages are quite similar suggesting (1) no district-wide zonation; (2) no variation in isotopic composition with amount of lead (the relative magnitude of the Bonne Terre, Leadwood-Desloge, and Federal Mines is roughly 1 : 2 : 6); (3) there is as much variation in isotopic composition within a single mine as there is in the entire district. When Doe and Delevaux's three samples from No. 8 mine (Old Lead Belt) are averaged and compared with the average of the three from Bixby area (New Lead Belt), the two averages are also similar, again suggesting no district-wide zonation (Old Lead Belt and New Lead Belt are approximately 50 miles apart.) This is in contrast to Heyl et al. (1966) who documented a decrease in the $^{206}Pb/^{204}Pb$ ratios from northeast to southwest across the upper Mississippi Valley district (see Fig. 17), a distance of about 70 miles.

Although a district-wide zoning is not yet apparent in southeast Missouri, Brown (1967) has documented a vertical zoning in one mine, the Bonne Terre, relative to the Precambrian Basement knob underlying the deposit. This isotopic zoning, expressed as $\%^{204}Pb$ in Fig. 16A shows an upward and outward decrease in radiogenicity away from the basement high. In reference to this pattern, Brown concluded (1967, p. 422) that "the lead ... was derived mainly from the sedimentary rocks but the radiogenic contaminant came directly from the Precambrian. ... the sediments probably supplied two-thirds of the Lead Belt lead and the basement one third." It is interesting to note that, while the Pb-isotopic composition produces contours which essentially cross-cut stratigraphy, sulphur isotopes of the same galena samples (Fig. 16B), produce a contour pattern essentially parallel to stratigraphy, centering on the so-called "reef-rock" (Brown, 1967, and personal communication, 1968). Brown (p. 421) had also earlier noted the correlation between sulphur isotopes and stratigraphy but had not contoured the sulphur-isotope data as in Fig. 16B. The differing contour patterns of Pb and S suggest that the Pb source was principally in the basement, or at least came from that direction, whereas the sulphur probably had its source within the enclosing strata (as a result of bacterial reduction of sulphate?).

The decreasing radiogenicity of Pb-isotopes away from a basement high in southeast Missouri finds its counterpart, albeit in a different scale, in Upper Mississippi Valley (Fig. 17) where the $^{206}Pb/^{204}Pb$ ratios decrease southwestward away from Precambrian outcrops north-northeast of the Upper Mississippi Valley district. This basement-related zoning is in concert with the conclusions of Doe and Delevaux (1972) and Richards et al. (1972) regarding the similarity in "age" of source rocks in southeast Missouri and Upper Mississippi Valley respectively. A basement source for the lead in southeast Missouri is at variance with Doe and Delevaux (1972) who ruled out the basement in favour of the Lamotte sandstone as the more likely source. However, their conclusions may be premature on two counts: (1) the lead isotopic composition of the basement rocks was based on only two samples of felsite in spite of the fact that large volumes of granitic intrusions are also known to be present; (2) the three samples of Lamotte analyzed by

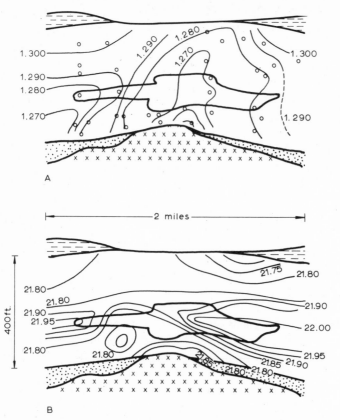

Fig. 16. A. Vertical section of Bonne Terre Mine, southeast Missouri, showing vertical zonation of % ^{204}Pb. Stippled pattern at top is Davis shale, dotted pattern is Lamotte sandstone, crosses are Precambrian basement, and area without pattern is Bonneterre Formation (dolomite). Area outlined in centre is "reef rock" (from Brown, 1967). B. Same section as above but showing contours of ^{32}S/^{34}S (from Brown, 1967, and personal communication, 1968).

the authors were of the upper orthoquartzite unit (which just as likely could have had as its provenance Precambrian rocks far to the north); the lower, locally derived, arkosic unit was not analyzed. With these reservations in mind and considering the isotopic zoning patterns shown in Figs. 16 and 17 together with the similarity in "age" of source rock for the two areas, a basement source cannot yet be ruled out. In fairness, however, the zoning of Pb-isotopes shown in Fig. 16A could just as well be the result of formation waters flowing updip through the Lamotte sandstone and thence into the Bonneterre dolomite. In view of the common substitution of Pb$^+$ for K$^+$ in potash feldspar, derivation of lead from a locally-derived sandstone as the immediate source (with basement as an ultimate source) is, nevertheless, an alternative possibility. It would help to explain the fact that the southeast Missouri district, unlike most other carbonate-hosted lead–zinc deposits, is lead, rather than zinc, dominated.

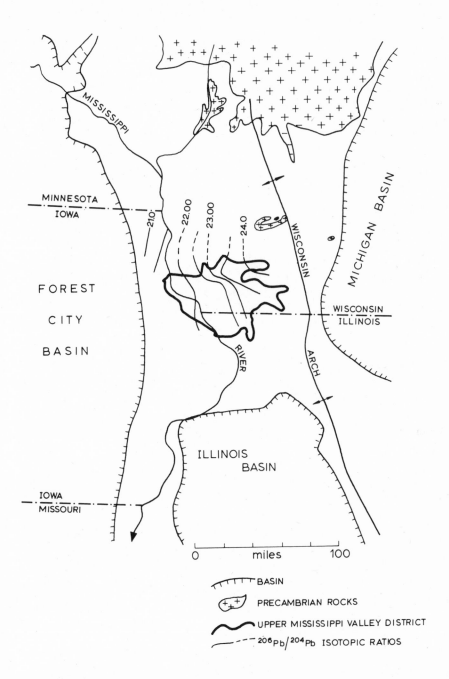

Fig. 17. Simplified geologic map of Upper Mississippi Valley district, U.S.A., showing zonation of $^{206}Pb/^{204}Pb$ ratios (from Heyl et al., 1973).

TABLE IX

Lead isotope ratios in galenas from selected localities in the Appalachian Valley zinc district, U.S.A. (data from Heyl et al., 1966, p. 939)

Locality	206/204	207/204	208/204
Birmingham Mines, Pa.	18.48	15.60	38.52
German Valley, N.J.	18.55	15.57	38.11
Bamford Mine, Pa.	18.61	15.64	38.42
Flat Gap Mine, Tenn.	19.04	15.65	39.17
Jackson Mine, Tenn.	19.33	15.70	38.98

No modern Pb-isotope values appear to be available for the largest Mississippi Valley district of all — the so-called Tri-State district of Missouri, Kansas, and Oklahoma. The only data are those of Russell and Farquhar (1960) shown in Fig. 14. The data are considered to be too scattered to reliably indicate whether they define an anomalous trend or not. They are, however, radiogenic with $^{206}Pb/^{204}Pb$ values greater than 20.00, similar in this respect to Upper Mississippi Valley and southeast Missouri. It is perhaps interesting to note that, according to the compilation of basement ages in central United States by Muehlburger et al. (1967), all three districts are floored by rocks of about 1350 m.y. age.

(b) *Appalachian Valley, U.S.A.* Data are available from only a few scattered localities in the Appalachian Valley region of eastern United States, the fourth district to contain deposits of the so-called "Mississippi Valley" type. The deposits represented by data in Fig. 18A are so scattered, both geographically and geologically that it is unlikely that the few data available can be considered representative of the Appalachian Valley region. Within this region, the Mascot—Jefferson City district in eastern Tennessee is the largest, yet lead is so rare that no modern isotopic analyses are available. Friedensville, in Pennsylvania, is also only represented by two nearby deposits reportedly "similar" to the Friedensville (Heyl et al., 1966, p. 954) and the Austinville—Ivanhoe district in Virginia is not represented at all. Nevertheless, in spite of these deficiencies, the data available are presented here in order to contrast their lead isotopic composition with those of the three major Mississippi Valley districts discussed previously. The main difference is, of

TABLE X

Lead isotope ratios in galenas from Pine Point, Canada (data from Cumming and Robertson, 1969)

No.	206/204	207/204	208/204
6009a	18.16	15.62	38.14
6009a	18.19	15.62	38.28
6009b	18.18	15.62	38.25
8	18.24	15.69	38.29
9	18.18	15.64	38.26
11	18.17	15.61	37.96

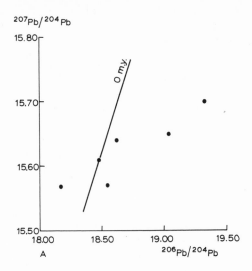

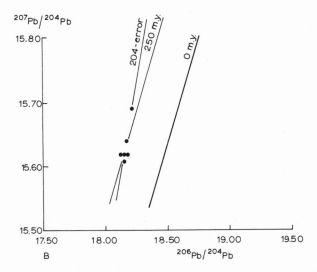

Fig. 18. A. Lead isotope ratios in galenas from Appalachian Valley district, U.S.A. (data presented in Table IX). B. Lead isotope ratios in galenas from Pine Point district, Canada (data presented in Table X).

course, as pointed out by Heyl et al. (1966, p. 942) and by Cannon et al. (1961), that leads from the three districts contained in the drainage basin of the Mississippi Valley are markedly more radiogenic than those in the Appalachian Valley.

In discussing the significance of the Appalachian Valley lead isotopes Heyl et al. (1966, p. 954, 956) conclude:

"Although the results obtained from the Appalachian Valley zinc districts make it possible to differentiate these leads from those of the Mississippi Valley districts, they provide us with little insight into the mechanism of ore generation and deposition ... galenas in carbonate rocks in the Appalachian Valley zinc districts contain leads essentially of ordinary isotopic composition, and similar isotopically to those found elsewhere in the Appalachian Mountains. Thus, even though geologic features of the deposits in the folded Appalachian Mountains are markedly similar in many respects to those in the Mississippi Valley, the lead isotopic composition suggests that at least the galenas have a different origin than the J-type leads of the Mississippi Valley, and that the two groups of deposits may be unrelated genetically."

It may be worthwhile to note here that the Appalachian Valley district, occurring as it does on the eastern seaboard of the United States and contained in Cambro–Ordovician carbonate rocks, could represent the miogeosynclinal equivalent of the coeval volcanogenic Cu–Pb–Zn massive sulphide deposits of Maine, Quebec, New Brunswick (Bathurst) and Newfoundland regions to the north. These latter deposits are considered to be the result of volcanic exhalations into Cambro–Ordovician seas. Lead in volcanogenic deposits is notably non-radiogenic and some of this "volcanic" lead, perhaps derived from similar island arc activity now buried beneath younger sediments east of Appalachian Valley district, may have been co-precipitated with the shelf carbonates now represented in Appalachian Valley district and later concentrated, with only minor contamination, into the deposits of this famous region. In this respect, these leads would have their origin in contemporanous volcanism and would not be "basement-derived" in the same manner as postulated for the three Mississippi Valley districts with their distinctly radiogenic leads.

(c) *Pine Point, Canada.* Hosted in Middle Devonian carbonates (Skall, 1975), the Pine Point ore-field is Canada's largest Mississippi Valley-type lead–zinc district. Unfortunately, lead isotopic data from Pine Point are fragmentary and may not be representative of the district as a whole. Of the six analyses shown in Fig. 18B three are from the same specimen and the remaining three are from weathered specimens collected from the surface long before mining began. Because the Pine Point ore-field is comprised of about 40 known ore bodies within an area of about 30 X 6 miles, of which only two or three deposits at the east side of the district originally outcropped, the data of Cumming and Robertson (1969), cannot yet be considered representative of the district as a whole. The analyses (Fig. 18B), though possessing considerable 204 error, are nevertheless markedly less radiogenic than those from any of the U.S. Mississippi Valley-type lead–zinc districts. Until more isotopic data are available from Pine Point to determine the extent to which available data are representative, interpretations are speculative. There is little or no contemporaneous known volcanic exhalative activity in the Pine Point area, either in the subsurface immediately to the west, or in the Cordillera far to the west, to account for the non-radiogenic character of the lead. The deposit is situated close to the Precambrian Shield and might be expected to yield radiogenic lead by the "basement-source" model described earlier for the major U.S. Mississippi Valley districts. More analyses of the Pine Point ore-leads might produce an anomalous lead line similar to Upper Mississippi Valley (Fig. 13) in which case a basement source could be a possibility. If, as postulated by Jackson and Beales (1967), the metals originated, as their immediate source, in an adjacent shale basin, the lead may have become sufficiently homogenized (either during shale sedimentation and/or during transport to its present site in the carbonates) so as to lose its anomalous character and approach single-stage conditions. The data, however, do not permit further speculation on the origin of Pine Point lead; suffice it to say that the data available are among the least radiogenic of the Mississippi Valley-type deposits.

(d) *Silesian district, Poland.* The Silesian—Cracovian lead—zinc deposits of the Middle Triassic in southern Poland are among the largest zinc reserves in Europe. The ore-bearing Triassic carbonates rest unconformably on Paleozoic basement rock peripheral to the Bohemian massif to the west. According to Ridge and Smolarska (1972) the ores are found in major part in solution cavities, localized largely by major to minor fractures.

Lead isotope ratios from seven deposits in the district (36 X 18 miles) are extremely uniform (Fig. 19A) and bespeak of a source of similar uniform composition. The area, being adjacent to the crystalline Bohemian massif, might be expected to have anomalous leads and produce a typical two-stage pattern similar to Upper Mississippi Valley. However, it does not; neither is there evidence of contemporaneous volcanic activity to explain the uniform composition. In these respects, the isotope pattern is similar to that of Pine Point except that there is no adjacent shale basin in the Triassic situation in which to conveniently mix and homogenize the lead before introducing it into the existing orebodies. Ridge and Smolarska (1972, p. 225) concluded that:

". . . the uniformity of lead isotope ratios in the Polish deposits eliminates as a reasonable explanation the concept of syngenetic precipitation of lead derived from the surrounding land surface. These ratios do not, however, distinguish between deposition from hydrothermal solutions entering solid rock and emplacing the ores by replacement and open-space filling and from volcanic-hydrothermal fluids reaching the sea floor."

Whatever the source of the lead, it must have been extremely uniform in isotopic composition, or else the leads became well mixed enroute to the site of deposition,

TABLE XI

Lead isotope ratios of galenas from the Silesia-Cracow district, Poland (data from Ridge and Smolarska, 1972)

Mine	206/204	207/204	208/204
Warynski	18.41	15.62	38.41
Nowy Dwor	18.44	15.62	38.43
Marchlewski	18.41	15.60	38.41
Trzebionka	18.43	15.61	38.43
Matylda	18.42	15.61	38.42
Olkusz	18.42	15.62	38.43
Boleslaw	18.42	15.61	38.42

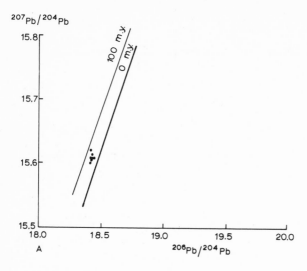

A

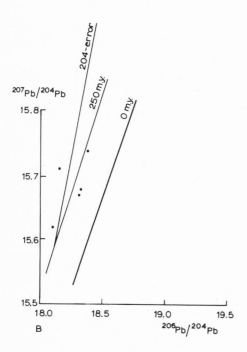

B

Fig. 19. A. Lead isotope ratios in galenas from mines in the Silesia-Cracow district, Poland (data presented in Table XI). B. Lead isotope ratios in galenas from deposits of the Central Plain, Ireland (data presented in Table XII).

TABLE XII

Lead isotope ratios of galenas from selected deposits, Central Plain, Ireland (data from Greig et al., 1971)

Deposit	206/204	207/204	208/204	Remarks
Riofinex	18.14	15.71	38.37	
Tynagh	18.10	15.62	38.09	average of 2
Silvermines (Mogul)	18.38	15.74	38.50	average of 5
Magcobar	18.32	15.68	38.38	
Gortdrum	18.29	15.67	38.41	

to produce such an extremely narrow range in isotopic composition over such a large area.

(e) *Central Plain, Ireland.* The relatively recent discovery of several significant lead—zinc deposits in the Carboniferous of Ireland has resulted in the delineation of a new lead—zinc province in northwestern Europe. Most of the deposits are spatially related to basement highs consisting for the most part of Middle Paleozoic sediments (Greig et al., 1971; Morrissey et al., 1971). Three centres of Lower Carboniferous volcanism are known in the Central Irish Plain and occur within 30 miles of all major sulphide deposits. Lead isotopic analyses from five of the major deposits are shown in Fig. 19B. Notwithstanding the large 204-error indicated, the data clearly indicate a population pattern similar to that for Pine Point (Fig. 18B) and Poland (Fig. 19A) and, in common with these, can be explained in terms of a simple single-stage evolution of lead as suggested by Greig et al. (1971) for Ireland.

Discussion. The foregoing illustrates that, as a group, carbonate-hosted, strata-bound lead—zinc deposits exhibit no uniformity in their lead isotopic patterns. Within the classical Mississippi Valley drainage basin, the three major districts (Figs. 12—14) show a tendency toward anomalous lead distribution (particularly the Upper Mississippi Valley district) although the Tri-State data (Fig. 14) are somewhat conjectural since they are analyses from instruments several generations removed from those of Doe and Delevaux's for southeast Missouri (Fig. 12). Nevertheless, if the three Mississippi Valley districts were plotted on a common diagram, they would define a distinct population with $^{206}Pb/^{204}Pb$ ratios greater than 20 and together define a composite anomalous lead isotope distribution for the central U.S.A. as a whole. In contrast, the fourth classical U.S.A. district, the Appalachian Valley, yields leads considerably less radiogenic than those in central U.S.A. Whether or not they define an anomalous lead line is conjectural due to insufficient data from the major deposits in the district but, nevertheless, appear to constitute a separate population.

The three remaining examples, Pine Point, Silesia, and Ireland are completely different from previous districts in their lead isotopic pattern. All are significantly less radiogenic than the classical central U.S.A. districts in spite of the fact that both the Silesian and Pine Point ore-fields exhibit many geological features of the so-called Mississippi Valley-type of deposit and both are spatially related to adjacent Precambrian highs. Taking into account the effect of apparently large 204-errors in the Pine Point and Irish analyses, they appear to define single-stage populations similar to the Silesian deposits. Assuming a single stage evolution, the three districts, Pine Point, Silesia and Ireland, give average model lead ages of 250, 50, and 250 m.y. respectively. Although these mineralization ages, which are considerably younger than the host rocks in each of the three districts, could perhaps be considered geologically acceptable in the sense that there is little or no evidence to suggest the deposits could *not* be as young as the model ages indicate, such a genetic model would require massive replacement of carbonate to produce the several concordant deposits known to occur in these districts. These model ages are not impossible of course, but are at variance with other lines of evidence which suggest mineralization in deposits of this type, if not syngenetic or diagenetic, occurs soon after deposition of host rocks (see Sangster, Vol. 6, Chapter 9).

In any case, regardless of the interpretation and genetic modelling one cares to derive from the isotopic patterns presented here, it is apparent that carbonate-hosted strata-bound lead—zinc deposits, spatially unrelated to igneous intrusions, can give rise to unpredictable lead isotopic compositions.

REFERENCES

Anger, G., Nielsen, H., Puchelt, H. and Rickie, W., 1966. Sulfur isotopes in the Rammelsberg ore deposit (Germany). *Econ. Geol.,* 61: 511—536.

Ault, W.V. and Kulp, J.L., 1959. The isotopic geochemistry of sulfur. *Geochim. Cosmochim. Acta.,* 16: 201—233.

Ault, W.V. and Kulp, J.L., 1960. Sulphur isotopes and ore deposits. *Econ. Geol.,* 55: 73—100.

Austin, S.R., 1970a. Some patterns of sulfur isotope distribution in uranium deposits. *Earth-Sci. Bull.,* 3: 5—22.

Austin, S.R., 1970b. Revised tabulation of sulfur isotope analyses. *U.S.A.E.C., Tech. Mem.,* AEC-RD-8.

Bachinski, D.J., 1969. Bond strength and sulfur isotopic fractionation in coexisting sulfides. *Econ. Geol.,* 64: 56—65.

Bastin, E.S., 1926. A hypothesis of bacterial influence in the genesis of certain sulphide deposits. *J. Geol.,* 34: 773—792.

Berner, R.A., 1971. *Principles of Chemical Sedimentology.* McGraw-Hill, N.Y., 240 pp.

Bodenlos, A.J., 1973. Sulphur. In: D.A. Brobst and W.P. Pratt (Editors), *United States Mineral Resources.* U.S., Geol. Surv., Prof. Pap., 820: 605—618.

Bowen, H.J.M., 1966. *Trace Elements in Biochemistry.* Academic Press, London, 239 pp.

Brown, J.S., 1967. Isotopic zoning of lead and sulfur in southeast Missouri. *Econ. Geol., Monogr.,* 3: 410—426.

Cannon, R.S., Pierce, A.P., Antweiler, T.C. and Buck, K.L., 1961. The data of lead isotope geology related to problems of ore genesis. *Econ. Geol.,* 56: 1—38.

Cheney, E.S. and Jensen, M.L., 1966. Stable isotopic geology of the Gas Hills, Wyoming, uranium district. *Econ. Geol.*, 61: 44–71.

Chukhrov, F.V., 1971. Isotopic composition of sulfur and genesis of copper ores of Zdfrezkazan and Udokan. *Int. Geol. Rev.*, 13: 1429–1434.

Claypool, G.E., Holser, W.T., Kaplan, I.R., Sakai, H., Zak, I., 1972. Sulfur and oxygen isotope geochemistry of evaporite sulphates. *Geol. Soc. Am., Abstr.*, 4 (7): 473 (abstr.).

Cumming, G.L. and Robertson, D.K., 1969. Isotopic composition of lead from the Pine Point deposit. *Econ. Geol.*, 64: 731–732.

Czamanske, G.K. and Rye, R.O., 1974. Experimentally determined sulfur isotope fractionations between sphalerite and galena in the temperature range 600°C to 275°C. *Econ. Geol.*, 69: 17–25.

Dobbin, C.E., 1968. Geology of natural gases rich in helium, nitrogen, carbon dioxide, and hydrogen sulfide. In: *Natural Gases of North America, Mem. Am. Assoc. Pet. Geol.*, 9: 1957–1969.

Doe, B.R., 1970. Lead isotopes. In: *Minerals, Rocks, and Inorganic Materials*, 3. Springer, Berlin–Heidelberg–New York, 137 pp.

Doe, B.R. and Delevaux, M.H., 1972. Source of lead in southeast Missouri galena ores. *Econ. Geol.*, 67: 409–425.

Doe, B.R. and Stacey, J.S., 1974. The application of lead isotopes to the problems of ore genesis and ore prospect evaluation: A review. *Econ. Geol.*, 69: 757–776.

Evans, T.L., Campbell, F.A. and Krouse, H.R., 1968. A reconnaissance study of some Western Canadian lead–zinc deposits. *Econ. Geol.*, 63: 349–359.

Farquhar, R.M., 1954. *The Lead Isotope Methods of Geological Age Determination*. Unpubl. Ph.D. Thesis, Univ. of Toronto, 81 pp.

Granger, H.C. and Warren, C.G., 1969. Unstable sulfur compounds and the origin of roll-type uranium deposits. *Econ. Geol.*, 64: 160–171.

Greig, J.A., Baadsgaard, H., Cumming, G.L., Folinsbee, R.E., Krouse, H.R., Ohmoto, H., Sasaki, A. and Smejkal, V., 1971. Lead and sulphur isotopes of the Irish base metal mines in Carboniferous carbonate host rocks. *Soc. Min. Geol. Japan, Spec. Issue*, 2: 84–92.

Grootenboer, J. and Schwarcz, H.P., 1969. Experimentally determined sulfur isotope fractionations between sulfide minerals. *Earth Planet. Sci. Lett.*, 7: 162–166.

Hall, W.E. and Friedman, I., 1963. Composition of fluid inclusions, Cave-in-Rock fluorite district, Illinois, and Upper Mississippi Valley zinc–lead districts. *Econ. Geol.*, 58: 886–911.

Harrison, A.G. and Thode, H.G., 1957. The kinetic isotope effect in the chemical reduction of sulfate. *Trans. Faraday Soc.*, 53: 1648–1651.

Heyl, A.V. and King, E.R., 1966. Aeromagnetic and tectonic analysis of the Upper Mississippi Valley zinc–lead district. *U.S. Geol. Surv. Bull.*, 1242A.

Heyl, A.V., Delevaux, M.H., Zartman, R.E. and Brock, M.R., 1966. Isotopic study of galenas from the Upper Mississippi Valley, the Illinois–Kentucky, and some Appalachian Valley mineral districts. *Econ. Geol.*, 61: 933–961.

Heyl, A.V., Broughton, W.A. and West, W.S., 1973. Guidebook to the Upper Mississippi Valley base-metal district. *Wisc. Geol. Nat. Hist. Surv. Circ.*, 16, 50 pp.

Heyl, A.V., Landis, G.P. and Zartman, R.E., 1974. Isotopic evidence for the origin of Mississippi Valley-type mineral deposits: A review. *Econ. Geol.*, 69: 992–1006.

Hoefs, J., 1973. *Stable Isotope Geochemistry*. Springer, New York, N.Y., 140 pp.

Hoefs, J., Nielsen, H. and Schidlowski, M., 1968. Sulfur isotope abundances in pyrite from the Witwatersrand conglomerates. *Econ. Geol.*, 63: 975–977.

Holser, W.T. and Kaplan, I.R., 1966. Isotope geochemistry of sedimentary sulphates. *Chem. Geol.*, 1: 93–135.

Jackson, S.A. and Beales, F.W., 1967. An aspect of sedimentary basin evolution. The concentration of Mississippi Valley-type ores during late stages of diagenesis. *Can. Pet. Geol., Bull.*, 15: 383–433.

Jensen, M.L., 1958. Sulfur isotopes and the origin of sandstone-type uranium deposits. *Econ. Geol.*, 53: 598–616.

Jensen, M.L., 1959. Sulphur isotopes and hydrothermal mineral deposits. *Econ. Geol.*, 54: 374–394.

Jensen, M.L. (Editor), 1962. Biogeochemistry of sulfur isotopes. *Symp., Yale Univ.*, 193 pp.

Jensen, M.L., 1967. Sulphur isotopes and mineral genesis. In: H.L. Barnes (Editor), *Geochemistry of Hydrothermal Ore Deposits*. Holt, Rinehart and Winston, N.Y., pp. 143–165.

Jensen, M.L. and Nakai, N., 1963. Sulphur isotope meteorite standards: results and recommendations. In: Biogeochemistry of Sulphur Isotopes. *Proc. Nat. Sci. Found. Symp., Yale Univ., April, 1962*, pp. 1–15.

Kajiwara, Y., 1971. Sulfur isotope study of the Kuroko-ores of the Shakanai No. 1 deposits, Akita Prefecture, Japan. *Geochem. J.,* 41: 157–181.

Kajiwara, Y. and Date, J., 1971. Sulfur isotope study of Kuroko-type and Kieslager-type stratabound massive sulfide deposits in Japan. *Geochem. J.,* 5: 133–150.

Kajiwara, Y. and Krouse, H.R., 1971. Sulfur isotope partitioning in metallic sulfide systems. *Can. J. Earth Sci.,* 8: 1397–1408.

Kajiwara, Y. and Sasaki, A., 1969. Experimental study of sulfur isotopic fractionation between coexistent sulfide minerals. *Earth Planet. Sci. Lett.,* 7: 271–277.

Kanasewich, E.R., 1962. Approximate age of tectonic activity using anomalous lead isotopes. *R. Astron. Soc. Geophys. J.,* 7: 158–168.

Kanasewich, E.R., 1968. The interpretation of lead isotopes and their geological significance. In: E.I. Hamilton and R.M. Farquhar (Editors), Radiometric dating for geologists. Interscience, London, pp. 147–223.

Kanasewich, E.R. and Farquhar, R.M., 1965. Lead isotope ratios from the Cobalt-Noranda area, Canada. *Can. J. Earth Sci.,* 2: 361–384.

Kanehira, K. and Tatsumi, T., 1970. Bedded cupriferous iron sulphide deposits in Japan, a review. In: *Volcanism and Ore Genesis*. Univ. of Tokyo Press, Tokyo, pp. 51–76.

Kaplan, I.R., Emery, K.D. and Rittenberg, S.C., 1963. The distribution and isotopic abundance of sulphur in recent marine sediments off southern California. *Geochim. Cosmochim. Acta,* 27: 297–331.

Krogh, T.E. and Davis, G.L., 1971. Zircon U–Pb ages of Archean metavolcanic rocks in the Canadian Shield. *Carnegie Inst. Wash., Geophys. Lab., Year Book, 70,* 1970–71: 241–242.

Lusk, J., 1969. Base metal zoning in the Heath Steele B-1 Orebody, New Brunswick, Canada. *Econ. Geol.,* 64: 509–518.

Lusk, J., 1972. Examination of volcanic-exhalative and biogenic origin for sulfur in the stratiform massive sulfide deposits of New Brunswick. *Econ. Geol.,* 67: 169–183.

Lusk, J. and Crocket, J.H., 1969. Sulfur isotope fractionation in coexisting sulfides from the Heath Steele B-1 orebody, New Brunswick, Canada. *Econ. Geol.,* 64: 147–155.

Macnamara, J. and Thode, H.G., 1950. Comparison of the isotopic constitution of terrestrial and meteoritic sulphur. *Phys. Rev.,* 78: 307.

McAllister, A.L. and Lamarche, R.Y., 1972. Mineral deposits of southern Quebec and New Brunswick. *Int. Geol. Congr., 24th, Montreal, Guideb. Excursion* A58.

Morrissey, C.J., Davis, G.R. and Steed, B.M., 1971. Mineralization in the Lower Carboniferous of central Ireland. *Trans. Inst. Min. Metall., Sect. B,* 80: 174–185.

Morton, R.D., Goble, R.J. and Fritz, P., 1974. The mineralogy, sulfur-isotope composition and origin of some copper deposits in the Belt Supergroup, southwest Alberta, Canada. *Miner. Deposita,* 9: 223–241.

Muehlberger, W.R., Denison, R.E. and Lidiak, E.G., 1967. Basement rocks in continental interior of United States. *Bull. Am. Assoc. Pet. Geol.,* 51: 2351–2380.

Myers, J.C., 1968. Sulfur – its occurrence, production, and economics. In: Natural Gases of North America. *Mem. Am. Assoc. Pet. Geol.,* 9: 1948–1956.

Nakai, N. and Jensen, M.L., 1967. Sources of atmospheric sulfur compounds. *Geochem. J.,* 1: 199–210.

Nielson, H., 1966. Sulphur isotopes in the marine cycle and the $\delta^{34}S$ of earlier oceans (in German). *Geol. Rev.,* 55: 160–172.

Ohmoto, H., 1970. Influence of pH and fO_2 of hydrothermal fluids on the isotopic composition of sulfur species. *Geol. Soc. Am., Abs. Progr.,* 2: 640.

Ohmoto, H., 1972. Systematics of sulfur and carbon isotopes in hydrothermal ore deposits. *Econ. Geol.*, 67: 551–578.

Ostic, R.G., Russell, R.D. and Stanton, R.L., 1967. Additional measurements of the isotopic composition of lead from stratiform deposits. *Can. J. Earth Sci.*, 4: 245–270.

Phillips, J.S., 1960. *Sandstone-Type Copper Deposits of the Western United States*. Ph.D. Thesis, Harvard Univ.

Rafter, T.A., Kaplan, I.R. and Hulston, J.R., 1958a. Sulphur isotope variation in nature. 6. Sulphur isotopic measurements on the discharge from fumaroles on White Island. *N. Z. J. Sci.*, 1: 154–171.

Rafter, T.A., Kaplan, I.R. and Hulston, J.R., 1958b. Sulphur isotopic variations in nature. 7. Sulphur isotope measurements on sulphur and sulphates in New Zealand geothermal and volcanic areas. *N. Z. J. Sci.*, 3: 209–218.

Rees, C.E., 1973. A steady-state model for sulphur isotope fractionation in bacterial reduction processes. *Geochim. Cosmochim. Acta*, 37: 1141–1162.

Richards, J.R., Yonk, A.K. and Keighin, C.W., 1972. A re-assessment of the Upper Mississippi Valley lead isotope data. *Miner. Deposit*, 7: 285–291.

Ridge, J.D. and Smolarska, I., 1972. Factors bearing on the genesis of the Silesian–Cracovian lead–zinc deposits in southern Poland. *Int. Geol. Congr., 24th, Sect. 6*, pp. 216–229.

Roscoe, S.M., 1965. Geochemical and isotopic studies, Noranda and Matagami areas. *Can. Inst. Min. Metall. Trans.*, 68: 279–285.

Roscoe, S.M., 1968. Huronian rocks and uraniferous conglomerates in the Canadian Shield. *Geol. Surv. Can., Pap.*, 68–40.

Russell, R.D. and Farquhar, R.M., 1960. *Lead Isotopes in Geology*. Interscience, New York, N.Y.

Russell, R.D., Ulrych, T.J. and Kollar, F., 1961. Anomalous leads from Broken Hill, Australia. *J. Geophys. Res.*, 66: 1495–1498.

Rye, R.O. and Czamanske, G.K., 1969. Experimental determination of sphalerite-galena sulfur isotope fractionation and application to the ores at Providencia, Mexico. *Geol. Soc. Am. Ann. Meet., Milwaukee*, 195–196 (abstr.).

Rye, R.O. and Ohmoto, H., 1974. Sulfur and carbon isotopes and ore genesis: a review. *Econ. Geol.*, 69: 826–842.

Sakai, H., 1968. Isotopic properties of sulfur compounds in hydrothermal processes. *Geochem. J.*, 2: 29–49.

Sakai, H. and Matsubaya, O., 1974. Isotopic geochemistry of the thermal waters of Japan and its bearing on the Kuroko ore solutions. *Econ. Geol.*, 69: 974–991.

Sakai, H. and Nagasawa, H., 1958. Fractionation of sulfur isotopes in volcanic gases. *Geochim. Cosmochim. Acta*, 15: 32–39.

Sangster, D.F., 1968. Relative sulfur isotope abundances of ancient seas and strata-bound sulphide deposits. *Geol. Assoc. Can., Proc.*, 19: 79–91.

Sangster, D.F., 1971. Sulphur isotopes, stratabound sulphide deposits, and ancient seas. *Soc. Min. Geol. Jap., Spec. Issue*, 3: 295–299.

Sangster, D.F., 1972a. Isotopic studies of ore-leads in the Hanson Lake-Flin Flon-Snow Lake mineral belt, Saskatchewan and Manitoba. *Can. J. Earth Sci.*, 9: 500–513.

Sangster, D.F., 1972b. Precambrian volcanogenic massive sulphide deposits in Canada: a review. *Geol. Surv. Can., Pap.*, 72–22: 44 pp.

Sasaki, A., 1970. Seawater sulfate as a possible determinant for sulfur isotopic compositions of some strata-bound sulfide ores. *Geochem. J.*, 4: 41–51.

Sasaki, A., 1974. Isotopic data of Kuroko deposits. In: K. Ishihara (Editor), *Geology of Kuroko Deposits Soc. Min. Geol. Jap., Spec. Issue*, 6: 389–397.

Sasaki, A. and Kajiwara, Y., 1971. Evidence of isotopic exchange between seawater sulfate and some syngenetic sulfide ores. *Soc. Min. Geol. Jap., Spec. Issue*, 3: 289–294.

Sasaki, A. and Krouse, H.R., 1969. Sulphur isotopes and the Pine Point lead-zinc mineralization. *Econ. Geol.*, 64: 718–730.

Sato, K. and Sasaki, A., 1973. Lead isotopes of the black ore (Kuroko) deposits from Japan. *Econ. Geol.*, 61: 547–552.

Sato, T., 1972. Behaviours of ore-forming solutions in seawater. *Min. Geol.*, 22: 31–42.

Sato, T., 1974. Distribution and geological setting of the Kuroko deposits. In: K. Ishihara (Editor), *Geology of Kuroko Deposits. Soc. Min. Geol. Jap., Spec. Issue*, 6: pp. 1–18.

Schiller, W.R., Gehlen, K.V. and Nelsen, H., 1970. Hydrothermal exchange and fractionation of sulfur isotopes in synthesized ZnS and PbS. *Econ. Geol.*, 65: 350–351.

Schwarcz, H.P. and Burnie, S.W., 1973. Influence of sedimentary environments on sulfur isotope ratios in clastic rocks: a review. *Miner. Deposita*, 8: 264–277.

Skall, H., 1975. The paleoenvironment of the Pine Point lead–zinc district. *Econ. Geol.*, 70: 22–47.

Solomon, P.J., 1965. Investigations into sulphide mineralization at Mt. Isa, Queensland. *Econ. Geol.*, 60: 737–765.

Stacey, J.S., Delevaux, M.H. and Ulrych, T.J., 1969. Some triple-filament lead isotope ratio measurements and an absolute growth curve for single-stage leads. *Earth Planet. Sci. Lett.*, 6: 15–25.

Stanton, R.L., 1960. The application of sulphur isotope studies in ore genesis theory, a suggested model. *N. Z. J. Geol. Geophys.*, 3: 375–389.

Stanton, R.L., 1972. *Ore Petrology*. McGraw-Hill, New York, N.Y., 713 pp.

Stanton, R.L. and Rafter, T.A., 1966. The isotopic constitution of sulphur in some stratiform lead–zinc sulphide ores. *Miner. Deposita*, 2: 16–29.

Temple, K.L., 1964. Syngenesis of sulfide ores: an evaluation of biochemical aspects. *Econ. Geol.*, 59: 1473–1491.

Temple, K.L. and Le Roux, N.W., 1964. Syngenesis of sulfide ores: Sulphate-reducing bacteria and copper toxicity. *Econ. Geol.*, 59: 271–278.

Thode, H.G. and Monster, J., 1965. Sulphur isotope geochemistry of petroleum evaporite and ancient seas. In: *Fluids in Subsurface Environments. Mem. Am. Assoc. Pet. Geol.*, 4: 367–377.

Thode, H.G. and Monster, J., 1970. Sulfur isotope abundances and genetic relations of oil accumulations in Middle East basin. *Bull. Am. Assoc. Pet. Geol.*, 54: 627–637.

Thode, H.G. and Rees, C.E., 1970. Sulphur isotope geochemistry and Middle East oil studies. *Endeavour*, 29: 24–28.

Thode, H.G., Kleerekoper, H. and McElcheran, D.E., 1951. Isotopic fractionation of sulphate. *Research (Lond.)*, 4: 581–582.

Thode, H.G., Monster, J. and Dunford, H.B., 1958. Sulphur isotope abundances in petroleum and associated materials. *Bull. Am. Assoc. Pet. Geol.*, 42: 2619–2641.

Thode, H.G., Wanless, R.K. and Wallouch, H.R., 1954. The origin of native sulphur deposits from isotope fractionation studies. *Geochim. Cosmochim. Acta*, 5: 286–298.

Tokunaga, M. and Honma, H., 1974. Fluid inclusions in the minerals from some Kuroko deposits. In: K. Ishihara (Editor), *Geology of Kuroko Deposits. Soc. Min. Geol. Jap., Spec. Issue*, 6: 385–388.

Tudge, A.P. and Thode, H.G., 1950. Thermodynamic properties of isotopic compounds of sulphur. *Can. J. Res.*, 28: 567–579.

Vinogradov, W.I. and Schanin, L.L., 1969. Variations in sulphur isotope composition in old oceans. *J. Appl. Geol.*, 15: 33–36 (in German).

Warren, C.G., 1971. A method for discriminating between biogenic and chemical origins of the ore-stage pyrite in a roll-type uranium deposit. *Econ. Geol.*, 66: 919–928.

Warren, C.G., 1972. Sulfur isotopes as a clue to the genetic geochemistry of a roll-type uranium deposit. *Econ. Geol.*, 67: 759–767.

Woodward, L.A., Kaufman, W.H., Schumacher, O.L. and Talbott, L.W., 1974. Strata-bound copper deposits in Triassic sandstones of Sierra Nacimiento, New Mexico. *Econ. Geol.*, 69: 108–120.

York, D., 1969. Least-squares fittings of a straight line with correlated errors. *Earth Planet. Sci. Lett.*, 5: 320–324.

Chapter 9

URANIUM-, THORIUM- AND LEAD-ISOTOPE STUDIES OF STRATA-BOUND ORES

V. KÖPPEL and R. SAAGER

INTRODUCTION

The radioactive decay of uranium and thorium to lead as well as the isotopic variations of lead in uranium-free ores are of great importance for an understanding of the genesis of ore deposits, as:

(1) Not only may the time of ore deposition be determined, but also the subsequent history of the deposit may be elucidated.

(2) The variations of the lead-isotope abundances are indicative of the source of the lead in ore deposits.

(3) The isotopic variations of crustal lead in ore deposits and those of mantle-derived lead in oceanic basalts yield important information on the geochemistry of uranium, thorium and lead in the crust and in the mantle.

(4) Systematic variations of lead-isotope ratios around ore bodies may be used as a tool for prospecting (see Cannon Jr. et al., 1971; Doe and Stacey, 1974).

AGE DETERMINATIONS: RADIOACTIVE DECAY

The radioactive parent isotopes and the stable end products are listed in Table I, as well as the respective decay constants.

From Table I it may be seen that two methods exist to determine an age: (1) from the amounts of ^{4}He and of uranium and thorium, or (2) from the atomic ratios ^{206}Pb/^{238}U, ^{207}Pb/^{235}U and ^{208}Pb/^{232}Th.

The U—Th—He method has never been widely applied to the problem of age determination of ores. Fanale (1964) reported He-α ages of magnetites with trace amounts of uranium from ore deposits.

The U—Pb and Th—Pb method

The equation describing the radioactive decay is given by:

$$dN/dt = -\lambda N \tag{1}$$

TABLE I

The uranium—lead and thorium—lead radioactive systems

Radioactive parent isotope	^{238}U	^{235}U	^{232}Th
		Intermediate daughter products	
Decay constants	$\lambda_8 \begin{cases} 1.537 \cdot 10^{-10}/y *1 \\ 1.5513 \cdot 10^{-10}/y*2 \end{cases}$	$\lambda_5 \begin{cases} 9.722 \cdot 10^{-10}/y*1 \\ 9.849 \cdot 10^{-10}/y*2 \end{cases}$	$\lambda_2 \begin{cases} 4.881 \cdot 10^{-11}/y*1 \\ 4.987 \cdot 10^{-11}/y*3 \end{cases}$
	$8\,^4He$	$7\,^4He$	$6\,^4He$
Stable end products	^{206}Pb	^{207}Pb	^{208}Pb

[1] Stieff et al. (1959).
[2] Jaffey et al. (1971).
[3] Kulp et al. (1954).

where λ = decay constant, N = number of radioactive parent atoms.

Integrating eq. 1 yields:

$$N = N_0 e^{-\lambda t} \tag{2}$$

where N = number of remaining radioactive parent isotopes, N_0 = original number of radioactive parent isotopes.

The condition of a closed system is:

$$D = N_0 - N \tag{3}$$

where D = number of radiogenic daughter isotopes.

Combining eqs. 2 and 3 one obtains:

$$Ne^{\lambda t} = N_0 = D + N \tag{4}$$

$$D = N(e^{\lambda t} - 1) \tag{5}$$

$$D/N = e^{\lambda t} - 1 \tag{6}$$

$$t = 1/\lambda \ln(D/N + 1) \tag{7}$$

From the. ratio of parent to daughter isotopes one may calculate a number with the dimension of t. This number corresponds to a meaningful age only if certain conditions are fulfilled:

(1) The element containing the radiogenic daughter isotope should not be incorporated into the mineral at the time of crystallization. This condition is seldom fulfilled. Most of the uranium minerals contain a measurable quantity of a common-lead compo-

nent (see section on common lead). In the case of the $^{238}U-^{206}Pb$ system eq. 5 has to be modified:

$$^{206}Pb_{total} = {}^{206}Pb_f + {}^{238}U(e^{\lambda_8 t} - 1) \tag{8}$$

where $^{206}Pb_f$ is the number of ^{206}Pb isotopes incorporated at the time of the mineral formation. One therefore has to apply a so-called "common-lead correction" before calculating an age. There fortunately exists a fourth lead isotope, the ^{204}Pb, of which no radioactive parent is known. This isotope is therefore the basis of the common-lead correction. If there is no thorium present in the sample, then the ^{208}Pb may also serve as a basis for the correction. The isotopic composition of the common lead has to be obtained from uranium- and thorium-free phases that were formed by the same process and at the same time as the uranium- and thorium-bearing phase.

(2) Eq. 3 implies a closed-system behaviour, i.e. during the lifetime of the mineral neither lead nor uranium or thorium have left or entered the mineral. In the case of the Th–Pb system, one may test whether this assumption holds or not by analyzing a number of samples, preferably from one hand specimen, and plotting the data on a $^{208}Pb/^{204}Pb$ versus $^{232}Th/^{204}Pb$ curve, analogous to the diagram used to plot Rb–Sr data known as the Nicolaysen diagram (Nicolaysen, 1961). If the data points lie on a straight line then one is justified in assuming a closed-system behaviour since the time of crystallization or since the time of the last lead-isotope homogeneization. The great advantage of the U–Pb system is that there are two coupled systems with differing decay constants which thereby offer a simple method for testing the assumption of a closed U–Pb system in one

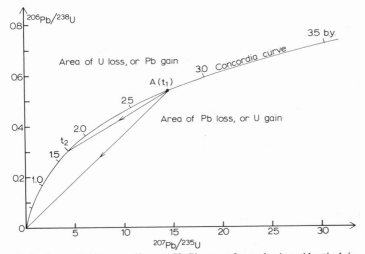

Fig. 1. Concordia diagram. The two U–Pb ages of sample A are identical, i.e. concordant, and the data point plots on the concordia curve. A recent lead loss, or uranium gain, shifts the data point along AO towards O, and a lead loss, or uranium gain, at time t_1 shifts it towards t_2 along $t_1 t_2$.

analysis. Wetherill (1956b) has introduced the so-called "concordia diagram" (Fig. 1). The atomic ratios of $^{206}Pb/^{238}U$ are plotted against $^{207}Pb/^{235}U$. The concordia curve is the locus of points the isotope ratios of which yield identical ages (eq. 7).

Provided the appropriate common-lead correction has been applied, the data points of samples that have acted as closed systems for uranium and lead since the time of their formation (t_1) will lie on the concordia curve. A recent lead loss, for example, with no fractionation of the lead isotopes will shift the data point A (Fig. 1) towards the origin O along the line AO. The distance from A is directly related to the amount of lead lost. The two ages calculated from eq. 9 and 10 will no longer be identical, $t_{235} > t_{238}$. The ages are discordant. Wetherill (1956b) has shown that an episodic lead loss in the past will shift the data point A towards t_2 along the line $t_1 t_2$ (Fig. 1). "Episodic" means that the disturbance of the U—Pb system lasted a short time compared to the lifetime of the mineral. (See also Gale and Musset, 1973, for episodic U—Pb models.)

A lead gain, or uranium loss, would move the data points into the field above the concordia curve. Such cases are rarely observed. Data points below the concordia curve are generally interpreted as indicating lead loss(es) and less frequently as pointing to a uranium gain.

The two equations:

$$^{206}Pb = {}^{238}U(e^{\lambda_8 t} - 1)$$
(9)

and

$$^{207}Pb = {}^{235}U(e^{\lambda_5 t} - 1)$$
(10)

can be combined to:

$$\frac{^{207}Pb}{^{206}Pb} = \frac{^{235}U}{^{238}U} \frac{(e^{\lambda_5 t} - 1)}{(e^{\lambda_8 t} - 1)}$$
(11)

where $^{235}U/^{238}U = 1/137.88$, the present-day ratio.

The age calculated from eq. 11 is referred to as the lead—lead age. It is calculated from the lead-isotopic ratios alone after having applied the common-lead correction.

The lead—lead age may also be derived graphically from the two U—Pb ages on the concordia diagram. A straight line drawn through the origin of the concordia curve and the data point intercepts the concordia curve at the time mark which is identical to the lead—lead age obtained from eq. 11.

Experience has shown that if a suite of uranium-bearing minerals from a rock specimen is analyzed, one often observes a linear array of data points on the concordia diagram (Fig. 2). Such a linear array may be the product of:

(1) Different degrees of episodic lead loss, or uranium gain, at t_2 of minerals formed at t_1. The event at t_2 may correspond either to a metamorphic or a tectonic event.

(2) A composite sample consisting of minerals formed at t_1 as well as minerals formed at t_2.

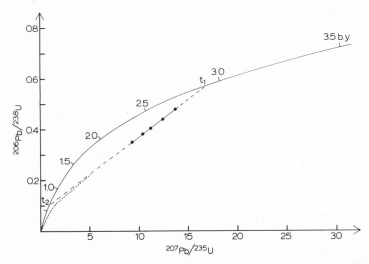

Fig. 2. A suite of cogenetic uranium-bearing minerals formed at time t_1 exhibit discordant age patterns ($t_{235} > t_{238}$). The broken line is the discordia corresponding to the cases 1 and 2, the dotted curve is the discordia corresponding to case 3 discussed in the text.

(3) Continuous loss of lead (Nicolaysen, 1957, Tilton, 1960, Wasserburg, 1963) or of intermediate daughter products (Wetherill, 1956b). The data points lie on a curve the upper part of which resembles a straight line. Only when 70% or more of the lead is continuously lost will the data points lie on the curved part of the discordia. The lower intercept of the extrapolated straight part of the discordia with the concordia has no geological significance, the upper intercept closely approaches t_1 (Fig. 2).

(4) Complex histories with more than one episodic lead loss, or episodic lead loss superimposed upon continuous lead loss. Such a behaviour of the U—Pb system can also lead to regular and even quasi-linear arrays of data points. The upper intercept with the concordia curve of such extrapolated quasi-linear arrays yields a minimum age, whereas the lower extrapolated intercept has no geological significance (Allègre et al., 1974).

One normally interprets the data according to the simplest model compatible with all other observations including careful mineralogical observations of the material analyzed.

Application. There are several factors that restrict the application of the U—Pb and Th—Pb methods of determining ages of ores. The problem that many uranium- and thorium-bearing minerals are susceptible to lead loss may be overcome by treating the data as outlined in the previous paragraphs. However, the problem remains when one deals with young samples of approximately 200 m.y. of age or less. In this age range the curvature of the concordia curve is extremely small and the extrapolated intercepts have normally large uncertainties.

Most of the uranium minerals are easily dissolved and redeposited, thus requiring a strict mineralogical and mineragraphic control of the sample.

Dooley Jr. et al. (1964, 1966) and Rosholt et al. (1964) reported significant fractionations between uranium and its intermediate decay products in roll features of sandstone-type uranium deposits in the Shirley Basin, Wyoming. Their investigations allow of drawing conclusions regarding the migration of uranium in permeable sandstones.

One also has to bear in mind that high-grade uranium ores of great age may have acted as a natural reactor, thereby increasing the $^{238}U/^{235}U$ ratio. Such a natural fossil reactor has recently been discovered in a sedimentary uranium deposit at Oklo, Gabon (see CEA, 1974).

In spite of all these restrictions, successful attempts have been made to establish the age of sedimentary uranium deposits. Dooley Jr. et al. (1974) obtained Early Miocene ages of 22 ± 3 m.y. on massive pitchblende from Gas Hills and Shirley Basins, Wyoming. The samples were in radioactive equilibrium. Less reliable, discordant U–Pb ages were obtained on ore-grade samples. The authors attribute the discordancy to lack of radioactive equilibrium and migration of elements in recent times.

Extensive age work has been carried out on uranium-bearing minerals from the Witwatersrand, South Africa (see section on case histories).

Common lead

The term "common lead" refers to lead in minerals, or whole-rock samples, with low U/Pb ratios, the $^{206}Pb/^{204}Pb$ ratio usually being less then 20. It should, however, be noted that even uranium-free minerals such as galenas may have exceptionally high $^{206}Pb/^{204}Pb$ ratios exceeding 100. Such ratios have been observed in occurrences in the neighbourhood of uranium deposits, e.g., Blind River (Mair et al., 1960), Witwatersrand (Burger et al., 1962), and Beaverlodge, Saskatchewan (Köppel, 1968).

The importance of the isotopic variations of common lead is based on the fact that the isotopic composition is frozen at the time when the lead was isolated from uranium and that therefore the composition is a function of the U/Pb ratios of the sources and the respective resident times of the lead in the sources. As different sources, such as the mantle, the lower or the upper crust have distinct U/Pb ratios, the isotopic composition of lead reflects its sources and thereby helps to solve genetic problems of ore formation.

Common lead has lost its importance as a tool for dating minerals, because it has been recognized that the evolution of the lead isotopes is hardly ever a simple one. Nevertheless, the discovery of systematic variations of the so-called "lead model ages" with ages obtained by other methods helped to shed light on processes affecting crust and mantle. Nevertheless, using models of evolution that account for such complications, it still is possible to establish an age for certain types of ore deposits (see below).

The earlier discussed U–Th–Pb method is based upon the disintegration of the radioactive parents from the time t at which a uranium- and thorium-bearing phase crystallizes

TABLE II

Symbols and parameters

	At time T	At time t	At present
$^{206}Pb/^{204}Pb$	a_0	x	a
$^{207}Pb/^{204}Pb$	b_0	y	b
$^{208}Pb/^{204}Pb$	c_0	z	c
$^{238}U/^{204}Pb$	$\mu e^{\lambda_8 t}$	$\mu e^{\lambda_8 t}$	μ
$^{235}U^{204}Pb$	$V e^{\lambda_5 T}$	$V e^{\lambda_5 t}$	V (137.88 $V = \mu$)
$^{232}Th/^{204}Pb$	$W e^{\lambda_2 T}$	$W e^{\lambda_2 t}$	W
$^{232}Th/^{238}U$			k

$a_0 = 9.307 \pm 0.003$*
$b_0 = 10.294 \pm 0.003$* $T = 4.57$ b.y. ** (age of the earth)
$c_0 = 29.476 \pm 0.009$*

* Tatsumoto et al. (1973).
** Tilton (1973).

up to the present. In contrast to this, the common-lead method is based upon the evolution of the lead since the time T when the earth started to act as a closed system for uranium, thorium and lead (the age of the earth) to the time t when the lead was isolated from uranium and thorium. The equations describing the growth of the lead-isotope ratios in a single and closed uranium–thorium–lead reservoir are (see Table II for parameters and symbols):

$$^{206}Pb/^{204}Pb = (^{206}Pb/^{204}Pb)_0 + (^{238}U/^{204}Pb)(e^{\lambda_8 T} - e^{\lambda_8 t}) \tag{12}$$

$$^{207}Pb/^{204}Pb = (^{207}Pb/^{204}Pb)_0 + (^{235}U/^{204}Pb)(e^{\lambda_5 T} - e^{\lambda_5 t}) \tag{13}$$

$$^{208}Pb/^{204}Pb = (^{208}Pb/^{204}Pb)_0 + (^{232}Th/^{204}Pb)(e^{\lambda_2 T} - e^{\lambda_2 t}) \tag{14}$$

The subscript "o" refers to the isotope ratios of the primordial lead as observed in the troilite phase of the Canyon Diablo meteorite. This lead is the least radiogenic so far observed. A very similar lead has been found by Tilton (1973) in the chondrite Mezö-Madaras after having applied a lead correction for the in-situ decay of uranium. The assumption that the solar system, including the meteorites, was created at the same time from the same material implies that the lead-isotopic composition of the primordial lead of the earth is equal to that of the most primitive lead observed in meteorites. It should be noted that in the $^{207}Pb/^{204}Pb$ versus $^{206}Pb/^{204}Pb$ and in the $^{208}Pb/^{204}Pb$ versus $^{206}Pb/^{204}Pb$ diagrams, all data points of terrestrial common lead form an array the lower extrapolation of which comprises the most primitive meteoritic lead. The value of 4.57 b.y. for the age of the earth was obtained by Tilton (1973) and by Tatsumoto et al.

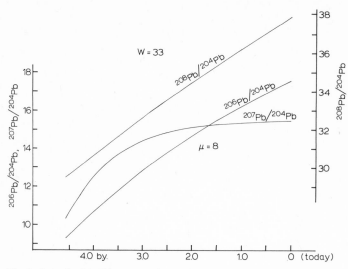

Fig. 3. Growth of lead-isotope ratios with time.

(1973) using the lead-isotope composition of meteorites and the new decay constants (Table I).

Eqs. 12—14 yield single-stage growth curves for the lead-isotope ratios. "Single stage" refers to the assumption that the μ value is not changed in any way except for the decay of uranium. Fig. 3 shows the growth of the ratios with time and Figs. 4a and 4b the conventional plots of $^{207}Pb/^{204}Pb$ versus $^{206}Pb/^{204}Pb$ and $^{208}Pb/^{204}Pb$ versus $^{206}Pb/^{204}Pb$.

In Fig. 4a two single-stage growth curves are plotted for different μ values and also the single-stage isochrons which are the loci of ratios that evolved between T and t in closed U—Pb reservoirs. The equation for the single-stage isochron is:

$$\frac{(^{207}Pb/^{204}Pb) - b_0}{(^{206}Pb/^{204}Pb) - a_0} = \frac{^{235}U(e^{\lambda_5 T} - e^{\lambda_5 t})}{^{238}U(e^{\lambda_8 T} - e^{\lambda_8 t})} \tag{15}$$

This equation represents a straight line:

$$\frac{^{207}Pb}{^{204}Pb} = b_0 + \frac{1}{137.88}\left(\frac{e^{\lambda_5 T} - e^{\lambda_5 t}}{e^{\lambda_8 T} - e^{\lambda_8 t}}\right)\left(\frac{^{206}Pb}{^{204}Pb}\right) - a_0 \tag{16}$$

Prior to the advent of the concept of plate tectonics, one of the major questions of lead-isotope geochemistry was whether single-stage leads exist on earth and if so, how they can be recognized. From Fig. 4a it is obvious that a single-stage lead cannot lie to the right of the 0 b.y. isochron, it also should plot in an area with a geochemically reasonable μ value (approximately 7—13).

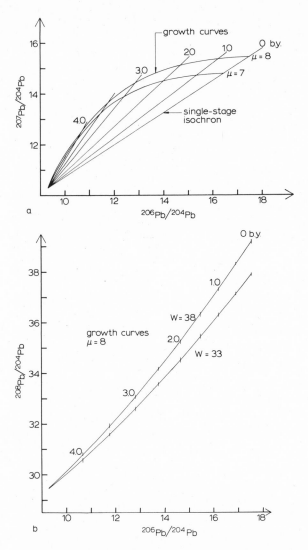

Fig. 4a,b. Conventional lead—lead diagrams showing single-stage growth curves and single-stage iso-chrons (Fig. 4a).

Houtermans (1946) envisaged an earth in which reservoirs with different U/Pb ratios (or μ values) persisted during the lifetime of the earth; single-stage leads will, therefore, lie on a family of growth curves, as shown for example in Fig. 4a, and leads extracted at the same time from different reservoirs will lie on primary isochrons. (For a full discussion of the so-called "Holmes—Houtermans method" see R.D. Russell and Farquhar, 1960.)

The lead growth curve of conformable ores[1]

Stanton and R.D. Russell (1959) observed that a number of conformable lead-bearing sulfide deposits, which were thought to have originated from island-arc as well as from submarine volcanism, exhibited each a uniform lead-isotopic composition that plots in the lead–lead diagrams on a single growth curve. Such leads were therefore taken as single-stage leads, and it was assumed that they could have derived only from the mantle as crustal rocks have widely differing U/Pb ratios.

In many cases the conformable ore leads contrasted with leads from epigenetic vein-type deposits which often exhibit a heterogeneous isotopic composition due to the admixture of crustal lead that developed in rocks with variable U/Pb ratios. During metamorphism the lead-isotopic ratios may be homogenized and further radiogenic lead exsolved from uranium-bearing phases may be added to minerals, leading to abnormally high $^{206}Pb/^{204}Pb$ ratios (Cumming and Gudjurgis, 1973).

The concept of single-stage leads in conformable ores associated with volcanism was supported by the observation that galenas from intramagmatic deposits of the noritic Sudbury complex, which probably is of mantle origin, also contained a lead that fitted the same growth curve (R.D. Russell and Farquhar, 1960).

The concept of a single-stage growth curve was subsequently maintained with only slight modifications. Some deposits were added to the list of those containing apparently a single-stage lead, others were removed as additional analyses revealed the possibility that they contain multi-stage leads. Improved mass-spectrometric techniques and the availability of absolute lead isotopic standards made a more precise definition of the growth curve possible (Stacey et al., 1969).

More complex models describing the evolution of ore leads have been proposed by many authors already early on in the art of lead-isotope geochemistry, assuming a continuous differentiation of mantle and crust (e.g. Bate and Kulp, 1955). At first these models received little attention. The possibility of crustal processes averaging the lead-isotopic composition in such a way that complex histories are masked, has already been proposed by D.M. Shaw (1957).

More precise data on oceanic volcanic rocks of mantle origin and, as already pointed out, the concept of plate tectonics placed serious doubts on the theory of the single-stage development of lead isotopes (e.g. J.R. Richards, 1971). For these reasons, models of differentiation of mantle and crust and/or mixing, both episodic and continuous, received more attention. Some of these accounted for the comparatively small age discrepancies between single-stage model ages of conformable ore deposits and the ages obtained by other methods, such as the stratigraphic age or the absolute age of the host rock. Furthermore, the high $^{207}Pb/^{204}Pb$ and low $^{206}Pb/^{204}Pb$ ratios of conformable ore leads, when compared with the mantle-derived lead from oceanic volcanics (Fig. 5a), were

---

[1] The term conformable is equivalent to the expression strata-bound.

explained. The reader is referred to R.D. Russell and Farquhar (1960), Kanasewich (1968), Armstrong (1968), R.D. Russell (1972), Armstrong and Hein (1973), Sinha and Tilton (1973), Robertson (1973), and R.D. Russell and Birnie (1974). In this context it should be noted that eqs. 12–14 must be replaced by equations containing terms accounting for the addition and loss of uranium, thorium and lead as proposed and evolved, among others, by Allègre and Michard (1973).

The isotopic ratios of the primordial lead of the troilite phase of the Canyon Diablo meteorite was newly determined by Tatsumoto et al. (1973). The single-stage growth curves were thereby seriously displaced, such that the model ages for the conformable ore deposits are considerably younger than the ages obtained by other methods. The new and higher decay constants for the two uranium isotopes have only slightly compensated this offset. *With these new parameters governing the evolution of lead, it is now clearly evident that the conformable ore deposits do not contain single-stage leads.*

To facilitate the comparison of model ages with the other ages, Doe and Stacey (1974) proposed a single-stage growth curve with T = 4.43 b.y., which minimizes the age differences. There are, however, no indications of such low lead–lead ages of meteorites and hence of T, the age of the earth. The authors state that the apparently single-stage leads are the results of mixing processes of lead that developed in sources with varying U/Pb and Th/Pb ratios.

Allsopp et al. (1974) developed a two-stage model based on the $^{87}Sr/^{86}Sr$ ratios of ancient South African crustal rocks and on the lead-isotopic data of the conformable ore deposits. They propose a major differentiation 3.75 b.y. ago of the initially homogeneous U/Pb and Th/Pb systems with a μ value of 7.19 and a W value of 31.21. Since the major differentiation 3.75 b.y. ago the lead of conformable ore deposits closely follows an evolution in a milieu with a μ value of 9.74 and a W value of 38.19[1]. An inspection of Fig. 5a, 5b and Table III reveals that the lead of some deposits still exhibits minor, but still significant deviations from such a two-stage model. It therefore is apparent that this model is a good first approximation describing the evolution of leads in conformable ores. The question remains open as to whether mixing processes of leads from different U/Pb and Th/Pb environments are only of minor importance or whether large-scale mixing always occurred to such a degree that a quasi two-stage is closely approached, without, however, obliterating the major event 3.75 b.y. ago. It should also be noted that this model alone does not explain the mantle-derived lead as is seen today in oceanic volcanic rocks. At least one more stage of evolution, i.e. an increase of the μ value of the mantle approximately 2000 m.y. ago is necessary to explain the high $^{206}Pb/^{204}Pb$ ratios.

In Table III, ore deposits containing leads that define the single, or two-stage growth curves are listed. The list follows approximately that of R.D. Russell (1972) and that of Allsopp et al. (1974). It could be expanded, however; most of the additional deposits are of comparatively young ages in the range of 0–500 m.y. and therefore do not help in defining the growth curve more precisely (see Ostic et al., 1967; Stacey et al., 1969; Doe,

---

[1] Personal communication J.D. Kramers. For the exact parameters, see also Stacey and Kramers (1975).

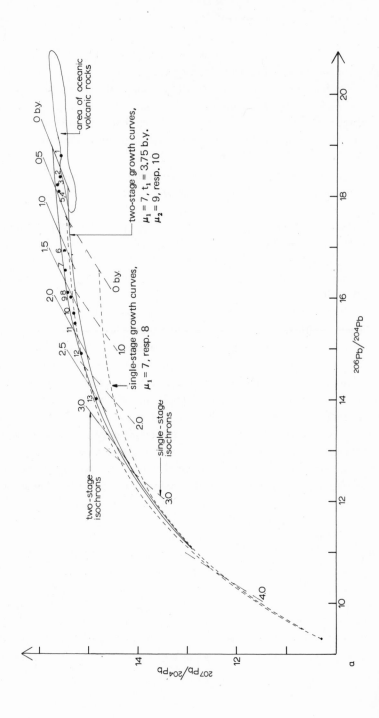

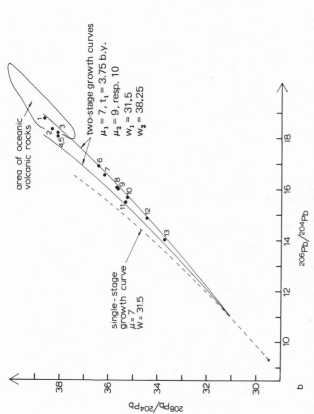

Fig. 5a,b. The data points (R.D. Russell, 1972) of galenas from conformable ores that were used to define the single-stage growth curve are plotted. The area of oceanic volcanic rocks contains the adjusted values given by Russell.

1 = Red Sea; *2* = Halls Peak, N.S.W.; *3* = Bathurst, N.B.; *4* = Cobar, N.S.W.; *5* = Captain's Flat, N.S.W.; *6* = Balmat, N.Y.; *7* = Sullivan, B.C.; *8* = Mount Isa, Qld.; *9* = Broken Hill, N.S.W.; *10* = Southwest Finland; *11* = Sudbury, Ont.; *12* = Cobalt, Ont.; *13* = Geneva Lake, Ont. (See also Table III).

TABLE III

List of ore deposits containing leads that define the single-, or two-stage growth curves.

Deposit/ locality	Ores				
	principal metals	relationship with country rocks	remarks on ore genesis	model lead age [1]	model lead age [2]
Red Sea	Zn, Cu, Ag, Pb, (Fe, Mn)	finely banded, conformable with unconsolidated normal Red Sea sediments	geothermally heated brines, discharged from submarine vents into a number of pools along median Red Sea rift valley	+100 m.y.	−600 to −700 m.y.
Halls Peak N.S.W.	Pb, Zn, (pyritic)	stratiform	ore formation probably related with acidic volcanics	+150 m.y.	−500 m.y.
Bathurst N.B.	Cu, Zn, Pb, (pyritic to pyrrhotitic)	stratabound	exhalative-sedimentary together with submarine rhyolitic volcanics.	+400 m.y.	−200 m.y.
Cobar N.S.W.	Cu, Au, Zn, Pb, (pyrrhotitic to pyritic)	stratabound	probably volcanogen	+400 m.y.	−100 m.y.
Captain's Flat N.S.W.	Zn, Pb, Cu, (pyritic)	stratiform	association with explosive volcanics and exhalites suggests syngenetic volcanic origin	+400 m.y.	− 50 m.y.
Balmat N.Y.	Zn, Pb, Cu	stratabound	formed during retrograde metamorphism of carbonate country rocks	+1100 m.y.	+600 m.y.
Sullivan B.C.	Zn, Pb, Ag, (pyritic)	stratabound	formed by exhalative sedimentary processes, probably in a Red Sea environment	+1400 m.y.	+900 m.y.
Mount Isa Queensland	Pb, Zn, Cu, Ag, (pyritic to pyrrhotitic)	stratiform	formed by precipitation contemporaneous with enclosing sediments	+1600 m.y.	+1200 m.y.
Broken Hill N.S.W.	Pb, Zn, Ag	stratabound	closely related with enclosing sedimentary rocks, modified by strong regional metamorphism	+1600 m.y.	+1300 m.y.
Southwest Finland	Pb, Zn, Cu, Sb, Bi	discordant skarn mineralizations, fracture and vein fillings	partly formed by contact metamorphic metasomatic processes	+1800 m.y.	+1400 m.y.
Sudbury Ont.	Ni, Cu, (Pb)	stratabound	ore deposited during intrusion of sulfide rich silicate magma along the base of Sudbury main felsic norite	+1900 m.y.	+1500 m.y.
Cobalt Ont.	Ni, Cu, Ag, Co	veins	veins considered to be thermal remobilization products from Archean volcano-sedimentary rocks due to intrusion of Nipissing diabase	+2150 m.y.	+1800 m.y.
Geneva Lake Ont.	Zn, Pb, Ag, (pyritic)	stratabound	ores associated with mafic to felsic extrusive rocks and sediments; the deposits probably have formed from submarine exhalations	+2600 m.y.	+2300 m.y.

[1] Two-stage evolution model (Allsopp et al., 1974), compare also Stacey and Kramers (1975).
[2] Singel-stage growth curve.

Ores	Country rocks			References
remarks	types	metamor-phic grade	geologic age	
extremely fine grained to amorphous in sedimentary black ooze	detrital and calcareous normal Red Sea sediments		Recent	Degens and Ross, 1969; Bischoff and Manheim, 1969; Holmes and Tooms, 1972; Hackett and Bischoff, 1973.
fine grained strongly crossfaulted, banded	greywackes, water-worked tuffs, carbonaceous shales	low ·	+240 m.y.	King, 1965; Ostic et al., 1969.
banding parallel to bedding of country rocks	greywackes, tuffs, andesitic-rhyolitic volcanics, cherts	low	+460 m.y.	Lea and Rancourt, 1958; Stanton, 1959, 1960; Helmstaedt, 1970
tabular loads in shear zones, in parts distinctly vein fillings	greywackes, tuffs, siltstones, impure carbonaceous shales	low	+410 m.y.	Thompson, 1953; King, 1965; Russell and Lewis, 1965; Ostic et al., 1969.
flat en echelon lenses, distinctly banded, fine grained	tuffaceous, calcareous shales, cherts, acid volcanics	low	+420 m.y.	Kenny and Mulholland, 1941; Glassone and Paine, 1965; King, 1965.
	siliceous magnesian marbles	medium?	+1100 to +1200 m.y.	Doe, 1962; Doe and Stacey, 1974.
banded, zonal distributed	argillites, siltstones, impure quartzites	low-medium	+1580 m.y.	Freeze, 1962; Leach and Wanless, 1962; Sinclair, 1966; Sangster, 1972.
banded, strongly folded, very fine grained	dolomitic siltstones, shales, tuffs, pyritic black shales	low ·	+1650 m.y.	Knight, 1953, 1958; King, 1965; Bennet, 1965; Farquhason and Wilson, 1971; Stanton, 1962.
coarse grained en echelon drag folded	metasediments, probably former, volcanic greywackes, shales argillaceous	high	+1600 m.y.	King and Thompson, 1953; Carruthers, 1965; Lewis et al., 1965; Stanton, 1955, 1973.
5 deposits: Attu, Korsnäs, Pakila, Orijärvi, Metsamouttu	biotite gneisses, leptites, skarns, amphibolites, serpentinites	medium-high	+1600 to +1800 m.y.	Pehrmann, 1931; Kuovo, 1958, Vassjoki, 1965.
partly modified by metamorphism	felsic norite, granophyre, quartz dioritic to noritic sublayer	low	+1720 m.y.	Stockwell, 1964; Souch and Podolsky, 1968; Naldrett and Kullerud, 1967; Ulrych and Russell, 1964.
vein fillings, in faults and fractures, complex paragenesis	basic and intermediate lavas, interflow sediments, greywackes, conglomerates, diabase	low	+2170 to +2230 m.y.	Thomson, 1965; van Schmus, 1965; Boyle and Dass, 1971; Kanasewich and Farquhar, 1965; Hutchinson et al., 1971; Thorpe, 1974.
fine grained, banded	basic and intermediate lavas and interflow sediments	low	+2540 m.y.	Gilmour, 1965, 1971; Hutchinson et al., 1971; Stacey et al., 1969.

1970; Tugarinov et al., 1972). The isotopic composition of the lead in some of the older conformable deposits does not fit the growth curve. Flin Flon, for example, was considered by Sangster (1972b) to contain a single-stage lead identical to that of other deposits in the area. Slawson and R.D. Russell (1973) demonstrated that the isotopic composition is not uniform and that the leads, therefore, exhibit a multi-stage history. If there is evidence showing that the lead with the lowest isotopic ratios is the starting point of an isochron, or the end member of a mixing line, then one may consider this lead as a "single"-stage lead. Barberton is not included because Saager and Köppel (1976) have shown that these leads lie on an isochron and are also two-stage or possibly higher-order leads (see section on case histories).

The recent Red Sea deposits have been added to the list, because the deposits possess many features characteristic of the conformable ore deposits (see Degens and Ross, 1969, and their contribution to the present publication: Chapter 4, Vol. 4). Some additional comments concerning their lead-isotopic compositions are required:

(1) In contrast to other conformable deposits, the lead-isotopic composition is not homogeneous. However, one has to bear in mind that the sediments are still unconsolidated and extremely rich in water. One may therefore envisage that future diagenetic and metamorphic processes will homogenize the isotopic composition. (2) The investigations of Cooper and J.R. Richards (1969), of Delevaux and Doe (1974) as well as of Reynolds and Dasch (1971), show that the lead of the Red Sea brine is a mixture of mantle-derived lead with lead that is typically found in oceanic sediments such as the manganese nodules (Fig. 6a,b).

Another example where metalliferous oceanic sediments contain mantle-derived lead was found near the East Pacific Rise (Dasch et al., 1971). Sato and Sasaki (1973) reported lead-isotopic compositions of Kuroko ores which demonstrate that they also contain a mantle-derived lead component. These examples show that part of the lead in conformable ore deposits may indeed contain mantle-derived lead.

(Anomalous) lead lines, multi-stage leads

Many lead-isotopic studies, especially from vein-type ore deposits, have revealed the presence of linear arrays of data points in the $^{207}Pb/^{204}Pb$ versus $^{206}Pb/^{204}Pb$ and also in the $^{208}Pb/^{204}Pb$ versus $^{206}Pb/^{204}Pb$ diagrams. Such linear arrays have been termed anomalous lead lines in contrast to the so-called ordinary leads which were thought to be single-stage leads. As it is very doubtful that single-stage leads still exist, the term anomalous should be abandoned.

Straight lines in both diagrams will be produced by incomplete mixing of two different leads. The slope of the line has no age meaning (Fig. 7).

A linear array, however, may also have the significance of an isochron, and the array may be obtained as follows:

(1) At some time t, prior to t_1, isotopic homogeneization occurred, for instance by magmatic processes. At time t_1 differentiation occurred leading to the formation of

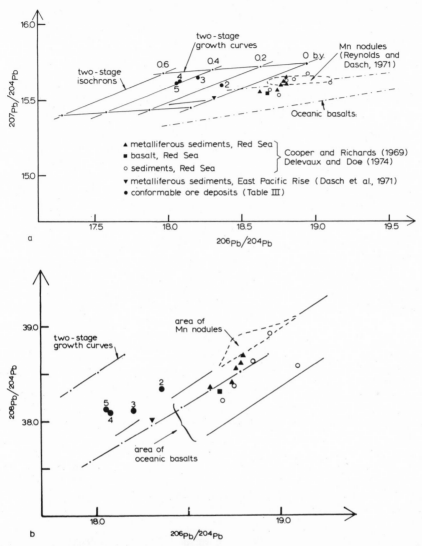

Fig. 6a,b. Data points of recently deposited metalliferous sediments (Red Sea and East Pacific Rise). The data points from the Red Sea scatter between the oceanic basalt lines (compare Delevaux and Doe, 1974, and R.D. Russell, 1972) and the area of Mn nodules (Reynolds and Dasch, 1971). The lead appears therefore to be a mixture of mantle lead and a lead typically found in oceans. The latter is probably also a composite of mantle-derived and crustal lead. The two-stage growth curves are those shown in Fig. 5a. The uncertainties of the $^{206}Pb/^{204}Pb$ and $^{207}Pb/^{204}Pb$ ratios should not exceed 0.3% and of the $^{208}Pb/^{204}Pb$ ratio 0.5%.

various subsystems, each having its own μ value ($\mu_{2,1}$ to $\mu_{2,n}$) (Fig. 8). At time t_2 lead was extracted from these subsystems and isolated from uranium. Eqs. 12 and 13 can then be rewritten:

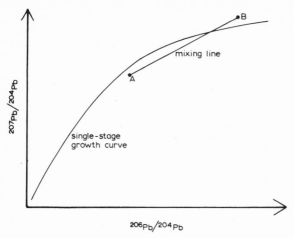

Fig. 7. Data points of a mixture of component A and B lie on a mixing line. The slope has no age significance.

$$(^{206}Pb/^{204}Pb)_{t_2} = (^{206}Pb/^{204}Pb)_{t_1} + \mu_{2,i}(e^{\lambda_8 t_1} - e^{\lambda_8 t_2}) \qquad (17)$$

$$(^{207}Pb/^{204}Pb)_{t_2} = (^{207}Pb/^{204}Pb)_{t_1} + (\mu_{2,i}/137.88)(e^{\lambda_5 t_1} - e^{\lambda_5 t_2}) \qquad (18)$$

These two equations may be combined similarly as eq. 15 yielding a straight-line equation (eq. 16; Fig. 9). Examples will be discussed in the sections on case histories.

(2) Radiogenic lead that developed between t_1 and t_2 in a system with a value of $\mu = \infty$, i.e. free of ^{204}Pb, is added to a homogeneous common lead and incompletely mixed (Fig. 10). The mixing line will also have a time meaning, the slope of the line being the $^{207}Pb/^{206}Pb$ ratio of the radiogenic lead:

$$s = \frac{\Delta(^{207}Pb/^{204}Pb)}{\Delta(^{206}Pb/^{204}Pb)} = \left(\frac{^{207}Pb}{^{206}Pb}\right)_{rad.} = \frac{1}{137.88}\frac{(e^{\lambda_5 t_1} - e^{\lambda_5 t_2})}{(e^{\lambda_8 t_1} - e^{\lambda_8 t_2})} \qquad (19)$$

It should be noted that in the $^{208}Pb/^{204}Pb$ versus $^{206}Pb/^{204}Pb$ diagram, a linear array is not necessarily observed in the cases (1) and (2). In case (1) the fractionation of uranium and lead is independent of the uranium and thorium fractionation, and in case

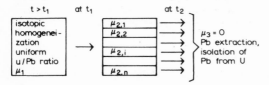

Fig. 8. Schematic diagram illustrating the development of U–Pb subsystems with $\mu_{2,1}$ to $\mu_{2,n}$ by fractionation of a homogeneous source with either a single-stage or multi-stage lead at time t_1.

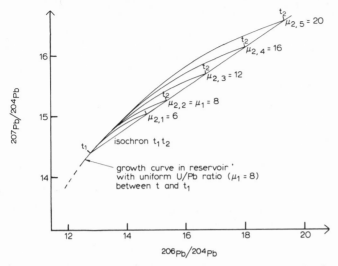

Fig. 9. The growth curves of various subsystems are shown for the time interval $t_1 - t_2$. At time t_2 the data points of the subsystems will lie on the isochron $t_1 t_2$.

(2) the U/Th ratio of the system with $\mu = \infty$ may not be the same in all parts of the system.

From the combination of the equations 17 and 18, which also leads to eq. 19, one can calculate either t_1 or t_2. In case (1), t_1 is the age of the source rock from which the lead was extracted at t_2 to form a mineralization (see case history of the Barberton Mountain Land). In case (2) which is realized for example in the Witwatersrand (see below), t_1 is the age of the source of the radiogenic component. In both cases t_2 corresponds to the time of mineralization if deposition occurred shortly after extraction.

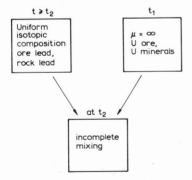

Fig. 10. Diagram illustrating how a lead—lead isochron can be obtained by incomplete mixing of a homogeneous common lead with a radiogenic lead free of ^{204}Pb.

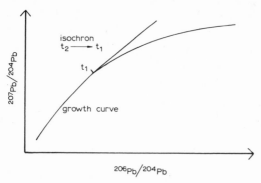

Fig. 11. A lead–lead isochron with a slope, i.e, a $^{207}Pb/^{206}Pb$ ratio, that corresponds to the instantaneous production rate of the two isotopes at time t_1. The isochron is tangential to the growth curve defined by the μ value of the homogeneous source prior to differentiation.

If neither t_1 nor t_2 are known, then one may let $t_1 \rightarrow t_2$ (= t_i) and the formula for the slope s (Fig. 11) is:

$$s = \frac{1}{137.88} \frac{\lambda_5 e^{\lambda_5 t_i}}{\lambda_8 e^{\lambda_8 t_i}} \tag{20}$$

where t_i is the maximum age of mineralization and a minimum age for the source in which the lead developped. For a rigorous treatment of multi-stage leads see Gale and Mussett (1973).

It must be emphasized that lead lines are a most helpful tool for solving genetic problems of ore deposition. In the following case histories a number of examples will be briefly described in which lead lines not only assisted in determining ages of mineralization but also periods of metamorphism as well as possible sources of the metals.

CASE HISTORIES

Mississippi Valley lead–zinc deposits

The important strata-bound lead–zinc–barite–fluorite ore deposits in the central United States all lie within the Mississippi Valley, and three districts are generally distinguished (Fig. 12): (a) the Southeast Missouri district, (b) the Tri-State district of Missouri, Kansas and Oklahoma, and (c) the Upper Mississippi Valley or Wisconsin–Illinois district.

According to Brown (1970) and Heyl et al. (1974) the following common characteristics can be distinguished in the deposits of the Mississippi Valley (see also the comparative table in Hagni's Chapter 10 in Vol. 6):

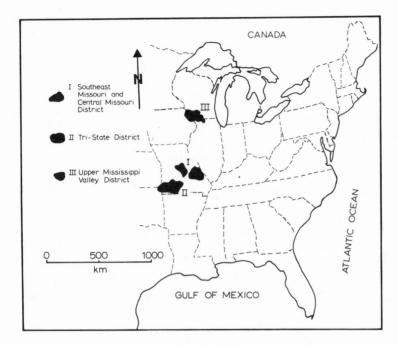

Fig. 12. The main districts of the Mississippi Valley-type lead–zinc ores.

(1) The ores occur primarily in Upper Cambrian carbonate rocks, originally either limestone or dolomite (Fig. 13). In the three principal districts the ores occur in certain preferred host-rock horizons and can be described as strata-bound, although minor fractures may impregnate associated sandstones in a distinctly discordant way.

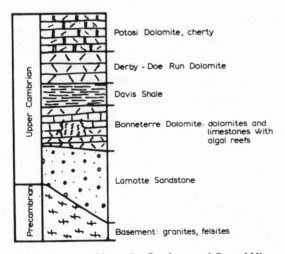

Fig. 13. Stratigraphic section Southeast and Central Missouri District.

(2) Most abundant minerals are galena, iron-poor sphalerite, barite, fluorite, iron sulphides and chalcopyrite.

(3) The country rocks are not metamorphosed and do not exhibit any signs of marked physical alteration since deposition except for weathering at outcrops.

(4) Post-Precambrian igneous activity is minor and where present shows little or no apparent connection with the mineralization.

(5) Geological evidence as well as oxygen- and sulphur-isotope data indicate a low-temperature ore formation (Hall and Friedman, 1969; Pinckney and Rafter, 1972; Pinckney and Rye, 1972; Heyl et al., 1974). This is further substantiated by fluid inclusion data which indicate an ore deposition at 200°C or less by basin brines (Roedder, 1967).

(6) The deposits contain markedly radiogenic lead, of J-type with $^{206}Pb/^{204}Pb$ ratios of 20 or greater.

(7) Each district has its distinct range of $^{206}Pb/^{204}Pb$ and $^{208}Pb/^{204}Pb$ ratios. Within a district one observes a regional zonation of the ratios which may be used as a guide for exploration. Finally, the isotopic ratios may also vary within single crystals. The genesis of the Mississippi Valley deposits has been investigated by numerous workers and various quite opposing interpretations of the deposits are advocated. The debates centre mainly around the questions of the provenance of the metal and the timing of deposition with respect to the enclosing sediments.

Ohle Jr. (1967) and McKnight (1967) suggested a deep-seated, probably alkaline or potassic igneous source as being responsible for the mineralizing emanations; similarly for the Hicks Dome area in the Kentucky–Illinois district, Grogan and Bradbury (1967), Zartman et al. (1967), and Hall and Heyl (1968) proposed that some of the metals derived from an alkalic cryptovolcano. A concentration of the metals from the surrounding sedimentary rocks by a lateral secretion mechanism has been favoured by many isotope and geochemical investigators, e.g. Brown (1967), Skinner (1967), Helgeson (1967), and Roedder (1967). In contrast to these *epigenetic* interpretations of the ore formation, Amstutz and Park (1967), Amstutz et al. (1961) and Park and Amstutz (1968), using primary sedimentary features, proposed that the ores may have been deposited during or after the formation of the sedimentary structures, i.e. during different periods of diagenesis. The authors, therefore, suggested a *syngenetic* synsedimentary or diagenic formation of the deposits. (See also Jackson and Beales, 1967, and Wolf, 1976).

Lead-isotopic investigations. The investigation of the lead-isotope characteristics of sulphide-ore deposits can offer valuable information concerning the history of the lead and, thus, can be of great assistance for the delineation of the source regions of the ores. A great number of lead-isotopic investigations, therefore, has been used to help in the interpretation of the processes which eventually led to the formation of the Mississippi Valley ore deposits, e.g. Cannon Jr. et al. (1961 and 1963), Heyl et al. (1966), Brown (1967), McKnight (1967), Cannon Jr. and Pierce (1967), Doe and Delevaux (1972), and J.R. Richards et al. (1972).

The presence of abnormally radiogenic lead in the ores of the Mississippi Valley, yielding negative or future model ages was discovered by Nier (1938) in a galena sample from Joplin Mine, Missouri. Houtermans (1953) proposed for such abnormally radiogenic leads the designation *Joplin-* or *J-type leads*.

Many lead-isotope investigators, especially in the *Upper Mississippi Valley district*, favoured a formation of the ores by lateral secretion with the original source of the lead being the Precambrian basement (Kanasewich, 1962; Heyl et al., 1966). The main reason for this conclusion was the conspicuous similarity between the source age of the lead, as determined from the slope of the lead–lead isochron, and the accepted age of the basement rocks (i.e. 1450 m.y.). Based on the lead-isotope pattern obtained from galenas of the *Southeast Missouri district*, Brown (1967), however, arrived at the conclusion that the ore-forming fluid was mainly connate water generated in the Paleozoic rocks overlying the basement. It circulated, highly concentrated, mainly in the Lamotte Sandstone, a permeable aquiver, and flowed upward into the overlying gently domed Bonneterre Dolomite, where it spread laterally causing the distinct lateral and vertical lead-isotopic zoning observed in the district. According to Brown (1967), two-thirds of the ore lead were leached from the permeable Paleozoic sediments and one-third was derived from the Precambrian basement, accounting for all the radiogenic anomaly of the ore leads. Cannon Jr. et al. (1963) found significant differences in the lead-isotopic composition in successive growth zones from a large galena crystal originating from the Picher field, Oklahoma, in the *Tri-State district*. They concluded that these isotopic variations imply a progressive leaching of lead, probably during a considerable time-span, from rocks containing appropriate amounts of lead, uranium and thorium. The data furthermore suggests that the source rocks were younger than Precambrian, presumably some of the Paleozoic carbonate sediments in which the Tri-State ores were deposited.

In a recent study, Doe and Delevaux (1972) proposed another interpretation for the source of the lead in the Southeast Missouri district. As all earlier conclusions from lead-isotopic studies were made without knowledge of the isotope composition of the lead in possible source rocks, Doe and Delevaux (1972) included samples from such rocks in their investigation. They analyzed drill-core samples from extensive felsites of the *Precambrian basement*, from arkosic and orthoquartzitic sandstones of the *Upper Cambrian Lamotte Sandstone*, from calcitic and dolomitic limestones of the *Upper Cambrian Bonneterre Dolomite*, which is the main ore-bearing horizon in the district, and from calcareous shales of the *Upper Cambrian Davis Formation* of the Elvins Group (for the stratigraphic column, see Fig. 13). From their study Doe and Delevaux (1972) concluded that the major supplier of lead was the Lamotte Sandstone. The host rock of the ores, the Bonneterre Dolomite, has possibly also contributed some of the lead.

In Fig. 14b, the $^{208}Pb/^{204}Pb$ ratios are plotted versus the $^{206}Pb/^{204}Pb$ ratios of the ore lead. Assuming that some mixing of the lead occurred in the metal transporting brines, the lead-isotopic compositions of possible source rocks should either lie on the same linear array extending along the lower and upper continuation of the ore-lead array,

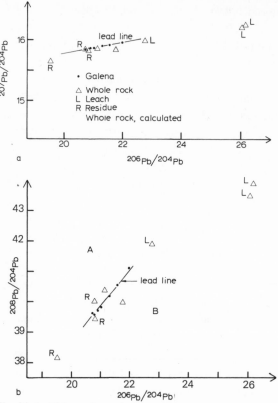

Fig. 14a,b. Galenas from the Upper Cambrian dolomite and whole-rock samples of the underlying Lamotte Sandstone of Southeast Missouri which is a possible source of the ore lead (Doe and Dele-vaux, 1972).

or else should plot in the fields designated *A* and *B* in Fig. 14b, such that the lines connecting data points cross the array of the ore lead. This latter condition is only fulfilled by the lead of the Lamotte Sandstone and the Bonneterre Dolomite.

Whereas U–Th fractionation in the $^{208}Pb/^{204}Pb$ versus $^{206}Pb/^{204}Pb$ diagrams commonly causes a scatter of the data points of otherwise undisturbed U–Th–Pb systems, the systematics of the coupled U–Pb systems are such that, in the $^{207}Pb/^{204}Pb$ versus $^{206}Pb/^{204}Pb$ diagrams, linear arrays are only observed if the following conditions are realized: (a) the isotopic composition of the lead was homogeneous at one time in the history of the rock; (b) the U/Pb ratio (μ value) differed in various subunits of the rock; and (c) no subsequent event disturbed the U–Pb systems of the subunits.

It should be noted that the lead–lead diagrams are relatively insensitive to small changes of the U/Pb and Th/Pb ratios in rocks which took place within the last several 100 m.y., and linear arrays will therefore not be destroyed by comparatively young events. Such disturbances, however, will be readily noticed in diagrams involving U/Pb ratios, as for instance in the $^{206}Pb/^{204}Pb$ versus $^{238}U/^{204}Pb$ plot or in the concordia diagram.

The $^{206}Pb/^{204}Pb$ and $^{207}Pb/^{204}Pb$ ratios of the ore lead are plotted in Fig. 14a. They exhibit a linear array with a slope of 0.0864 which is interpreted as an isochron. From the slope one may determine t_1, if t_2 the time of mineralization is known (eq. 19); t_1 is approximately 1300 m.y. if the mineralization took place 100 m.y. ago, which indicates that the ore lead contains a radiogenic portion that developed in a source rock during the time-span 1300–100 m.y. ago. If the mineralization occurred recently (i.e. t_2 = 0 m.y.), then t_1 is approximately 1350 m.y.

The possible source rocks must today contain a lead with similar $^{206}Pb/^{204}Pb$ and $^{207}Pb/^{204}Pb$ ratios which also show a considerable spread. The slope of the array should be equal to the slope of the array obtained from the ore-lead samples, or possibly slightly smaller if the mineralization occurred some time in the past. If two slightly different slopes are observed, they should intersect to the left of the ore array. Again, these conditions are only met by the lead of the Lamotte Sandstone and partly also by that of the Bonneterre Dolomite. The lead of the latter, however, normally has lower $^{206}Pb/^{204}Pb$ ratios which cannot explain the entire range of the values of the ore-lead samples.

The slope of the array of the lead from the Lamotte Sandstone indicates an age of 1200 to 1300 m.y., similar to that of the ore lead and also similar to that of the Cambrian Bonneterre Dolomite. *These ages thus do not indicate the time of deposition of these sediments, but rather represent an indication of the age of the source area which contributed detritus to them.* Therefore, t_1 signifies an integrated age of the source material, as stated by Doe and Delevaux (1972), rather than the age of the source itself.

Other lead–zinc deposits in carbonate rocks

Kesler and Ascarrunz (1973) reported on the isotopic composition of sulphur and lead from deposits in central Guatemala. The $\delta^{34}S‰$ values show similar variations to those in the Mississippi Valley deposits, and the differences in the $\delta^{34}S‰$ values of coexisting galena and sphalerite indicate similar temperatures of deposition to those in the Mississippi Valley ores, i.e. less than 100–250°C. The lead-isotopic composition varies in a similar manner to that in the Mississippi Valley deposits, the range of variation, however, being much smaller.

A similar pattern of lead-isotopic ratios has been observed in the lead–zinc ores of the western Balkan Mountains. Lead–zinc mineralizations occur mainly in Anisian carbonates, whereas a superimposed lead–copper mineralization also occurs in Jurassic and Paleozoic sediments (see Minčeva-Stefanova, 1967).

A similar pattern in the lead-isotopic ratios has been observed in the lead–zinc ores in Bulgaria (see Minčeva-Stefanova, 1967).

Not all lead–zinc deposits in carbonate rocks contain lead with a variable isotopic composition. The Pine Point deposit occurs in Middle Devonian limestones and dolomites with intercalated shales and evaporites. The series is underlain by a basal sandstone which rests upon the Precambrian basement (Jackson and Folinsbee, 1969). The lead-isotopic composition is remarkably uniform (Cumming and Robertson, 1969, Cumming et al., 1971). The mean isotopic ratios plot in the lead—lead diagram closely to the growth curve of conformable ore leads. The model age is approximately 270 m.y. (two-stage model by Allsopp et al., 1974) and is clearly younger than the age of the host rock (375 m.y.).

Although data on the cryptic lead in the host rock, as well as in the underlying rocks and in the sedimentary cover, are lacking one may tentatively conclude that the lead derived from a deep-seated source and was well mixed prior to deposition. Near-surface rocks would most likely contain a more radiogenic lead with a considerable range, especially in the $^{206}Pb/^{204}Pb$ ratio. N. Campbell (1967) drew attention to faults in the Precambrian basement the projection of which underly the reef and ore zones. This author favoured an epigenetic origin for the ore and a deep-seated origin for the ore solutions. The faults are considered as the channelways for the rising fluids.

The sulfur isotopes suggest that the evaporites were the source for this element (Sasaki and Krouse, 1969). The $\delta^{34}S‰$ values of galena and sphalerite show similar differences to those of the Mississippi Valley deposits (Pinckney and Rafter, 1972), indicating thereby similar temperatures of deposition of around 200°C.

The oxygen-isotope pattern of the Pine Point deposit (Fritz, 1969) is also similar to that of the Mississippi Valley deposits (Hall and Friedman, 1969). In both cases the decrease of the $\delta^{18}O‰$, towards the ore testifies to the influx of heated meteoric water.

Pine Point is, therefore, an example demonstrating how isotopic studies may be used to gain information on the origin of elements and how different elements of a deposit may have different origins. It should be furthermore noted here that whereas the lead-isotopic abundances are related to time and origin, the isotopic ratios of light elements such as sulphur and oxygen reflect the physico-chemical conditions of the mineral formation as well as their origin.

Broken Hill lode

The regional setting[1]. The famous sulphide deposit of Broken Hill in New South Wales, Australia, is known to contain one of the largest lead—zinc concentrations found anywhere on earth. Since the beginning of mining operations in 1884, the deposit has produced well over 14 million tons of lead and 12 million tons of zinc; and the entire lode is expected to yield over 120 million tons of ore. The main occurrence in the vicinity of Broken Hill extends along strike for over 6500 m, its greatest depth is approximately 850 m and its maximum width some 150 m. Additional minor occurrences of the so-

---

[1] Up-to-date information is presented by Both and Rutland in Chapter 6, Vol. 4.

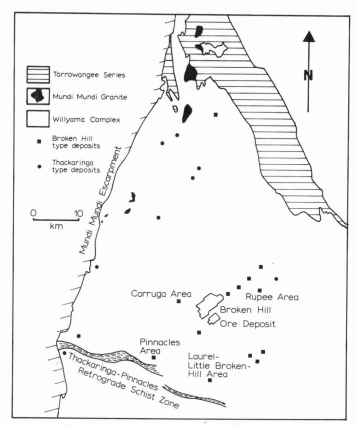

Fig. 15. Locality map and regional geology of the Willyama Complex (Carruthers, 1965; Both, 1973).

called Broken Hill type ores are numerous and they are found within a distance of more than 40 km from the main Broken Hill lode (Fig. 15).

All *Broken Hill type mineralizations* occur in high-grade metamorphic rocks of the *Willyama Complex* which covers an area of some 9000 km^2. To the northeast the metamorphic rocks of the Willyama Complex are unconformably overlain by the Torrowangee sediments and to the west, at the N–S striking Mundi Mundi escarpment, they disappear under recent alluvial sand deposits (Fig. 15).

In many cases the rocks of the Willyama Complex exhibit distinct lithological banding and it is agreed by most investigators that, in the main, they were originally a sedimentary sequence of argillaceous and arenaceous sediments with minor calcareous members (Carruthers, 1965; Lewis et al., 1965). All these rocks have been strongly folded and more or less intensively metamorphosed. In the environment of the Broken Hill lode, the metamorphism is that of the sillimanite–almandine subfacies of the almandine–amphibolite facies. The immediate host rocks of the ores now consist of K-feldspar–garnet–silli-

manite—biotite gneisses, quartz—plagioclase—biotite—garnet gneisses, granitic gneiss, amphibolite and iron oxide rich layers, the so-called "Banded Iron Formation".

The $^{87}Sr/^{86}Sr$ ratios yield a maximum age of 1820 ± 100 m.y. for the deposition of
the acidic members of the Willyama Complex (Pidgeon, 1967). The same author, using
the Rb—Sr method, obtained an age of 1640 ± 40 m.y. for the high-grade Willyama
metamorphic event during which an almost complete redistribution of the Sr isotopes
took place within the individual rock members. This high-grade metamorphism was followed at 1540 ± 40 m.y. by the extensive emplacement of muscovite pegmatites and the
intrusion of alkali-rich Mundi Mundi granites to the north of the zinc—lead mineralizations. A low-grade regional metamorphic event eventually is indicated by biotite ages of
500 ± 10 m.y.. It coincides with the emplacement of muscovite pegmatites at Thackaringa (Pidgeon, 1967).

The lead—zinc lode at Broken Hill consists of a number of complexly folded layers, of
which six constitute the main ore body (Fig. 16). However, no one layer is ore-bearing

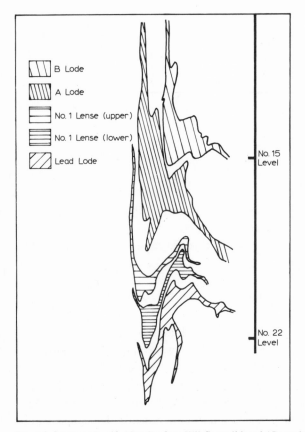

Fig. 16. Section No. 62, New Broken Hill Consolidated (Carruthers, 1965).

over the entire 6500-m extent of the mineable ore. All of the six lenses are broadly conformable with the foliation of the host rocks, although there are some local discordances and transgressions present (King and Thomson, 1953). Very often the lenses exhibit a conspicuous spatial relationship with the accompanying banded iron formations. The folding of the Broken Hill ore is en-échelon in type (J.D. Campbell, 1958), and recent work by Hobbs (1966), Anderson (1966), and Williams (1967) has shown that at least two major periods of folding, one at an acute angle to the other, can be recognized in the Broken Hill district.

The principal sulfide minerals in all the Broken Hill type ores are generally coarse-grained galena and sphalerite with minor amounts of pyrite and chalcopyrite. Wallrock alteration has in general been found to be insignificant.

In strong contrast to the broadly conformable Broken Hill type deposits are the lead–silver mineralizations of the *Thackaringa type* which occur to the west and northwest of Broken Hill. They usually are coarse-grained quartz–siderite mineralizations which occur as fracture fillings and, in the main, they possess only limited economic importance. King and Thomson (1953) and King (1958) — see also Chapter 6 by Both and Rutland, Vol. 4 — suggested that these deposits have been formed by metamorphic remobilizations of Broken Hill type ores and thus represent a classic example of epigenetic mineralization later than the main Willyama metamorphism.

Genesis of the Broken Hill lode. For a long time the Broken Hill lodes were considered to be of epigenetic origin, formed by hydrothermal solutions of presumed magmatic derivation which extensively replaced the country rocks (Stillwell, 1926, 1959; Gustafson, 1939; Stillwell and Edwards, 1956; Den Tex, 1958; etc.) This view was seriously challenged amongst others by Ramdohr (1951), King and Thompson (1953), King (1958), and Carruthers (1965) who suggested a syngenetic origin for the ore and later recrystallization and modification during the metamorphic events (see Chapter 5 by King, Vol. 1, on evolution of genetic hypotheses). Stanton (1955, 1960, and 1973), and Stanton and Richards (1961), further advancing these ideas, arrived at the assumption that the ore is of submarine volcano-sedimentary origin.

Since the extremely strong metamorphic overprint certainly has obliterated any primary sedimentary features of the ores as well as of the sediments, it still poses great difficulties to demonstrate in an unequivocal way that the ore experienced the same history as its metamorphic host rocks.

In a comprehensive review of the available data on the origin and metamorphic history of the Broken Hill ores, Hobbs et al. (1968) concluded, that: (a) a stratigraphic control of the ore is not clearly established; (b) the sulphides have been deformed by the retrograde regional metamorphism; and (c) the ores may have been involved in the Willyama metamorphic event. However, the authors could not find any clear evidence that the sulphides were present prior to the oldest known metamorphic event in the Willyama Complex.

Lawrence (1973) in a minerographic study of the sulphides of the Broken Hill lode showed that the ore minerals have been affected by two periods of metamorphism. The high-grade Willyama metamorphism caused intensive brecciation and recrystallization of ore and gangue minerals. The retrograde metamorphism (Thackaringa event) then super-imposed the earlier developed ore structures and again fragmented the gangue and spha-lerite and recrystallized galena.

Sulphur-isotope measurements of the isotope fractionation between coexisting galena and sphalerite from Broken Hill ore indicate a temperature of 540°C (Grootenboer and Schwarcz, 1969), or 600°C according to the equilibrium curve of Czamanske and Rye (1974); a temperature which is definitely too high for hydrothermal deposition, but reasonably consistent with the high-grade metamorphism of the Willyama event.

. A study of the minor-element geochemistry of sulphide minerals in the Broken Hill lode indicates a significant variation in the composition of galena and sphalerite between the various lenses (Both, 1973). This observation together with the distinct absence of systematic chemical variations within the lenses (Carruthers, 1965) seems to be difficult to explain by a replacement ore formation, but could indicate that each of the different lenses represents a distinct layer of metal-rich sediments which essentially was deposited at the same time as the enclosing sediments (Both, 1973).

Stanton (1973), who investigated the chemical relationship between the sulphide lenses and the "Banded Iron Formations" of Broken Hill, found close links in their chemistries. He concluded that the two rock types represent exhalative/hydrothermal material contributed to essentially similar sea-floor environments and, thus, merely reflect differences in the intensity of the submarine exhalative source.

The lead isotopes. Stanton and R.D. Russell (1959) reported lead-isotopic ratios of the Broken Hill ore and demonstrated its uniform isotopic composition. The data points plotted on the single-stage lead growth curve in the same way as the data of other conformable ore deposits. As mentioned in the earlier section on the single-stage lead growth curve of conformable ores, the model ages — approximately 1600 m.y. — agreed well with the time of metamorphism of the host rocks as determined by Pidgeon (1967), and S.E. Shaw (1968). The new values of the parameters for the single-stage growth curve, however, seriously lowered the model age (see Table III). By adopting the model proposed by Allsopp et al. (1974), the agreement between model age and age of meta-morphism is restored.

In spite of the agreement of the model ages, the lead-isotopic data, however, do not shed light on the controversy as to whether the metals were introduced during sedimenta-tion or during metamorphism. As pointed out by Pidgeon (1967), the sedimentation age of the Willyama Complex could be as high as 1820 m.y. If this maximum age is correct and if a synsedimentary deposition of the metals took place, then one might expect to observe a higher model age. However, if the metamorphism at 1640 m.y. was the first one to affect the Willyama Complex, then it could well be that a lead-isotopic homogeneiza-tion between ore and host rock occurred because of the high mobility of trace elements in water-rich sediments[1]. This implies that lead from the host rock was added to the ore lead and that the μ value of the host rock was similar to that of the source in which the ore lead developed. A calculation shows that a maximum of 2% of the lead would derive from the host rocks or, estimating the total lead content of the deposit to amount to approximately 20 million tons, some 400,000 tons of lead. Assuming that each ton of country rock contributed 10 g of lead, a volume of about 10—15 km^3 of rock would have been involved in the exchange. Such a volume appears reasonable in size, when compared to the size of the ore fields.

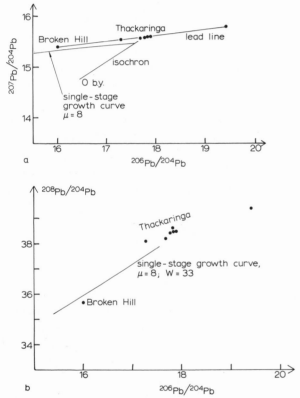

Fig. 17. a. The data points from the Thackaringa vein-type deposits define a lead line that passes through the data point of Broken Hill. b. The scatter of the data points indicates that the lead is not a two-component mixture, and therefore the lead line in Fig. 17a represents an isochron (Stanton and Russell, 1959; Russell et al., 1961).

---

[1] For compaction as an important process in modifying sediments and in the origin of ores, see the summary by Wolf (1976).

In the Thackaringa-type vein deposits, the lead-isotopic composition of galenas is clearly heterogeneous and shows a linear array of the data points in the $^{207}Pb/^{204}Pb$ versus $^{206}Pb/^{204}Pb$ diagram and a scatter in the $^{208}Pb/^{204}Pb$ versus $^{206}Pb/^{204}Pb$ diagram (Fig. 17a,b) (R.D. Russell et al., 1961). This contrasting behaviour in the two lead diagrams demonstrates that the array of the Thackaringa-type leads has the significance of an isochron. As pointed out by Kanasewich (1962), the slope of the isochron indicates the addition of radiogenic ^{206}Pb and ^{207}Pb which developed between $t_1 = 1620$ m.y. (the time of metamorphism in the Broken Hill lode) and $t_2 = 485 \pm 80$ m.y. to lead of the Broken Hill type. The latter event (t_2) coincides with the emplacement of muscovite pegmatites and a low-grade regional metamorphism which occurred approximately 500 m.y. ago (J.R. Richards and Pidgeon, 1963; Pidgeon, 1967). This metamorphic event triggered a loss of radiogenic lead in uranium-bearing minerals that mixed with lead such as is found in the Broken Hill area, and eventually it was deposited in veins, forming the Thackaringa-type deposits. The scatter of the data points in the $^{208}Pb/^{204}Pb$ versus $^{206}Pb/^{204}Pb$ diagram can be explained by the variable U/Th ratios of the minerals which lost part of their radiogenic lead.

The Witwatersrand deposits

The geological setting[1]. The sediments of the Witwatersrand and Dominion Reef Sequences in the Republic of South Africa contain the largest gold concentration known on earth. During 1973 the various mines working the Witwatersrand deposit, produced 847 tons of gold. Most of the economically exploitable gold occurs in conglomerate ("banket") and carbon reefs within the upper portions of the Witwatersrand Sequence which overlies the strata of the Dominion Reef Sequence. The latter contains in its lower portion economically important uraniferous conglomerates.

Together with the sediments and volcanics of the Ventersdorp Sequence, the argillaceous and arenaceous sequences are referred to as the Witwatersrand Triad. Allsopp (1964), investigating whole-rock samples by the Rb—Sr method on an underlying granite obtained a maximum age of 2820 ± 55 m.y. for the Dominion Reef Sequence. The Ventersdorp Sequence, at the top of the Witwatersrand Triad, has been dated by Van Niekerk and Burger (1964) at 2300 ± 100 m.y. using the U—Pb method on zircons from interbedded quartz-porphyry lavas. The Witwatersrand Sequence, some 7600 m thick, lies between the two.

The Dominion Reef and Witwatersrand sediments are confined to the central portion of the Kaapvaal Craton (Fig. 18) and occur in an elliptical depository — the Witwatersrand basin — which is framed by a number of domelike inliers of granite—greenstone basement.

---

[1] For details, see Chapters 1 and 2 by Pretorius, Vol. 7.

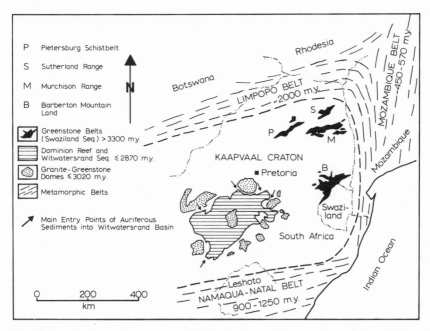

Fig. 18. Locality map and generalized regional geology of the greenstone belts and the primary and placer gold deposits in the Kaapvaal Craton.

Despite their great age, the sediments of the Witwatersrand Triad exhibit only the effects of a mild low-grade regional metamorphism. Argillaceous sediments have been converted to shales and slates, whereas arenaceous sediments show the effects of secondary silification and pyrophyllitization. Generally, the auriferous conglomerates are composed of distinctly water-worn vein quartz, quartzite and chert pebbles which are firmly cemented in a fine-grained matrix of quartz, sericite, phyllosilicates and opaques. The latter comprise among others pyrite, uraninite, arsenopyrite, chromite, rutile, galena, gold and chalcopyrite.

The genesis of the ore deposit. The problem of the origin of the Witwatersrand ores has been a controversial matter between placerists and hydrothermalists, since the discovery of the goldfields (see Mellor, 1916; Young, 1917; Graton, 1930; Sharpe, 1949; Davidson, 1953 and 1957; etc.). Minerographic studies by Liebenberg (1955), Ramdohr (1955), Saager (1969) and Schidlowski (1970) — see also Chapters 1 and 2 by Pretorius, Vol. 7 — led these authors to the conclusion that the gold and uraninite, as well as a large portion of the sulphides, must be of placer origin. Subsequent to their deposition some of the placer minerals — notably gold and a portion of the pyrite — were reconstituted during the low-grade metamorphism and thereby lost their detrital features.

Recently aquired geological and sedimentological data (Brock and Pretorius, 1964) indicate that almost all payable gold mineralizations occur along the northwestern arc of the Witwatersrand basin and are confined to five large fan-like areas or deltas (Fig. 18). High-energy environments close to the points of entry of sedimentary material into the basin, at the apex of the deltas, gave rise to typical conglomerate reefs in which gold and other heavy-mineral constituents were concentrated by various fluviatile and deltaic sedimentary processes. Low-energy environments at the far end of the deltas led to the formation of the carbon-seam type reefs (Pretorius, 1966).

With the *modified placer theory* as a concept for the metallogenesis of the Witwatersrand ores generally being accepted, the attention of students of the deposit has diverted to the problem of the original source of the gold and the sulfides within the sediments of the Triad. Viljoen et al. (1970) suggested that the primary source of the Witwatersrand gold has to be sought in the Archean greenstone belts of the Kaapvaal and Rhodesian cratons. Employing the greenstone terrain of the Barberton Mountain Land in the eastern Transvaal as a model source terrain for the Dominion Reef and Witwatersrand sediments, these authors were able to show a high degree of correlation between expected placer minerals derived from successive erosional levels of such a source terrain and the actual placer minerals encountered in the Witwatersrand depository.

Saager (1974) investigated the trace-element geochemistry of pyrites from the Witwatersrand ore horizons and from primary gold deposits of the greenstone terrains of the Archean Swaziland Sequence. He observed that detrital Witwatersrand pyrites and pyrites from primary deposits in the greenstone belts possess close similarities in their trace-element geochemistry, whereas "in-situ" formed Witwatersrand pyrites are markedly different in their trace-element composition. This observation underlines the possibility of an origin of the detrital Witwatersrand pyrites from an ancient greenstone terrain and, thus, supports the gold source concept proposed by Viljoen et al. (1970).

Isotope investigations. Nicolaysen et al. (1962) analysed U—Pb ages of detrital uraninites and monazites as well as of total — uraninite-bearing — conglomerate samples from the Dominion Reef Sequence (Fig. 19). One of the four analyses yielded concordant ages of 3100 m.y.. The discordant ages of two samples point to a recent lead loss, or uranium gain, whereas the discordant age pattern of the fourth sample indicates a uranium loss, or possible lead gain. All four samples lie, within the analytical uncertainties, on a line that intercepts the concordia curve at 0 m.y. to 200 m.y., the time of the recent disturbance of the U—Pb systems, and at 3100 m.y., the time of mineral formation. The crystallization age of 3100 m.y. agrees with the hypothesis of a detrital origin of the uraninites and monazites.

The data points of uranium-bearing minerals from the Witwatersrand, Ventersdorp and Transvaal Sequences, analyzed by Louw and Strelow (1954), are plotted on Fig. 19. They show a considerable scatter and lie to the left of the data points from the Dominion Reef Sequence. Wetherill (1956a) concluded that most of the uranium minerals were formed

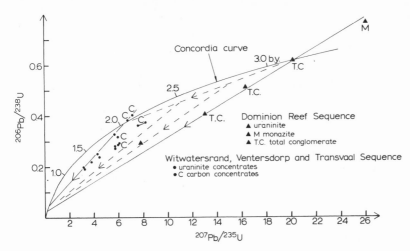

Fig. 19. Concordia diagram showing the U—Pb data points of minerals from the Witwatersrand Triad and the Transvaal Sequence (Nicolaysen et al., 1962; Louw and Strelow, 1954). The scale of the drawing did not require a recalculation of the ages with the new decay constants. The tie line connecting two data points indicates the uncertainty of the data from the Witwatersrand, Ventersdorp and Transvaal Sequences, due to the common-lead correction based on galena samples which show a considerable variation in their isotopic compositions (Fig. 20).

2040 ± 50 m.y. ago and suffered a recent lead loss 200 ± 150 m.y. ago. The data points to the right of the bounding line 2000—200 m.y., indicate that these samples are older, possibly also 3100 m.y. old. A first drastic lead loss has occurred 2000 m.y. ago, and a second about 200 m.y. ago. One may therefore interpret the samples lying on the 2000—200 m.y. line as being of the same age, but having suffered a complete lead loss 2000 m.y. ago, and a second partial lead loss in recent times.

Independent evidence for an event 2000 m.y. ago, respectively for approximately 3000 m.y. old components in the sediments of the Witwatersrand Sequence, was found by Burger et al. (1962) when analyzing galena samples from the sequence. The data points scatter on the $^{207}Pb/^{204}Pb$ versus $^{206}Pb/^{204}Pb$ diagram (Fig. 20) within a field which possesses a well-defined lead line as an upper boundary. The authors concluded that the lead line represents an isochron. From its slope of 0.37 ± 0.01 and from the time of lead loss 2040 m.y. ago, one may deduce the time of formation of the uranium-bearing minerals, or vice versa (eq. 19), which contributed lead to the galena. The time of crystallization of the uranium minerals is calculated at 2980 ± 100 m.y. and is in remarkable agreement with the time of formation as deduced from the U—Pb analyses on uraninites and monazites from the Dominion Reef Sequence (3100 ± 100 m.y.), as reported by Nicolaysen et al. (1962).

The period of alteration of the uranium minerals 2040 m.y. ago coincides in time with the formation of the Bushveld granite at Houtbeck as determined by the U—Pb method

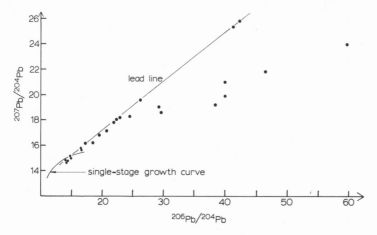

Fig. 20. Data points of galenas from the Witwatersrand Sequence (Burger et al., 1962).

on monazites (Nicolaysen et al., 1958). During this period the sediments of the Witwatersrand Triad were obviously mildly metamorphosed. Not all of the galena samples analyzed by Burger et al. (1962) fitted the lead line. Some of the data points fall below the isochron. The lead expelled during the second disturbance of the U–Pb systems (Fig. 19) had a ^{207}Pb/^{206}Pb ratio smaller than the lead exsolved 2040 m.y. ago. The authors, thus, attributed this scatter to the subsequent lead loss of the uranium minerals, approximately 200 m.y. ago. At the time of the second lead loss, either new galena crystals were formed or else new lead was added to existing galenas.

In a recent study Köppel and Saager (1974) tested the gold-source hypothesis of Viljoen et al. (1970) by measuring the lead-isotopic composition of the trace lead in detrital and in-situ formed, authigenic pyrites from the Witwatersrand Sequence and comparing it with the lead-isotopic composition of pyrites and galenas in primary gold deposits of the Archean greenstone belts of the Kaapvaal Craton. In order to obtain conclusive evidence, it was necessary to show that the authigenic minerals of the Witwatersrand Sequence have lead-isotopic compositions that plot in a field distinct from that of minerals accompanying the primary gold ores. A somewhat hypothetical field of lead compositions in authigenic minerals can be delineated in the following way: (a) the crystallization age of the uranium-bearing phase is 3100 m.y.; (b) only two periods of lead losses occurred: the first 2040 m.y. ago, and the second in recent times, between 200 and 0 m.y. ago. The first period of lead loss eventually produced the lead line observed by Burger et al. (1962). It constitutes an upper boundary for the data field. The lower boundary is set by the line starting at the lower end of the upper boundary and possessing a slope that corresponds to the present-day ^{207}Pb/^{206}Pb ratio of 2040 m.y. old uranium minerals (Fig. 21).

If more than one period of lead loss occurred after 2040 m.y., the slope of the lower boundary will be smaller, because the present-day ^{207}Pb/^{206}Pb ratio of an uranium mineral which suffered an additional lead loss, e.g. 1000 m.y. ago, will be smaller. It should be noted that all data points of Burger et al. (1962) and all data points of authigenic pyrites, reported by Köppel and Saager (1974), plot within the predicted field.

The experimentally determined area of data points of sulfides from primary gold deposits (Fig. 21) does not completely overlap the field of authigenic minerals of the Witwatersrand Sequence. One of the detrital Witwatersrand pyrites (No. 25, Fig. 21) had a lead-isotopic composition outside the area of authigenic minerals. This composition was identical with the lead observed in a galena from the Rosetta Mine in the Barberton Mountain Land, which is one of the greenstone terrains of the Kaapvaal Craton. The hypothesis of Viljoen et al. (1970) concerning the source of the gold in the Witwatersrand Sequence is, thus, supported by this observed identity of the lead-isotopic compositions. Some other features of the data obtained by Köppel and Saager (1974) are noteworthy:

(a) The isotopic composition of lead in pyrites is not homogeneous. Leaching of the pyrites with HCl extracts a lead that is usually enriched in ^{206}Pb, ^{207}Pb and ^{208}Pb. This excess is probably due to the presence of either small amounts of uranium and thorium, or else of admixed lead that may derive from interstitial uranium and thorium of the host rock. The lead was transported by water and eventually absorbed by sulphides.

(b) Systematic relationships may be observed in the lead-isotopic compositions of pyrites collected within a very restricted volume of rock. The samples 3.1, 3.2, 3.4, 3.6 and a leach form a well-defined lead line (line a in Fig. 22). The position of this line is remarkable in that it does not enter the field of normal common leads. The slope of the line is 0.117 ± 0.001 and thus very close to the present-day $^{207}Pb/^{206}Pb$ ratio (0.126 ± 0.004) of a 2040 ± 50 m.y. old uranium mineral. The presence of considerable amounts of uranium in these pyrites was verified by analyses. It therefore appears reasonable to

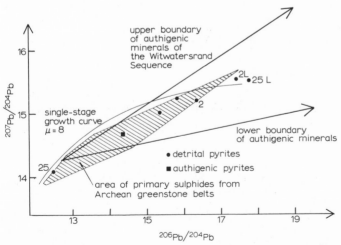

Fig. 21. Showing the relationship of the areas containing the data points of authigenic and detrital minerals of the Witwatersrand Sequence (Köppel and Saager, 1974; Saager and Köppel, 1975).

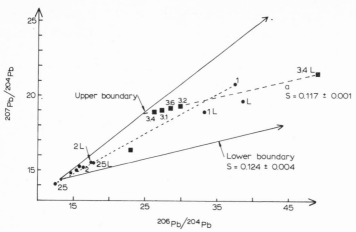

.Fig. 22. Authigenic (squares) and detrital (full circles) pyrites from the Witwatersrand Sequence, see text (Köppel and Saager, 1974).

assume that all these pyrites were formed approximately 2000 m.y. ago. They incorporated a lead with a composition given by the intersection of line a with the upper boundary of the field for the authigenic minerals of the Witwatersrand. At the time of their formation, the authigenic pyrites also incorporated uranium, which then caused the data points to lie on the isochron a.

Another remarkable lead line was obtained by three detrital pyrite samples which were collected at the same locality of a mine. The data points 25, 2 and 1 of the leached samples lie on a line with a slope of 0.271 ± 0.003 which corresponds to the present-day $^{207}Pb/^{206}Pb$ ratio of a 3300 m.y. old U–Pb system. Slopes yielding such ages were commonly observed in pyrite samples from greenstone belts (Saager and Köppel, 1976) and also in whole-rock samples (Sinha, 1972). The acid wash obviously removed all the lead that was added after the deposition of the pyrites in the Witwatersrand sediments.

Archean greenstone belts of the eastern Transvaal

Geological setting. The greenstone belts of the Kaapvaal Craton in the Republic of South Africa constitute some of the oldest and best preserved volcano-sedimentary assemblages known on earth and have been used as a model for other Archean greenstone terrains (Anhaeusser, 1971, 1973; Viljoen and Viljoen, 1970; and Chapter 7 by Anhaeusser and Button, Vol. 5). The greenstone belts have aroused considerable interest among earth scientists as they not only belong to the oldest rocks of the African continent, but also contain the oldest known remnants of life and at their base possess mafic volcanic rocks of an unusual chemical composition.

The tightly folded greenstone belts are scattered about in the complex granitic basement of the Kaapvaal Craton, and are collectively referred to as remnants of the *Swazi-*

land Sequence. Four major greenstone belts can be distinguished; they are the Barberton Mountain Land, the Pietersburg Schistbelt, the Murchison Range, and the Sutherland Range (Fig. 18).

In the Barberton Mountain Land, which is particularly well preserved and investigated, the Swaziland Sequence is subdivided into three lithological units (Viljoen and Viljoen, 1969; Anhaeusser 1971). The *Onverwacht Group* at the base of the sequence is characterized by mafic volcanics. They are conformably overlain by the argillaceous sediments of the *Fig Tree Group* which, in turn, is followed by the principally arenaceous sediments of the *Moodies Group*.

The rocks of the Swaziland Sequence are surrounded by a variety of granites which in the Barberton Mountain Land have been dated at 3260 ± 40 m.y. (U–Pb) to 2550 ± 70 m.y. (Rb–Sr) (Allsopp et al., 1968; Oosthuyzen, 1970). These granites deformed the assemblages to a varying degree and produced the low-grade regional metamorphism that can be observed throughout the greenstone belts of the craton.

In common with most other Early Precambrian greenstone terrains, the greenstone belts of the Kaapvaal Craton have been, and in the case of the Barberton Mountain Land still are, important gold producers. The gold occurrences are restricted to the volcano-sedimentary assemblages, the surrounding granitic basement being devoid of any significant gold or sulphide ore deposits (Viljoen et al., 1969).

Detailed geological and mineralogical descriptions of the gold and sulphide deposits in the greenstone belts of the eastern Transvaal are given amongst others by Van Eeden et al. (1939); De Villiers (1957); Schweigart and Liebenberg (1966); Viljoen et al. (1969), and Saager (1973 and 1974). Broadly speaking, the gold and sulphide occurrences can be divided[1] into the three following groups (Saager, 1973):

(a) *Strata-bound massive sulphide deposits* which often contain small gold values. The deposits usually occur in the Onverwacht Group and are associated with acid volcanics. The mineralizations are considered to be of an exhalative sedimentary origin and exhibit, in cases, a considerable metamorphic overprint. Sub-economic copper, zinc, and lead-bearing massive sulphides are only known from the northern flank of the Murchison greenstone belt, at Letaba. The ore bodies form a number of strata-bound, metamorphic deposits associated with extrusive quartz-prophyries which probably belong to the Onverwacht Group. A submarine exhalative sedimentary origin has also been postulated for the economically very important antimony ores near Gravelotte in the Murchison Range (Saager, 1974). The distinctly strata-bound ore lenses can be traced along a distance of some 50 km and are invariably associated with intermediary to acidic volcanics and dolomitic carbonates. Generally the ore has been intensively metamorphosed, which led to an extensive reconstitution and recrystallization of its constituents, namely the sulphides stibnite and berthierite.

---

[1] See Chapter 7 by Anhaeusser and Button, Vol. 5, for classification of economic deposits.

(b) *Complex disseminated sulfide ores* in porphyry bodies. The ore occurs as mineral veins and veinlets in the fracture system of the small porphyry stocks that intrude the Onverwacht Group rocks. The ores are considered to be subvolcanic equivalents to the stratiform massive sulphides which were formed during the emplacement of porphyry.

(c) *Gold–quartz veins* which form the most abundant and economically most important gold occurrences in the greenstone belts. Viljoen et al. (1969 and 1970), and Saager (1973 and 1974) showed that the gold and other elements of the veins were derived largely from the country rock by lateral secretion during the metamorphic period. The intruding granites, thus, played an important role in the initiation of the ore-forming processes, as they triggered the metamorphic mobilization, but they cannot be regarded as the ultimate sources of the gold and the sulphides present in the vein deposits.

Lead isotope data. In spite of many attempts, only minimum ages for these rocks have so far been established (see Saager and Köppel, 1976). Work on lead isotopes has been carried out by Ulrych et al. (1967) who found a least radiogenic ore lead. Sinha (1972) analyzed whole-rock samples by the U–Th–Pb method. Saager and Köppel (1976) reported lead-isotopic compositions of sulphides from primary gold deposits and massive sulphide occurrences of the Barberton Mountain Land and the Murchison greenstone belt.

The most important finding of Saager and Köppel (1976) was the existence of a lead line (Fig. 23) defined by galena, pyrite and antimonite samples from various types of deposits of the Barberton Mountain Land. The spread of the $^{208}Pb/^{204}Pb$ ratios is small compared to the analytical uncertainty. Therefore, the $^{208}Pb/^{204}Pb$ versus $^{206}Pb/^{204}Pb$ plot offers no opportunity of deciding whether the line represents an isochron or possibly a mixing line. From the metallogenetic concepts concerning the origin of the mineralizations, it follows that a mixing line may be envisaged if the metals were partly derived from the surrounding granites. The lead is then a mixture of a granitic lead with lead possibly mobilized from the host rock of the occurrences. If, however, the metals were derived by lateral secretion from the host rocks (Viljoen et al., 1969; Saager, 1973, 1974), then the lead line represents rather an isochron than a mixing line. Lead from feldspars of the surrounding granites was analyzed by Sinha and Tilton (1973) and by Oosthuyzen (1970). As seen in Fig. 23, the data points plot to the right of the sulphide lead line which, therefore, cannot be interpreted as a mixing line in the manner outlined above. Rather the line represents an isochron and further underlines the metallogenetic concept proposed by Viljoen et al. (1969) and Saager (1973 and 1974).

Rb–Sr data of a pegmatite crosscutting a primary gold ore at New Consort in the Barberton Mountain Land indicates a minimum age of 3030 ± 40 m.y. for the mineralization (Allsopp et al., 1968). If one sets t_2 = 3030 m.y. (eq. 19) one obtains a t_1 of 3800 m.y.. *This age is a maximum age for the Onverwacht Group of the Swaziland Sequence.* It should be noted that this maximum age closely agrees with the age of the major differentiation as proposed by Allsopp et al. (1974) in their two-stage lead evolution model.

Saager and Köppel (1976) pointed to the existence of a conformable ore deposit at

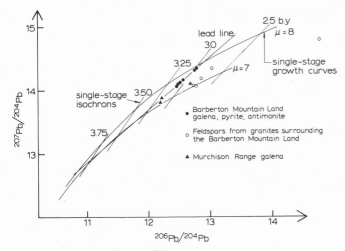

Fig. 23. Primary mineralizations in Archean greenstone belts of the Kaapvaal craton (Saager and Köppel, 1976). The position of the lead line with respect to the feldspar data (Sinha and Tilton, 1973; Oosthuyzen, 1970) suggests that the lead line is an isochron and not a mixing line of a lead component derived from the granites and another from the rocks of the greenstone belt.

Letaba in the Murchison Range (see previous section and Fig. 18). The lead-isotopic data, however, does not fit the model proposed by Allsopp et al. (1974), the apparent μ value being too small. Further analyses are needed to verify homogeneity of the isotopic composition of the lead from this deposit. Furthermore, data on the age of the host rock, a mildly metamorphosed extrusive quartz prophyry, and on the age of the metamorphic event are needed before placing the deposit on the list of conformable ores defining a lead-growth curve.

Old Star, a vein-type gold deposit in the Murchison Range, is remarkable because the lead is the least radiogenic so far observed in galenas. Yet the isotopic composition varies in different hand specimens, thereby verifying once again the experience that the vein-type deposits often contain multistage leads.

SUMMARY AND CONCLUSIONS

The isotopic systems of uranium, thorium and lead are a powerful tool in helping to solve some of the genetic problems of ore deposition.

Knowing the age of formation of an ore deposit is a prerequisite for the understanding of its genesis. Although obstacles do exist for using the U—Th—Pb methods for establishing the age of deposition, numerous successful attempts have been made especially in vein-type deposits containing uranium-bearing minerals. In the absence of such minerals, differing lead-isotopic ratios forming a lead line in the lead—lead diagram usually allow of establishing the age of deposition, or, at least, they allow of stating maximum and minimum ages.

Detailed isotopic investigations usually reveal a complex history of the ores demonstrating that after their formation reworking, due to tectonic activities, action of groundwater or heating during metamorphism of the mineral constituents, occurred. Such information may be gathered from investigations covering entire mineral belts but also from detailed studies of a single hand specimen. To obtain a maximum of information, both types of approaches are needed. Furthermore, such studies have to be accompanied by careful macroscopic and microscopic examinations of the geological setting of the sample and its composition.

Knowing the age of an ore body is also of importance in the search of new occurrences. It is worth remembering that the gold deposits of the Witwatersrand were at first considered to occur in recent alluvial deposits which, of course, had consequences in the search for the primary deposits.

Perhaps the most surprising aspect of lead-isotope geochemistry is that the leads of many conformable deposits form an evolution curve that can be explained by a simple two-stage model which allows for slight variations of the μ value of the second stage. For practical purposes, it is noteworthy that the ages obtained by the two-stage model agree closely with the ages established by other methods.

It is evident that additional data points along the growth curve in the age range of 1000 m.y. and higher are desirable. Before adding new deposits to the list of those defining the curve, it is, however, necessary to prove the homogeneity of the lead-isotopic composition and to know the age of the host rock.

It is not yet well understood whether the lead of the conformable deposits derives from a source with a rather homogeneous U/Pb ratio maintained over a substantial period of the earth's history, or whether extensive mixing occurred prior to the formation of the ore and possibly also, to a slighter extent, after its deposition during diagenesis or subsequent metamorphism.

A comparison of the lead of modern conformable ore deposits with that of recently formed oceanic volcanic rocks reveals that mantle-derived lead plays a significant role in some ores, whereas in others such a component, although it may be present, cannot be detected.

It would be worthwhile attempting to establish the original lead-isotopic composition of ancient basic rocks which are possibly of mantle origin and associated with sulphide deposits. The difficulties of such a project are considerable. One would have to establish the age of the rocks by independent means. One has to remember that the mobility of uranium is high and that therefore knowledge of the present-day uranium concentration generally does not allow for an adequate lead correction to arrive at the original composition. Furthermore, even mild metamorphism may change the isotopic composition of lead also in mineral phases virtually free of uranium.

Comparison of the lead-isotopic composition of ore lead with lead from associated rocks can be used to establish the source of the ore lead, or at least to exclude certain rock types as a source. Also the origin of ore minerals in placer deposits may be elucidated with the help of their lead-isotopic composition when compared with that of ores in primary deposits.

Finally, the isotopic composition of galenas may show a regional pattern, such as in the Mississippi Valley ores, which may be used to delineate areas susceptible of containing economic occurrences.

Analyzing mineral phases that contain only trace amounts of lead, such as pyrite and others, poses additional problems on the one hand, but on the other hand may yield information that can otherwise not be readily obtained from lead minerals. The smaller the lead content of a mineral is, the more susceptible the isotopic composition becomes for additions of lead from uranium- and thorium-bearing phases or from interstitial matter containing uranium. Lead lines obtained from minerals containing traces of lead will therefore in general be more difficult to interpret correctly than lead lines obtained from galena samples. In the former case, a lead line may be the result of small quantities of uranium incorporated at the time of their formation. One may also envisage the case that migrating solutions leached radiogenic lead from uranium- and thorium-bearing phases or from interstitial matter of the surrounding rock and that this lead eventually was absorbed by the sulphides. Finally, radiogenic and a common lead may not have been thouroughly mixed at the time of ore formation, leading thereby to variable isotopic compositions in the sulphides. Such studies should therefore be accompanied by a careful search for uranium in the sample material. Eventually investigations of the isotopic composition of trace lead with the aid of an ion microprobe mass analyzer may further help to evaluate the possibilities mentioned above.

REFERENCES

Allegre, C.J. and Richard, G., 1973. *Introduction a la geochimie*. Presses Universitaires de France, Paris, 220 pp.

Allegre, C.J., Albarede, F., Grünenfelder, M. and Köppel, V., 1974. $^{238}U/^{206}Pb - ^{235}U/^{207}Pb - ^{233}Th/^{208}Pb$ zircon geochronology in alpine and non-alpine environments. *Contrib. Mineral. Petrol.*, 43: 163–194.

Allsopp, H.L., 1964. Rb–Sr ages from the Western Transvaal. *Nature (London)*, 204: 361–363.

Allsopp, H.L., Ulrych, T.J. and Nicolaysen, L.O., 1968. Dating some significant events in the history of the Swaziland System by the Rb–Sr isochron method. *Can. J. Earth Sci.*, 5: 605–619.

Allsopp, H.L., Davies, R.D., Kramers, J., Nicolaysen, L.O. and Stacey, J.S., 1974. Isotopic evidence for a new model for the formation of the early crust. In: *2nd Annu. Meet. Eur. Geophys. Soc.*, University of Trieste. (Abstract.)

Amstutz, G.C. and Park, W.C., 1967. Stylolites of diagenetic age and their role in the interpretation of the southern Illinois fluorspar deposits. *Miner. Deposita*, 2: 44–53.

Amstutz, G.C., Uhley, R.P. and El Baz, F., 1961. Sedimentary features in the layered sulfide deposits of Frederiktown, Mo. *Geol. Soc. Am. Spec. Pap.*, 68: 128 pp.

Anderson, D.E., 1966. *Structural and Metamorphic Studies in the Mount Robe Area, Broken Hill, N.S.W.* Thesis, University of Sydney, Sydney, N.S.W., unpublished.

Anhaeusser, C.R., 1971. The Barberton Mountain Land, South Africa — A guide to the understanding of the Archean geology of Western Australia. *Geol. Soc. Aust. Spec. Publ.*, 3: 103–119.

Anhaeusser, C.R., 1973. The evolution of the Early Precambrian crust of southern Africa. *Philos. Trans. R. Soc. London*, A 273: 359–388.

Armstrong, R.L., 1968. A model for the evolution of Sr and Pb in a dynamic Earth. *Rev. Geophys.*, 6 (2): 175–200.

Armstrong, R.L. and Hein, S.M., 1973. Computer simulation of Pb and Sr isotope evolution of the Earth's crust and upper mantle. *Geochim. Cosmochim. Acta*, 37: 1–18.

Bate, G.L. and Kulp, J.L., 1955. Variations in the isotopic compositions of common lead and the history of the earth's crust. *Trans. Am. Geophys. Union*, abstract.

Bennet, E.M., 1965. Lead–zinc–silver and copper deposits of Mount Isa. In: *Commonw. Min. Metall. Congr., 8th*, 1: 233–246.

Bischoff, J.L. and Manheim, F.T., 1969. Economic potential of the Red Sea heavy metal deposit. In: E.T. Degens and D.A. Ross (Editors), *Hot Brines and Recent Heavy Metal Deposits in the Red Sea*. Springer, New York, N.Y., pp. 368–401.

Both, R.A., 1973. Minor element geochemistry of sulfide minerals in the Broken Hill Lode (N.S.W.) in relation to the origin of the ore. *Miner. Deposita*, 8: 349–370.

Boyle, R.W. and Dass, A.S., 1971. Origin of the native silver veins at Cobalt, Ontario. *Can. Mineral.*, 11: 414–417.

Brock, B.B. and Pretorius, D.A., 1964. Rand basin sedimentation and tectonics. In: S.H. Haughton (Editor), *The Geology of Some Ore Deposits in Southern Africa*. *Geol. Soc. S. Afr.*, 1: 549–599.

Brown, J.S., 1967. Isotopic zoning of lead and sulfur in Southeast Missouri. *Econ. Geol. Monogr.*, 3: 410–426.

Brown, J.S., 1970. Mississippi Valley type lead–zinc ores. *Miner. Deposita*, 5: 103–119.

Burger, A.J., Nicolaysen, L.O. and De Villiers, J.W.L., 1962. Lead isotopic compositions of galenas from the Witwatersrand and Orange Free State, and their relation to the Witwatersrand and Dominion Reef uraninites. *Geochim. Cosmochim. Acta*, 26: 25–59.

Campbell, J.D., 1958. En echelon folding. *Econ. Geol.*, 53: 448–472.

Campbell, N., 1967. Tectonics, reefs and stratiform lead–zinc deposits of the Pine Point area, Canada. In: J.S. Brown (Editor), *Genesis of Stratiform Lead–Zinc–Barite–Fluorite Deposits, a Symposium. Econ. Geol. Monogr.*, 3: 59–70.

Cannon Jr., R.S. and Pierce, A.P., 1967. Isotopic varieties of lead in stratiform deposits. *Econ. Geol. Monogr.*, 3: 427–434.

Cannon Jr., R.S., Pierce, A.P., Antweiler, J.L. and Buck, K.L., 1961. The data of lead isotope geology related to problems of ore genesis. *Econ. Geol.*, 56: 1–38.

Cannon Jr., R.S., Pierce, A.P. and Delevaux, M.H., 1963. Lead isotope variation with growth zoning in a galena crystal. *Science*, 142: 574–576.

Cannon Jr., R.S., Pierce, A.P. and Antweiler, J.C., 1971. Suggested uses of lead isotopes in exploration. *Can. Inst. Min. Metall., Spec. Vol.*, 11: 457–463.

Carruthers, D.S., 1965. An environmental view of Broken Hill ore occurrence. In: *Commonw. Min. Metall. Congr., 8th*, 1: 339–351.

Commissariat à l'Energie Atomique, 1974. Le phénomène d'Oklo. *Bull. Inf. Sci. Tech.*, 193.

Cooper, J.A. and Richards, J.R., 1969. Lead isotope measurements in sediments from Atlantis II and Discovery Deep areas. In: E.T. Degens and D.A. Ross (Editors), *Hot Brines and Recent Heavy Metal Deposits in the Red Sea*. Springer, New York, N.Y.

Cumming, G.L. and Gudjurgis, P.J., 1973. Alteration of trace lead isotopic ratios by post-ore metamorphic and hydrothermal activity. *Can. J. Earth Sci.*, 10: 1782–1789.

Cumming, G.L. and Robertson, D.K., 1969. Isotopic composition of lead from the Pine Point deposit. *Econ. Geol.*, 64: 731–732.

Cumming, G.L., Burke, M.D., Tsong, F. and McCullough, M., 1971. A digital mass spectrometer. *Can. J. Phys.* 49: 956–965.

Czamanske, G.K. and Rye, R.O., 1974. Experimentally determined sulfur isotope fractionations between sphalerite and galena in the temperature range 600°C–275°C. *Econ. Geol.*, 69: 17–25.

Dasch, E.J., Dymond, J.R. and Heath, G.R., 1971. Isotopic analysis of metalliferous sediments from the East Pacific Rise. *Earth Planet. Sci. Lett.*, 13: 175–180.

Davidson, C.F., 1953. The gold–uranium ores of the Witwatersrand. *Min. Mag.*, 88: 73–85.

Davidson, C.F., 1957. On the occurrence of uranium in ancient conglomerates. *Econ. Geol.*, 52: 668–693.

Degens, E.T. and Ross, D.A., 1969. *Hot Brines and Recent Heavy Metal Deposits in the Red Sea.* Springer, New York, N.Y., 600 pp.

Delevaux, M.H. and Doe, B.R., 1974. Preliminary report on uranium, thorium and lead contents and lead isotopic composition in sediment samples from the Red Sea. In: R.B. Whitmarsh, O.E. Weser and D.A. Ross (Editors), *Initial Reports of the Deep See Drilling Project*, 23: 943–946.

Den Tex, E., 1958. Studies in comparative petrofabric analysis: the Broken Hill Lode and its immediate wall rock. *Proc. Australas. Inst. Min. Metall.* (Stillwell Anniv. Vol.), pp. 77–104.

De Villiers, J.E., 1957. The mineralogy of the Barberton gold deposits. *Geol. Surv. S. Afr. Bull.*, 24: 60 pp.

Doe, B.R., 1962. Relationships of lead isotopes among granites, pegmatites, and sulphide ores near Balmat, New York. *J. Geophys. Res.*, 67: 28–95.

Doe, B.R., 1970. *Lead Isotopes*. Springer, Berlin, 137 pp.

Doe, B.R. and Delevaux, M.H., 1972. Source of lead in southeast Missouri galena ores. *Econ. Geol.*, 67: 409–425.

Doe, B.R. and Stacey, J.S., 1974. The application of lead isotopes to the problem of ore genesis and ore prospect evaluation: a review. *Econ. Geol.*, 69: 757–776.

Dooley Jr., J.R., Tatsumoto, M. and Rosholt, J.N., 1964. Radioactive disequilibrium studies of roll features, Shirley Basin, Wyoming. *Econ. Geol.*, 59: 586–595.

Dooley Jr., J.R., Granger, H.C. and Rosholt, J.N., 1966. Uranium-234 fractionation in the sandstone type uranium deposits of the Ambrosia Lake district, New Mexico. *Econ. Geol.*, 61: 1362–1382.

Dooley Jr., J.R., Harshman, E.N. and Rosholt, J.N., 1974. Uranium–lead ages of the uranium deposits of the Gas Hills and Shirley Basin, Wyoming. *Econ. Geol.*, 69: 527–531.

Fanale, E.P., 1964. *Helium in Magnetite*. Thesis, Columbia University, New York, N.Y., unpublished.

Farquharson, R.B. and Wilson, C.J.L., 1971. Rationalization of geochronology and structure at Mount Isa. *Econ. Geol.*, 66: 574–582.

Freeze, A.C., 1966. On the origin of the Sullivan orebody, Kimberley, B.C. *Can. Inst. Min. Metall. Spec. Vol. 8*: 263–294.

Fritz, P., 1969. The oxygen and carbon isotopic composition of carbonates from the Pine Point lead–zinc ore deposits. *Econ. Geol.*, 64: 733–742.

Gale, N.H. and Mussett, A.E., 1973. Episodic uranium–lead models and the interpretation of variations in the isotopic composition of lead in rocks. *Rev. Geophys. Space Phys.*, 11: 37–86.

Gilmour, P., 1965. The origin of the massive sulphide mineralization in the Noranda District, Northwestern Quebec. *Proc. Geol. Assoc. Can.*, 16: 63–81.

Gilmour, P., 1971. Strata-bound massive pyritic sulfide deposits – a review. *Econ. Geol.*, 66: 1239–1243.

Glasson, K.R. and Paine, V.R., 1965. Lead–zinc–copper ore deposits of Lake George Mines, Captain's Flat. In: *Commonw. Min. Metall. Congr., 8th*, pp. 423–431.

Graton, L.C., 1930. Hydrothermal origin of the gold in the Rand Gold Deposits, part 1. *Econ. Geol.*, 25: 185 pp. (supplement).

Grogan, R.M. and Bradbury, J.C., 1967. Origin of the stratiform fluorite deposits in southern Illinois. *Econ. Geol. Mem.*, 3: 40–51.

Grootenboer, J. and Schwarcz, H.P., 1969. Experimentally determined sulfur isotope fractionations between sulfide minerals. *Earth Planet. Sci. Lett.*, 7: 162–166.

Gustafson, J.K., 1939. *Geological investigation in Broken Hill, Final Report*. Central Geological Survey, unpublished.

Hackett Jr., J.P. and Bischoff, J.L., 1973. New data on the stratigraphy, extent, and geologic history of the Red Sea geothermal deposits. *Econ. Geol.*, 68: 553–564.

Hall, W.E. and Friedmann, I., 1969. Oxygen and carbon isotopic composition of ore and host rock of selected Mississippi Valley deposits. *U.S. Geol. Surv. Prof. Pap.*, 650 (C): C140–C148.

Hall, W.E. and Heyl, A.V., 1968. Distribution of minor elements in ore and host rock, Illinois–Kentucky fluorite district and Upper Mississippi Valley zinc–lead district. *Econ. Geol.*, 63: 655–670.

Helgeson, H.C., 1967. Silicate metamorphism in sediments and the genesis of hydrothermal solution. *Econ. Geol. Mem.*, 3: 333–342.

Helmstaedt, H., 1970. Structural geology of Portage Lake area, Bathurst–Newcastle District, N.B. *Geol. Surv. Can. Pap.*, 70 (28): 52 pp.

Heyl, A.V., 1967. Some aspects of genesis of stratiform lead–zinc–barite–fluorite deposits in the United States. *Econ. Geol. Mem.*, 3: 20–31.

Heyl, A.V., Delevaux, M.H., Zartman, R.E. and Brock, M.R., 1966. Isotopic study of galenas from the Upper Mississippi Valley, the Illinois–Kentucky and some Appalachian Valley mineral districts. *Econ. Geol.*, 61: 933–961.

Heyl, A.V., Landis, G.P. and Zartman, R.E., 1974. Isotopic evidence for the origin of Mississippi Valley-type mineral deposits: a review. *Econ. Geol.*, 69: 992–1006.

Hobbs, B.E., 1966. The structural environment of the northern part of the Broken Hill orebody. *J. Geol. Soc. Aust.*, 13: 315–338.

Hobbs, B.E., Ransom, D.M., Vernon, R.H. and Williams, P.F., 1968. The Broken Hill ore body, Australia – a review of recent work. *Miner. Deposita*, 3: 293–316.

Holmes, R. and Tooms, J.S., 1972. Dispersion from a submarine exhalative orebody. In: *Proc. Int. Geochem. Expl. Symp. London, 4th*, pp. 193–202.

Houtermans, F.G., 1946. The isotope frequency in natural lead and the age of the uranium. *Naturwissenschaften*, 33: 185–186.

Houtermans, F.G., 1953. Determination of age of the earth from the isotopic composition of meteoric lead. *Nuovo Cimento*, 10: 1623–1633.

Hutchinson, R.W., Ridler, R.H. and Suffel, G.G., 1971. Metallogenic relationships in the Abitibi Belt, Canada: a model for Archean metallogeny. *Can. Inst. Min. Metall. Bull.*, 64: 106–115.

Jackson, S.A. and Beales, F.W., 1967. An aspect of sedimentary basin evolution: the concentration of Mississippi Valley-type ores during late stages of diagenesis. *Bull. Can. Pet. Geol.*, 15: 383–433.

Jackson, S.A. and Folinsbee, R.E., 1969. The Pine Point lead–zinc deposits, N.W.T., Canada. Introduction and paleoecology of the Presqu'ile Reef. *Econ. Geol.*, 64: 711–717.

Jaffey, A.H., Flynn, K.F., Glendenin, I.E., Bentley, W.C. and Essling, A.M., 1971, Precision measurements of half-lives and specific activities of ^{235}U and ^{238}U. *Phys. Rev.*, C: 4/5.

Kanasewich, E.R., 1962. Approximate age of tectonic activity using anomalous lead isotopes. *Geophys. J. R. Astron. Soc.*, 7: 158–168.

Kanasewich, E.R., 1968. The interpretation of lead isotopes and their geological significance. In: E.I. Hamilton and R.M. Farquhar (Editors), *Radiometric Dating for Geologists*. Interscience, New York, N.Y., 506 pp.

Kanasewich, E.R. and Farquhar, R.M., 1965. Lead isotope ratios from the Cobalt–Noranda area, Canada. *Can. J. Earth. Sci.*, 2: 361–384.

Kenny, E.J. and Mulholland, C.St.J., 1941. The ore deposit of Captain's Flat, N.S.W.. *Proc. Aust. Inst. Min. Metall.*, 122: 45–64.

Kesler, S.E. and Ascarrunz, R., 1973. Lead–zinc mineralization in carbonate rocks, central Guatemala. *Econ. Geol.*, 68: 1263–1274.

King, H.F., 1958. Notes on ore occurrences in highly metamorphosed Precambrian rocks. *Proc. Australas. Inst. Min. Metall.* (Stillwell Anniv. Vol.), pp. 143–168.

King, H.F., 1965. Lead–zinc ore deposits of Australia. In: *Commonw. Min. Metall. Congr., 8th*, 1: 24–30.

King, H.F. and Thomson, B.P., 1953. Geology of the Broken Hill District. In: A.B. Edwards (Editor), *Geology of Australian Ore Deposits (5th Emp. Min. Met. Congr., Melbourne)*, pp. 533–577.

Knight, C.L., 1953. Regional geology of Mount Isa. In: A.B. Edwards (Editor), *Geology of Australian Ore Deposits (5th Emp. Min. Met. Congr., Melbourne)*, pp. 352–360.

Knight, C.L., 1958. Ore genesis – the source bed concept. *Econ. Geol.*, 53: 808–817.

Köppel, V., 1968. Age and history of the uranium mineralization of the Beaverlodge area, Saskatchewan. *Geol. Surv. Can. Pap.*, 67 (31): 111 pp.

Köppel, V. and Saager, R., 1974. Lead isotope evidence on the detrital origin of Witwatersrand pyrites and its bearing on the provenance of the Witwatersrand gold. *Econ. Geol.*, 69: 318–331.

Kulp, J.L., Bate, G.L. and Broecker, W.S., 1954. Present status of the lead method of age determination. *Am. J. Sci.*, 252: 345–365.

Kuovo, O., 1958. Radioactive age of some Finish Precambrian minerals. *Bull. Com. Geol. Finl.*, 182: 1–70.

Lawrence, L.J., 1973. Polymetamorphism of the sulphide ores of Broken Hill, N.S.W., Australia. *Miner. Deposita*, 8: 211–236.

Lea, E.R. and Rancourt, C., 1958. Geology of the Brunswick Mining and Smelting orebodies, Gloucester County, N.B. *Can. Inst. Min. Met. Trans.*, 61: 95–105.

Leach, G.B. and Wanless, R.K., 1962. Lead isotope and potassium studies in the East Kootenay district. *Geol. Soc. Am., Buddington Vol.*, pp. 241–279.

Lewis, B.R., Forward, P.S. and Roberts, J.B., 1965. Geology of the Broken Hill Lode, reinterpreted. In: *Commonw. Mining and Met. Congr., 8th*, 1: 319–335.

Liebenberg, W.R., 1955. The occurrence and origin of gold and radioactive minerals in the Witwatersrand System, the Dominion Reef, the Ventersdorp Contact Reef and the Black Reef. *Geol. Soc. S. Afr. Trans.*, 58: 101–223.

Louw, J.D. and Strelow, F.W.E., 1954. Geological age determinations on Witwatersrand uraninites using the lead isotope method. *Geol. Soc. S. Afr. Trans.*, 57: 209–230.

Mair, J.A., Maynes, A.D., Patchett, J.E. and Russell, R.D., 1960. Isotopic evidence on the origin and age of the Blind River uranium deposits. *J. Geophys. Res.*, 65: 341–348.

McKnight, E.T., 1967. Bearing of isotopic composition of contained lead on the genesis of Mississippi Valley ore deposits. *Econ. Geol. Monogr.*, 3: 392–398.

Mellor, E.T., 1916. The conglomerates of the Witwatersrand. *Trans. Inst. Min. Met.*, 25: 226–348.

Minčeva-Stefanova, J., 1967. The genesis of the stratiform lead–zinc ore deposits of the "Sedmochislenitsi" type in Bulgaria. In: J.S. Brown (Editor), *Genesis of Stratiform Lead–Zinc–Barite–Fluorite Deposits, a Symposium. Econ. Geol. Monogr.*, 3: 147–155.

Naldrett, A.J. and Kullerud, G., 1967. A study of the Strathcona mine and its bearing on the origin of the nickel–copper ores of the Sudbury district, Ontario. *J. Petrol.*, 8: 433–451.

Nicolaysen, L.O., 1957. Solid diffusion in radioactive minerals and the measurement of the absolute age. *Geochim. Cosmochim. Acta*, 11: 41–59.

Nicolaysen, L.O., 1961. Graphic interpretation of discordant age measurements on metamorphic rocks. *Ann. N.Y. Acad. Sci.*, 91: 198–206.

Nicolaysen, L.O., De Villiers, J.W.L., Burger, A.J. and Strelow, F.W.E., 1958. New measurements relating to the absolute age of the Transvaal System and of the Bushveld Igneous Complex. *Geol. Soc. S. Afr. Trans.*, 61: 137–163.

Nicolaysen, L.O., Burger, A.J. and Liebenberg, L.R., 1962. Evidence of the extreme age of certain minerals from the Dominion Reef conglomerates and the underlying granite in the Western Transvaal. *Geochim. Cosmochim. Acta*, 26: 15–23.

Nier, A.O., 1938. Variations in the relative abundances of the isotopes of common lead from various sources. *J. Am. Chem. Soc.*, 60: p. 1571.

Ohle Jr., E.L., 1967. The origin of the ore deposits of the Mississippi Valley type. *Econ. Geol. Mem.*, 3: 33–39.

Oosthuyzen, E.J., 1970. *The Geochronology of a Suite of Rocks from the Granitic Terrain surrounding the Barberton Mountain Land*. Thesis, University of the Witwatersrand, Johannesburg, 94 pp., unpublished.

Ostic, R.G., Russell, R.D. and Stanton, R.L., 1967. Additional measurements of the isotopic composition of lead from stratiform deposits. *Can. J. Earth. Sci.*, 4: 243–269.

Park, W.C. and Amstutz, G.C., 1968. Primary "cut and fill" channels and gravitational diagenetic features. *Miner. Deposita*, 3: 66–80.

Pehrman, G., 1931. Über eine Sulfidlagerstätte auf der Insel Attu im südwestlichen Finnland. *Acta Acad. Abo. Ser. B*, VI: 6 pp.

Pidgeon, R.T., 1967. A rubidium–strontium geochemical study of the Willyama Complex, Broken Hill, Australia. *J. Petrol.*, 8: 283–324.

Pinckney, D.M. and Rafter, T.A., 1972. Fractionation of sulfur isotopes during ore deposition in the Upper Mississippi Valley zinc–lead district. *Econ. Geol.*, 67: 315–328.

Pinckney, D.M. and Rye, R.O., 1972. Variation of $^{18}O/^{16}O$, $^{13}C/^{12}C$, texture and mineralogy in altered limestone in the Hill mine, Cave-in-Rock district, Illinois. *Econ. Geol.*, 67: 1–18.

Pretorius, D.A., 1966. Conceptual geological models in the exploration for gold mineralizations in the Witwatersrand basin. In: *Symposium on Matematical Statistics and Computer Application in Ore Valuation. S. Afr. Inst. Min. Metall.*, pp. 255–266.

Ramdohr, P., 1950. Die Lagerstätte von Broken Hill in New South Wales im Lichte der neuen geologischen Erkenntnisse und erzmikroskopischen Untersuchungen. *Heidelb. Beitr. Mineral. Petrogr.*, 2: 291–333.

Ramdohr, P., 1955. Neue Beobachtungen an den Erzen des Witwatersrandes in Südafrika und ihre genetische Bedeutung. *Abh. Dtsch. Akad. Wiss. Berlin, Kl. Math. Naturwiss.*, 5: 1–43.

Reynolds, P.H., and Dasch, E., 1971. Lead isotopes in marine manganese nodules and the ore lead growth curve. *J. Geophys. Res.*, 76: 5124–5129.

Richards, J.R., 1971. Major lead orebodies – mantle origin? *Econ. Geol.*, 66: 425–434.

Richards, J.R. and Pidgeon, R.T., 1963. Some age measurements on micas from Broken Hill, Australia. *J. Geol. Soc. Aust.*, 10: 243–259.

Richards, J.R., Yonk, A.K. and Keighn, C.W., 1972. Upper Mississippi Valley lead isotopes re-examined. *Miner. Deposita*, 7: 285–291.

Richards, S.M., 1966. The banded iron formations at Broken Hill, Australia, and their relationship to the lead–zinc orebodies. *Econ. Geol.*, 61: 72–96.

Robertson, D.K., 1973. A model discussing the early history of the Earth based on a study of lead isotope ratios from veins in some Archean cratons of Africa. *Geochim. Cosmochim. Acta*, 37: 2099–2124.

Roedder, E., 1967. Environment of deposition of stratiform (Mississippi Valley type) ore deposits, from studies of fluid inclusions. *Econ. Geol. Mem.*, 3: 349–362.

Rosholt, J.N., Harshman, E.N., Shields, W.R. and Garner, E.L., 1964. Isotopic fractionation of uranium related to roll features in sandstone, Shirley Basin, Wyoming. *Econ. Geol.*, 59: 570–585.

Russell, R.D., 1972. Evolutionary model for lead isotopes in conformable ores and in oceanic volcanics. *Rev. Geophys. Space Phys.*, 10: 529–549.

Russell, R.D. and Birnie, D.J., 1974. A bi-directional mixing model for lead isotope evolution. *Phys. Earth Planet. Inter.*, 8: 158–166.

Russell, R.D. and Farquhar, R.M., 1960. *Lead Isotopes in Geology*. Interscience, New York, N.Y., 243 pp.

Russell, R.D., Ulrych, T.J. and Kollar, F., 1961. Anomalous leads from Broken Hill. *J. Geophys. Res.*, 66: 1495–1498.

Russell, R.T. and Lewis, B.R., 1965. Gold and copper deposits of the Cobar district. In: *Commonw. Min. Metall. Congr.*, 8th, 1: 411–419.

Saager, R., 1969. The relationship of silver and gold in the Basal Reef of the Witwatersrand System, South Africa. *Miner. Deposita*, 4: 93–113.

Saager, R., 1973. Metallogenese präkambrischer Goldvorkommen in den vulkano-sedimentären Gesteinskomplexen (greenstone belts) der Swaziland-Sequenz in Südafrika. *Geol. Rundsch.*, 62: 888–901.

Saager, R., 1974. *Geologische und geochemische Untersuchungen an primären und sekundären Goldvorkommen im frühen Präkambrium Südafrikas: Ein Beitrag zur Deutung der primären Herkunft des Goldes in der Witwatersrand Lagerstätte*. Thesis, University of Heidelberg, 150 pp., unpublished.

Saager, R. and Köppel, V., 1976. Lead isotopes and trace elements from sulfides of Archean greenstone belts in South Africa – a contribution to the knowledge of the oldest known mineralizations. *Econ. Geol.*, 71, in press.

Sangster, D.F., 1972a. Precambrian volcanogenic massive sulphide deposits in Canada: a review. *Geol. Surv. Can. Pap.*, 72 (22): 44 pp.

Sangster, D.F., 1972b. Isotopic studies of ore-leads in the Hanson Lake–Flin Flon–Snow Lake mineral belt, Saskatchewan and Manitoba. *Can. J. Earth Sci.*, 9: 500–513.

Sasaki, A. and Krouse, H.R., 1969. Sulfur isotopes and the Pine Point lead–zinc mineralization. *Econ. Geol.*, 64: 718–730.

Sato, K. and Sasaki, A., 1973. Lead isotopes of the Black Ore ("Kuroko") deposits from Japan. *Econ. Geol.*, 68: 547–552.

Schidlowski, M., 1970. Untersuchungen zur Metallogenese im südwestlichen Witwatersrand-Becken (Oranje-Freistaat-Goldfeld, Südafrika). *Beih. Geol. Jahrb.*, 85: 80 pp.

Schweigart, H. and Liebenberg, W.R., 1966. Mineralogy and chemical behaviour of some refractory gold ores from the Barberton Mountain Land. *Natl. Inst. Metall. Johannesburg Res. Rep.*, 8: 72 pp.

Sharpe, J.W.N., 1949. The economic auriferous banket of upper Witwatersrand beds and their relationship to sedimentation features. *Geol. Soc. S. Afr. Trans.*, 52: 265–300.

Shaw, D.M., 1957. Comments on the geochemical implications of lead isotope dating of galena deposits. *Econ. Geol.*, 52: 570–573.

Shaw, S.E., 1968. Rb–Sr isotopic studies of the Mine Sequence rocks at Broken Hill mines. In: M. Radmanovich and J.T. Woodcock (Editors). *Australas. Inst. Min. Metall. Monogr. Ser.*, 3: 185–198.

Sinclair, A.J., 1966. Anomalous leads from the Kootenay arc, British Columbia. *Can. Inst. Min. Metall. Spec. Vol.*, 8: 1709–1717.

Sinha, A.K., 1972. U–Th–P systematics and the age of the Onverwacht Series, South Africa. *Earth Planet. Sci. Lett.*, 16: 219–227.

Sinha, A.K. and Tilton, G.R., 1973. Isotopic evolution of common lead. *Geochim. Cosmochim. Acta*, 37: 1823–1849.

Skinner, B.J., 1967. Precipitation of Mississippi Valley type ores: a possible mechanism. *Econ. Geol. Mem.*, 3: 363–370.

Slawson, W.F. and Russell, R.D., 1973. A multistage history for Flin Flon lead. *Can. J. Earth Sci.*, 10: 582–583.

Souch, B.E. and Podolsky, T., 1968. The sulfide ores of Sudbury: their particular relation to a distinctive inclusion-bearing facies of the nickel irruptive. *Econ. Geol. Monogr.*, 4: 138–165.

Stacey, J.S. and Kramers, J.D., 1975. *Earth Planet. Sci. Lett.*, 26: 207–221.

Stacey, J.S., Delevaux, M.H. and Ulrych, T.J., 1969. Some triple-filament lead isotope ratio measurements and an absolute growth curve for single stage leads. *Earth Planet. Sci. Lett.*, 6: 15–25.

Stanton, R.L., 1955. Lower Palaeozoic mineralization near Bathurst, N.S.W.. *Econ. Geol.*, 50: 681–714.

Stanton, R.L., 1959. Mineral features and possible mode of emplacement of the Brunswick Mining and Smelting orebodies, Gloucester County, N.B.. *Can. Inst. Min. Metall. Bull.*, 52: 631–642.

Stanton, R.L., 1960. General features of the conformable "pyritic" orebodies. *Can. Inst. Min. Metall. Trans.*, 63: 22–27.

Stanton, R.L., 1962. Elemental constitution of the Black Star orebodies, Mount Isa, and its interpretation. *Inst. Min. Metall. Trans.*, 72: 69–124.

Stanton, R.L., 1973. A preliminary account of chemical relationships between sulfide lode and "Banded Iron Formation" at Broken Hill, New South Wales. *Econ. Geol.*, 67: 1128–1145.

Stanton, R.L. and Richards, S.M., 1961. The abundance of lead, zinc, copper and silver at Broken Hill. *Proc. Aust. Inst. Min. Metall.*, 198: 309–367.

Stanton, R.L. and Russell, R.D., 1959. Anomalous leads and the emplacement of lead sulfide ores. *Econ. Geol.*, 54: 588–607.

Stillwell, F.L., 1922. The rocks in the immediate neighbourhood of the Broken Hill Lode and their bearing on its origin. *Geol. Surv. N.S.W. Mem.*, 8. Appendix II.

Stillwell, F.L., 1959. Petrology of the Broken Hill Lode and its bearing on ore genesis. *Proc. Australas. Inst. Min. Metall.*, 190: 1–84.

Stillwell, F.L. and Edwards, A.B., 1956. Uralite dolerite dykes in relation to the Broken Hill Lode. *Proc. Australas. Inst. Min. Metall.*, 178: 213–228.

Stief, L.R., Stern, T.W., Seiki Oshiro and Senftle, F.E., 1959. Tables for the calculation of lead isotope ages. *U.S. Geol. Surv. Prof. Pap.*, 334 (A).

Stockwell, C.H., 1964. Fourth report on structural provinces, orogenies and time-classification of rocks of the Canadian Shield. *Geol. Surv. Can. Pap.*, 64 (17).

Tatsumoto, M., Knight, R.J. and Allègre, C.J., 1973. Time difference in the formation of meteorites as determined from the ratio of lead-207 to lead-206. *Science*, 180: 1279–1293.

Thomson, B.P., 1956. *Tectonics and Archean Sedimentation of the Barrier Ranges, N.S.W..* Thesis, Adelaide University, N.S.W., unpublished.

Thomson, R., 1965. Casey and Harris townships. *Ont. Dep. Min. Geol. Rep.*, No. 26.

Thorpe, R., 1974. Lead isotope evidence on the genesis of the silver–arsenide vein deposits of the Cobalt and Great Bear Lake areas, Canada. *Econ. Geol.*, 69: 777–791.

Tilton, G.R., 1960. Volume diffusion as a mechanism for discordant Pb ages. *J. Geophys. Res.*, 65: 2933–2945.

Tilton, G.R., 1973. Isotopic lead ages of chondritic meteorites. *Earth Planet. Sci. Lett.*, 19: 321–329.

Tugarinov, A.I., Mityayaeva, N.M., Zamyatin, N.I., Shilov, L.I., Lebedev, V.P., Miyasishchev, V.V. and Shilov, V.I., 1972. Lead and sulfur isotopic composition and the process of ore deposition in the deposits of the Atasuy region. *Geochem. Int.*, 9: 336–350.

Ulrych, T.J. and Russell, R.D., 1964. Gas source mass spectrometry of leads from Sudbury, Ontario. *Geochim. Cosmochim. Acta*, 28: 455–469.

Ulrych, T.J., Burger, A. and Nicolaysen, L.O., 1967. Least radiogenic terrestrial leads. *Earth Planet. Sci. Lett.*, 3: 179–189.

Vaasjoki, O., 1956. A comparison of the minor base metal contents of some Finish galenas. *Bull Comm. Geol. Finl.*, 172: 45–53.

Van Eeden, O.R., Partridge, F.C., Kent, L.E. and Brandt, J.W., 1939. The mineral deposits of the Murchison Range, east of Leydsdorp. *Geol. Surv. S. Afr. Mem.*, 36: 152 pp.

Van Niekerk, C.B. and Burger, A.J., 1964. The age of the Ventersdorp System. *Geol. Surv. S. Afr. Ann.*, 3: 75–86.

Van Schmus, W.R., 1965. The geochronology of the Blind River-Bruce mines area, Ontario, Canada. *J. Geol.*, 73: 755–780.

Viljoen, M.J. and Viljoen, R.P., 1969. An introduction to the geology of the Barberton granite–greenstone terrain. *Geol. Soc. S. Afr. Spec. Publ.*, 2: 9–27.

Viljoen, M.J. and Viljoen, R.P., 1970. The geology and geochemistry of the layered ultramafic bodies of the Kaapmuiden area. *Geol. Soc. S. Afr. Spec. Publ.*, 1: 661–688.

Viljoen, R.P., Saager, R. and Viljoen, M.J., 1969. Metallogenesis and ore control in the Steynsdorp Goldfield, Barberton Mountain Land, South Africa. *Econ. Geol.*, 64: 778–797.

Viljoen, R.P., Saager, R. and Viljoen, M.J., 1970. Some thoughts on the origin and processes responsible for the concentration of gold in the Early Precambrian of southern Africa. *Miner. Deposita*, 5: 164–180.

Wasserburg, G.J., 1963. Diffusion processes in lead–uranium systems. *J. Geophys. Res.*, 68: 4823–4846.

Wetherill, G.W., 1956a. An interpretation of the Rhodesia and Witwatersrand age pattern. *Geochim. Cosmochim. Acta*, 9: 290–292.

Wetherill, G.W., 1956b. Discordant uranium–lead ages. *Trans. Geophys. Union*, 37: 320–326.

Williams, P.F., 1967. Structural analysis of the Little Broken Hill area, New South Wales. *J. Geol. Soc. Aust.*, 14: 317–331.

Wolf, K.H., 1976. Ore genesis influenced by compaction. In: G.V. Chilingar and K.H. Wolf (Editors) *Compaction of Coarse-Grained Sediments, 2.* Elsevier, Amsterdam, in press.

Young, R.B., 1917. *The Banket of the South African Goldfields.* Gurney and Jackson, London.

Zartman, R.E., Brock, M.R., Heyl, A.V. and Thomas, H.H., 1967. K–AR and Rb–Sr ages of some alkalic intrusive rock from central and eastern United States. *Am. J. Sci.*, 265: 848–870.

Chapter 10

SEDIMENTARY GEOCHEMISTRY AND MINERALOGY OF THE SULFIDES OF LEAD, ZINC, COPPER AND IRON AND THEIR OCCURRENCE IN SEDIMENTARY ORE DEPOSITS

DAVID J. VAUGHAN

INTRODUCTION

The sulfide minerals of Pb, Zn, Cu and Fe are the most common sulfides in sedimentary rocks. They form sufficiently important sulfide ore deposits of sedimentary affiliation to justify separate consideration in this chapter.

Initially, the fundamental aspects of the chemistry and geochemistry of each element are considered. Lead and zinc are discussed before the more complex transition metals. Then, the sulfide minerals of each are examined in terms of their crystal chemistry, composition, important textures and associations. The stability relations among these sulfide minerals and the associated oxides and oxysalts are considered as a function of Eh, pH, f_{S_2}, f_{O_2} in the aqueous environment. In addition, problems associated with the dissolution, transport and precipitation of the elements as sulfides are considered. Finally, some key examples of Pb-, Zn-, Cu-, Fe-sulfide formation in recent and ancient sediments are discussed.

Throughout, the emphasis is on the application of physicochemical principles to the understanding of the nature and genesis of Pb-, Zn-, Cu-, and Fe-sulfide deposits of sedimentary affiliation. The term "sedimentary ores" is not intended to necessarily imply a syngenetic origin for the deposits mentioned, but is used to include all stratabound ores regardless of origin. Stratabound ores of volcanic affiliation would, for example, be considered "sedimentary" in this context.

ASPECTS OF Pb, Zn, Cu AND Fe CHEMISTRY AND GEOCHEMISTRY

The atomic number, atomic weight, and electronic configuration of each element is given in Table I, together with the radii of the geochemically important ions of these four elements. Lead, a much heavier element with completely filled *4f* and *5d* shells, is distant in the periodic table from the three other elements which are all in the second period. Although Zn, Cu, and Fe are closely associated in the second period, Fe is a true transi-

TABLE 1

Atomic weights, atomic numbers, electronic configurations and ionic radii for Pb, Zn, Cu, and Fe

Element	Atomic no.[*]	Atomic wt.[*]	Electronic configuration[*]	Ionic radii[**]	
				M^{2+}	M^{3+}
Pb	82	207.2	[Xe] $4f^{14}5d^{10}6s^26p^2$	1.18	
Zn	30	65.38	[Ar] $3d^{10}4s^2$	0.745	
Cu	29	63.54	[Ar] $3d^{10}4s^1$	0.73	(Cu^{1+}=0.96)
Fe	26	55.84	[Ar] $3d^64s^2$	0.77	0.645

[*] Data from Cotton and Wilkinson (1972).
[**] Data from Shannon and Prewitt (1969) using values for six-fold coordination (all high-spin states).

tion element (i.e., it has a partially filled d shell in all of its common oxidation states), whereas Zn is definitely not a transition element. Copper occupies an unusual position since it is a true transition element in the Cu^{2+} oxidation state, but not in the Cu^+ state. The distinction between transition and non-transition metals is important because the two differ markedly in aspects of their crystal chemistry, solution chemistry, and thermodynamic properties. This led to the formulation of the crystal field and ligand field theories — topics briefly treated in a later section.

Each of the four elements will now be briefly discussed. Further information may be obtained from such standard reference works as Goldschmidt (1958), Krauskopf (1967), Wedepohl (1969), Cotton and Wilkinson (1972).

Lead

Lead commonly occurs in the Pb^{2+} oxidation state and more rarely as Pb^{4+}. Only the divalent cation is known in the aqueous chemistry of lead. Except for the nitrate, most Pb^{2+} salts are insoluble in water and all Pb^{2+} salts, except PbS, can be dissolved in excess OH^-. Pb^{2+} forms complexes in aqueous solution, particularly in the presence of halide ions (e.g., members of the series $PbX^+ \rightarrow PbX_3^-$ where $X = F$, Cl, Br, I, are known). Concentrated brines may be important transporting agents for lead, a subject further considered in later sections.

Goldschmidt (1958) describes lead as being dominated by chalcophile and lithophile properties. The first assertion is evident from the dominance of galena as the major lead ore mineral and the extensive sulfosalt mineralogy of lead. The lithophile properties result from the large ionic radius of lead which enables it to replace K, Sr, Ba, and even Ca in certain minerals, such as feldspars, augites, and apatites. The sulfates and carbonates are common oxidation products from the weathering of galena, but the simple oxides are less common.

Zinc

Zinc is a much smaller atom than lead (Table I) with a chemistry dominated by the Zn^{2+} ion. Zinc also forms halide complexes in aqueous solution and there is an important chemistry of organo-zinc compounds. Although dominantly chalcophile and frequently occurring as sphalerite (β-ZnS), some lithophile character results from the similarity of radii between Zn^{2+} and ions such as Mg^{2+}, Mn^{2+}, Fe^{2+}, Co^{2+}, and Ni^{2+}. This permits their replacement by Zn^{2+} in oxides and silicates, such as magnetite, ilmenite, pyroxenes, amphiboles, and biotites. In sulfide and oxide environments, the small Zn^{2+} ion shows preference for the more compact tetrahedral coordination. Simple oxides and silicates of zinc are rare, and the common oxidation products of primary zinc sulfides are mainly sulfates and carbonates.

Copper

Copper, an element of dual nature, is a true transition metal in the divalent state (Cu^{2+}), but not in the monovalent state (Cu^+) when the $3d$ shell is completely filled with electrons. In the monovalent state, it resembles elements of a more noble character like silver and gold, whereas divalent copper has properties similar to other first series transition elements (Fe, Co, Ni, etc.). The Cu^{2+} state is more important in the aqueous chemistry of copper, partly due to the greater lattice and solvation energies of the Cu^{2+} ion compared to Cu^+, which is stable only in very low concentrations. There are a large number of Cu^{2+} salts, most of which dissolve in water to give the cupric ion $[Cu(H_2O)_6]^{2+}$. Complexes can be formed by successive displacement of water molecules from the cupric ion – e.g., $[Cu(NH_3)_4(H_2O)_2]^{2+}$ – and the extensive "complex" chemistry of Cu^{2+} involves halides, organic molecules, sulfides and other types of ligand.

Copper is a strongly chalcophile element, hence, little or no copper is found in the common rock-forming silicate minerals. However, alteration of the primary sulfides does result in a rich array of secondary minerals including oxides, hydroxy-carbonates, hydroxy-silicates and sulfides.

Iron

Iron is by far the most geologically important transition element, with a geochemistry dominated by the Fe^{2+} and Fe^{3+} oxidation states. It occurs in a wide variety of minerals, constituting a major component of many pyroxenes, olivines, amphiboles, micas, garnets and other silicates, and exhibits important oxide and sulfide mineralogies. The nature of iron-oxygen and iron-sulfur chemical bonding has recently been discussed by Vaughan et al. (1974). The existence of Fe^{3+} in sulfides has been a matter of debate but has been demonstrated in the mineral greigite (Fe_3S_4) by Coey et al. (1970) and by Vaughan and Ridout (1971).

In aqueous solutions, Fe^{2+} forms the ion $Fe(H_2O)_6{}^{2+}$ and Fe^{3+} also shows a strong tendency towards hydrolysis or complex formation with halide, cyanide, oxalate and other ligands. The rapid oxidation of Fe^{2+} to Fe^{3+} and the very low solubility of the Fe^{3+} hydroxides often severely limits the concentration of iron in surface waters.

Crystal field theory and the geochemistry of Fe and Cu

Any discussion of the geochemistry of iron or copper requires consideration of crystal field or ligand field theory. Crystal field theory and its application in the earth sciences has been discussed in detail by Burns (1970). Ligand field theory is merely a modification of crystal field theory, designed to take into account covalent interactions between cation and surrounding anions (ligands). Since crystal field theory has important consequences for the crystal chemistry and solution chemistry of Fe and Cu, certain aspects will be briefly reviewed.

The five $3d$ orbitals of first-series transition metals are equal in energy (degenerate) in the free atoms or ions. However, when surrounded by a group of anions or ligands, the "crystal field" of the anions causes those orbitals that are proximal to them, to be destabilized relative to the orbitals that are further from the anions. The nature of this splitting depends on the symmetry of the surrounding ions, and the situation for the common cases of octahedral and tetrahedral coordination are shown in Fig. 1. In octahedral coordination, two orbitals (termed e_g) are destabilized by $3/5\Delta_0$ (where Δ_0 is the total splitting) and three (t_{2g}) by $2/5\Delta_0$. The value of Δ_0 depends on the anions involved and the M–X distances. The corresponding situation for tetrahedral coordination is again given in Fig. 1. Also in Fig. 1 are shown the alternative ways in which the electrons of the

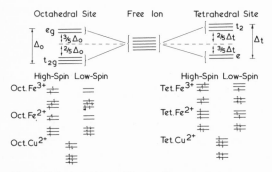

Fig. 1. Splitting of the $3d$ orbitals in octahedral and tetrahedral crystal fields and the high- and low-spin occupancy schemes of Fe^{3+}, Fe^{2+}, and Cu^{2+}.

ions Fe^{3+}, Fe^{2+}, and Cu^{2+} can occupy the t_{2g} and e_g, or e and t_2 orbitals. In general, there are two alternatives: one in which as few of the electrons as possible are paired in the same orbital (high-spin), and one in which the low-energy orbitals are as completely filled as possible (low-spin).

Only some of the configurations shown in Fig. 1 actually occur in minerals. Octahedral and tetrahedral Fe^{3+} are usually high-spin, whereas octahedral Fe^{2+} is high-spin in oxides, but in FeS_2 occurs in the low-spin state. Whether the high- or low-spin configuration occurs is largely due to whether the energy gained in occupying the t_{2g} or e orbitals (the "crystal field stabilization energy") overcomes the energy required to pair the electrons. The crystal field stabilization energy, although only a small part of the total energy of a system, can be critical in determining the distribution of transition metal cations between coordination positions in a mineral structure or between coexisting phases. Only one type of configuration is possible for Cu^{2+}, but here another phenomenon, the "Jahn-Teller Effect", becomes important. In this case, the single hole in the $3d$ shell results in asymmetry of the ion which thus favors distorted or low symmetry sites in a crystal structure.

Distribution of Pb, Zn, Cu, and Fe in the earth

In Table II, the average distributions of Pb, Zn, Cu, and Fe in various igneous and sedimentary rocks and in sea water and streams are listed. The data, which in some cases can only be regarded as crude estimates, are taken from compilations by Turekian and Wedepohl (1961), Vinogradov (1962), and Turekian (1969).

TABLE II

Average distribution of Pb, Zn, Cu, and Fe in rocks, ocean, and streams

	Lead (ppm)	Zinc (ppm)	Copper (ppm)	Iron (wt.%)	References
Ultramafics	1	50	10	9.43	Turekian and Wedepohl (1961)
	0.1	30	20	9.85	Vinogradov (1962)
Basalts	6	105	87	8.65	Turekian and Wedepohl (1961)
	8	130	100	8.56	Vinogradov (1962)
Granites					
high Ca	15	60	30	2.96	
low Ca	19	39	10	1.42	
Clays + shales	20	80	57	3.33	Vinogradov (1962)
Shales	20	95	45	4.72	Turekian and Wedepohl (1961)
Sandstones	7	16	1%	0.98	Turekian and Wedepohl (1961)
Carbonate rocks	9	20	4	0.38	Turekian and Wedepohl (1961)
Deep-sea					
carbonate rocks	9	35	30	0.90	Turekian and Wedepohl (1961)
clays	80	165	250	6.50	Turekian and Wedepohl (1961)
	(ppb)	(ppb)	(ppb)	(ppb)	
Sea water	0.03	5	0.9	3.4	Turekian (1969)
Streams	3	20	7	~0.67	Turekian (1969)

TABLE III

Sulfide minerals important in sedimentary associations

Mineral	Composition	Structure type[1]
Galena	PbS	NaCl (cb)
Sphalerite	$\sim$ZnS	sphal. type (cb)
Wurtzite	$\sim$ZnS	wurtz. type (hx)
Covellite	CuS	cov. type (hx)
Chalcocite	$\sim$Cu$_2$S	
Djurleite	Cu$_{1.96}$S	
Digenite	Cu$_9$S$_5$	
Anilite	Cu$_{1.75}$S	
Blaubleibender		
covellite	Cu$_{1.1}$S	(hx)
Chalcopyrite	CuFeS$_2$	sphal. type (te)
Cubanite	CuFe$_2$S$_3$	wurtz. der. (O)
Bornite	Cu$_5$FeS$_4$	sphal. der. (te)
Pyrite	$\sim$FeS$_2$	pyrite-type (cb)
Marcasite	$\sim$FeS$_2$	pyrite-der. (cb)
Pyrrhotites:		
troilite	FeS	NiAs-der. (hx)
monoclinic pyrr.	$\sim$Fe$_7$S$_8$	NiAs-der. (M)
hexagonal pyrr.	$\sim$Fe$_9$S$_{10}$'$\sim$Fe$_{10}$S$_{11}$?	NiAs-der. (hx)
	$\sim$Fe$_{11}$S$_{12}$?	
Mackinawite	Fe$_{1+x}$S	PbO-type (te)
Greigite	Fe$_3$S$_4$	spinel-type (cb)
Smythite	Fe$_{3.25}$S$_4$?	NiAs-der. (hx)

[1] cb = cubic, hx = hexagonal, te = tetragonal, O = orthorhombic, M = monoclinic.

Pb, Zn, Cu, AND Fe SULFIDE MINERALS OF SEDIMENTARY ASSOCIATION

The occurrence of Pb, Zn, Cu, and Fe as sulfides in sedimentary rocks is dominated by the small group of minerals listed in Table III. Diagnostic properties for the identification of these minerals are recorded elsewhere (Berry and Thompson, 1962; Ramdohr, 1969; Uytenbogaart and Burke, 1971), as are data on isotopic studies (this Vol., Chapters 7–9). The complex sulfide, sulfosalt, and oxysalt mineralogy resulting from their weathering and supergene enrichment will not be considered. (See also Chapter 1, this Vol. on trace elements.)

Structures and compositions of the major minerals

Galena, the most common and important lead mineral, has the NaCl (rocksalt) structure (Fig. 2) in which Pb^{2+} ions occupy PbS$_6$ octahedra which share corners through the common sulfur atoms. The importance of galena as a host mineral which can contain appreciable quantities of other elements (notably Ag, Sb, As, Bi, Cu, Zn, Cd, Fe, Se, and

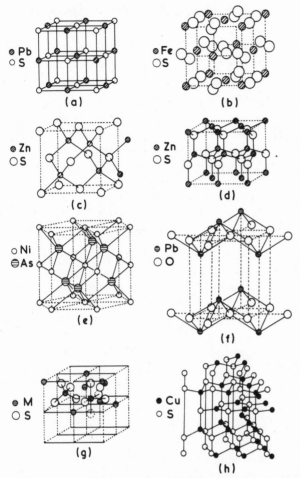

Fig. 2. Important structure types occurring in Pb, Zn, Cu, and Fe sulfides: (a) NaCl type; (b) pyrite type; (c) sphalerite type; (d) wurtzite type; (e) NiAs type; (f) tet PbO type; (g) spinel; (h) covellite.

Te) in the crystal lattice, has been discussed by a number of authors (e.g., Goldschmidt, 1958; Ramdohr, 1969). It seems likely that a limited amount of direct substitution (frequently as coupled substitutions, e.g., $Ag^+ + Bi^{3+}$ for $2Pb^{2+}$) occurs, but that the other elements occur as very fine, though discrete, particles of other minerals, such as freibergite, proustite, tetrahedrite, sphalerite, and even native silver. The presence of these silver-bearing minerals in the galena is generally the result of the breakdown of high-temperature solid solutions like that between galena and matildite ($AgBiS_2$) at $> 215°C$ (Craig, 1967). Silver contents can reach several percent and make galena a valuable ore of silver. Although no such solid solution appears to occur between PbS and $CuBiS_2$ (Karup-Møller, 1971), cuprian galena ($< 13\%$ Cu) has been reported from supergene alteration of copper ores (Clark and Sillitoe, 1971).

The structure of sphalerite (Fig. 2), in which regular zinc—sulfur tetrahedra are again linked by shared sulfur atoms at the corners of the tetrahedra, is also a fundamental structure type. Sphalerite is rarely pure, and the zinc can be isomorphously replaced by substantial amounts of other elements, including Fe, Mn, Co, Cd, Hg, and the sulfur by Se (Nickel, 1965). Of all these elements, iron is the principal substituent for zinc in the structure of natural sphalerites. It is frequently present in concentrations of up to 15 wt.% and as much as 26 wt.% has been recorded. Cadmium and manganese rarely exceed a maximum of 1 wt.%, except in meteorites, and the other substituents are rare.

Wurtzite, the other polymorph of ZnS, has a structure closely related to sphalerite with zinc again tetrahedrally coordinated, but the stacking of the tetrahedra differs so that the lattice has hexagonal symmetry (Fig. 2). Also, the tetrahedra are slightly elongated along the crystal c-axis direction. Wurtzite was formerly thought to be a high-temperature polymorph of sphalerite, but Scott (1968) has shown that wurtzite is, in fact, sulfur deficient relative to sphalerite. Electrical measurements show zinc vacancies in sphalerite and sulfur vacancies in wurtzite, with a total range in nonstoichiometry of ~1 atom %. The sphalerite—wurtzite phase change is, therefore, a function of f_{S_2} as well as temperature. Examination of the stability field of wurtzite in aqueous systems as a function of f_{O_2}, f_{S_2} and pH (see p. 337) are consistent with the occurrence of wurtzite in low f_{S_2} environments in nature (Scott, 1968). Examples include wurtzite—siderite concretions in western Pennsylvania and eastern Ohio (Seaman and Hamilton, 1950). Furthermore, the suggestion of Behre (1939), that radiating sphalerite blades in Upper Mississippi Valley-type deposits (see p. 356) originally precipitated as wurtzite, has been upheld.

The binary copper sulfides are of less importance as primary minerals, and also constitute a complex and incompletely characterized group. Early studies of the Cu—S system suggested there were only two phases of intermediate composition, covellite and chalcocite. However, studies by Roseboom (1966) indicated three phases with compositions close to Cu_2S which are stable at 25°C (Table III). Chalcocite has a composition close to Cu_2S, but with increasing temperature to ~105°C it exhibits a wider range of composition ($<Cu_2S \sim Cu_{1.998}S$) and inverts to hexagonal Cu_2S at about this temperature. Djurleite ($Cu_{1.96}S$) is a more copper-deficient phase with an apparently constant composition up to its inversion temperature of ~93°C when it breaks down to a mixture of the high-temperature (hexagonal) chalcocite and a high temperature form of the third phase — digenite. Digenite or "digenite solid solution" shows a range of composition at 25°C which increases with temperature. Inversion to high digenite, which has an even larger range of composition, takes place at ~80°C. The phase relations in the Cu—S system given in Fig. 3A are as shown by Roseboom (1966). However, subsequent studies by Morimoto and Koto (1970), following the discovery of yet another phase — anilite —

Fig. 3. Phase relations in the Cu—S system. A. From Roseboom (1966). Reproduced from *Econ. Geol.*, 1966, vol. 61, p. 648, with the publisher's permission. B. From Barton (1973). Reproduced from *Econ. Geol.*, 1973, vol. 68, p. 465, with the publisher's permission.
al = anilite; *bb-cv* = blaubleibender covellite; *cc* = chalcocite; *cv* = covellite; *co* = copper; *dg* = digenite; *dj* = djurleite; *su* = sulfur.

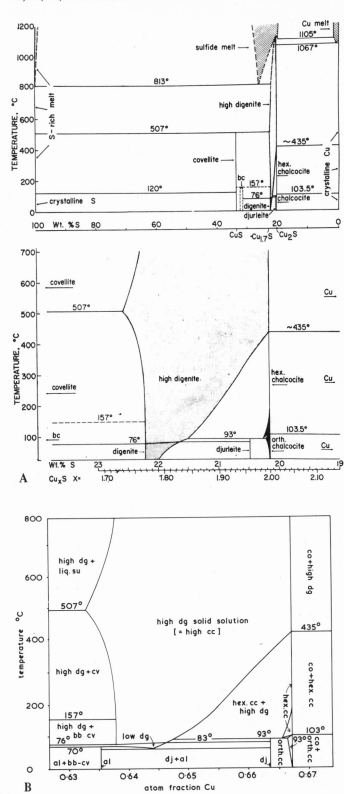

suggest the digenite solid solution is metastable at room temperature and decomposes to anilite + djurleite. A recently published diagram showing the phase relations in the chalcocite, djurleite, anilite portions is given in Fig. 3B. This is taken from Barton (1973) who incorporates data published before 1971 (see Barton, p. 455). Yet another complexity is the field of "blaubleibender[1] covellite" (Moh, 1964), which is included in this diagram, although its stability has not been totally established.

A detailed discussion of the structures of these phases and their interrelationships is not appropriate in a review of this type, but the example of covellite is included in Fig. 2. Covellite has a complex layer structure containing Cu^+ and Cu^{2+} ions and both S^{2-} anions and $(S-S)^{2-}$ polyanions. The Cu^+ is tetrahedrally coordinated to the polyanions, and the Cu^{2+} occurs in a triangular coordination to the S^{2-} ions.

The relationships between copper minerals shown in Fig. 3A and 3B are derived from studies in dry systems. Very little is known about the formation of these minerals in the low-temperature aqueous environments that are the particular concern of this chapter. Rickard (1970a) has reviewed much of the available data, and has also reported the synthesis of djurleite (Rickard, 1970b) from the reaction of cuprous oxide and aqueous sodium sulfide solution at 25°C (1 atm). Covellite formation in low-temperature aqueous solutions has also been described (Rickard, 1972), from which it was suggested that blaubleibender covellite is a metastable intermediary in the reduction of covellite to form chalcocite or djurleite.

From the complexities described in the binary Cu–S system, it is not surprising that the Cu–Fe–S system is one of the least understood ternary systems despite extensive investigations (see Barton, 1973; Cabri, 1973). Since phase equilibria in the dry systems at temperatures ≥100°C form the basis of most of these studies, they will not be considered here in any detail. Furthermore, below 100°C, phase relations are extremely difficult to characterize because of slow reaction rates. However, it should be pointed out that in addition to the minerals chalcopyrite, cubanite, and bornite listed in Table III, a number of other phases have been reported in the central portion of the Cu–Fe–S system – most of them in the last 5 years. Fig. 4 shows the compositions of these phases taken from Cabri (1973), who provides further data on their structures, stabilities, and relationships. It has only been through careful X-ray and electron-microprobe studies that phases, such as talnakhite, mooihoekite, and haycockite, has been distinguished from the more commonly reported minerals (e.g., Cabri and Hall, 1972). What role such phases may have in sedimentary environments is unknown.

The structure of chalcopyrite is the same as that of sphalerite (Fig. 2) except that the c-axis length is doubled through alternate filling of tetrahedra by copper and iron atoms. The cubanite structure represents a modification of the structure of wurtzite. It can be

---

[1] "Blaubleibender" or "blue remaining" covellite, so-called because it is distinguished from normal covellite in polished section by not changing to purple/red colours in oil immersion (i.e., it remains blue in colour).

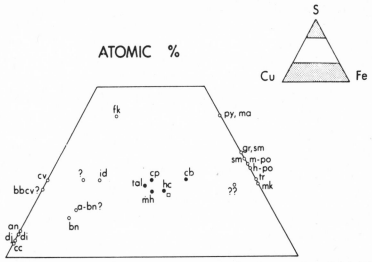

Fig. 4. Mineral phases occurring in part of the Cu−Fe−S system (from Cabri, 1973). *a-bn* = anomalous bornite; *an* = anilite; *bbcv* = blaubleibender covellite; *bn* = bornite; *cc* = chalcocite; *cp* = chalcopyrite; *cv* = covellite; *cb* = cubanite; *di* = digenite; *fk* = fukuchilite; *gr* = greigite; *hc* = haycockite Cu_3FeS_4; *mk* = mackinawite; *ma* = marcasite; *mh* = mooihoekite $Cu_9Fe_9S_{16}$; *m-po* = monoclinic pyrrhotite; *py* = pyrite; *sm* = smythite; *tal* = talnakite $Cu_9Fe_8S_{16}$; *tr* = troilite; *?* = probable new mineral Cu_5FeS_6; *?* = new mineral $Cu_{0.12}Fe_{0.94}S_1$; □ = synthetic $Cu_3Fe_4S_6$. Reproduced from *Econ. Geol.*, 1973, vol. 68, p. 444, with the publisher's permission.

considered as made up of slices of the wurtzite structure with adjacent pairs of FeS_4 tetrahedra sharing edges, and bringing pairs of iron atoms closer together. Bornite is structurally related to sphalerite and chalcopyrite, but only three quarters of the tetrahedral sites in the anion sublattice are filled. Bornite exists in three polymorphic forms (Morimoto and Kullerud, 1961), a complexity arising from the statistical distribution of metal atoms amongst slightly different positions within the sulfur tetrahedra. However, only the low-temperature tetragonal form occurs naturally. Bornite samples from red-bed type deposits are commonly sulfur-rich and exsolve lamellae of chalcopyrite, which increase in dimensions on heating between 75 and 400°C, but eventually homogenize at >500°C. Brett and Yund (1964) conclude that such bornites must have formed at temperatures below 75°C.

Pyrite and pyrrhotite are the most abundant natural sulfides. The structure of pyrite (Fig. 2) is related to the rocksalt structure, but $(S–S)^{2-}$ groups replace simple anions. The Fe^{2+} ions are, therefore, octahedrally coordinated, but because of the strong crystal field of the disulfide groups, ferrous iron is low-spin in pyrite. Pyrite is frequently reported as containing significant amounts of minor elements, although, as with galena, some material may be as other discrete phases. Fleischer (1955) reviewed early data in this field. Selenium and probably arsenic substitute for sulfur, and cobalt and nickel for iron. Since $FeS_2–CoS_2–NiS_2$ solid solution is limited at higher temperatures, many reported intermediate members must occur metastably at lower temperatures (Springer et al., 1964). The same is probably true of copper-bearing pyrites and such isostructural minerals as villamaninite, $(Cu,Fe,Ni,Co)S_2$ and fukuchilite, Cu_3FeS_8 (Ypma, 1968; Shimazaki and Clark, 1970).

The structure of pyrrhotite is based on the "NiAs" structure (Fig. 2) with metals again in octahedral coordination, but sulfur octahedra sharing faces perpendicular to the c-axis. The close approach of iron atoms along the c-axis results in magnetic interactions and the magnetic properties of pyrrhotite are complex (e.g., Schwarz and Vaughan, 1972). The "pyrrhotites" are really a whole group of minerals extending from FeS (troilite) to compositions close to Fe_7S_8 in composition ("monoclinic pyrrhotite"). The compositional range arises from vacancies in the metal atom sites and through ordering of these vacancies, the pyrrhotites have superstructures based on the "NiAs" structure. This complex subject has been reviewed by Ward (1971). Appreciable quantities of nickel ($\lesssim 1$ wt.%) are common in pyrrhotite and the partitioning of nickel and cobalt between pyrrhotite and pyrite is discussed on p. 331.

The minerals mackinawite and greigite, although only first described fairly recently (Evans et al., 1964; Skinner et al., 1964), are included because of their very important role as precursors of pyrite and pyrrhotite in sulfides precipitated from solution at low temperatures. Berner (1964) and Rickard (1969) have shown that the first iron sulfide formed from aqueous solution is the sulfur-deficient mackinawite (or an amorphous sulfide of the same composition). Sweeney and Kaplan (1973) have shown that in the presence of limited oxygen this can change into hexagonal pyrrhotite, and that on further reaction with elemental sulfur, either of these phases will produce greigite and finally pyrite (see Fig. 5).

Mackinawite has an unusual layer structure of the type characteristic of tetragonal PbO (Fig. 2), but with S in the Pb positions, and it can incorporate large amounts ($\sim$5–20 wt.%) of other transition metals (Cr, Co, Ni, Cu, etc.) into the lattice (Clark,

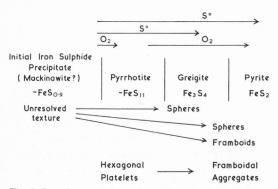

Fig. 5. Reactions of primary iron sulfide precipitate in aqueous solution. (Modified from Sweeney and Kaplan, *Econ. Geol.*, 1973, vol. 68, p.633, with the publisher's permission.)

1969; Vaughan, 1969; etc.). The thermal stability of natural specimens increases with increasing amounts of Fe, Ni, or Co (Takeno, 1965); pure mackinawite, near FeS composition, breaking down at $\sim 140°C$.

Greigite has a spinel structure (Fig. 2), and has been shown to be an inverse spinel — $Fe^{3+}[Fe^{2+},Fe^{3+}]S_4$ — by Mossbauer spectroscopy (Coey et al., 1970; Vaughan and Ridout, 1971). So little greigite has been observed in natural specimens that finds are still reported in the literature (e.g., Jedwab, 1967; Dell, 1972). Most reported occurrences are in reducing environments within recent sediments, in accordance with the low thermal stability of greigite, which breaks down at $\sim 180-200°C$ (Uda, 1967). Using arguments similar to those of crystal field theory, Vaughan et al. (1971) attribute the instability of greigite relative to other sulfospinels (linnaeite, violarite, etc.) to the occurrence of Fe^{2+} and Fe^{3+} in high-spin states. Analysis of a rare greigite occurrence in hydrothermal veins indicate 0.22% Ni (Radusinovic, 1966), but evidence of other substitutions in greigite is unavailable.

Smythite is another apparently rare iron sulfide, the role of which is still obscure. It was originally reported as being a polymorph of greigite (Erd et al., 1957), but subsequent work suggests a formula Fe_9S_{11} with some necessary nickel in the cation sites (Taylor and Williams, 1972). The structure is intermediate between the NiAs and $Cd(OH)_2$ types and appears, therefore, to be related to the pyrrhotites. Smythite has been reported in sedimentary iron ores (Chukrov et al., 1965), and like mackinawite and greigite, it is a low-temperature phase (Taylor, 1970). A metastable iron sulfide with the sphalerite structure has also been reported by Takeno et al. (1970) from low-temperature aqueous synthesis.

One other iron sulfide which must be mentioned is marcasite. Structurally, marcasite closely resembles pyrite and both contain $(S-S)^{2-}$ groups and have octahedrally coordinated low-spin Fe^{2+}. However, like mackinawite, greigite, and smythite, marcasite is a relatively low-temperature phase which has not been synthesized in the dry Fe–S system. This has led to suggestions that marcasite is a metastable phase relative to pyrite, or even that it contains $(OH)^-$ groups in the structure (Kullerud, 1967). A more appealing suggestion is that pyrite and marcasite are not monotropic polymorphs, but minerals of a slightly different composition, like sphalerite and wurtzite.

Forms and textures of the minerals

Galena frequently occurs as granular aggregates and as well developed crystals which are commonly zoned. "Reticulate" intergrowths of galena with colloform sphalerite or "schalenblende" (Ramdohr, 1969) are considered indicative of low temperatures of formation. However, the "colloform" textures of such ores, which contain wurtzite and often pyrite and marcasite in addition to sphalerite and galena, may not have originated in deposition from colloidal sulfide gels. Roedder (1968) has studied such assemblages and observed minerals which cannot have formed from gels, but grew as euhedral crystals on a surface in contact with an ore fluid. The banding of these ores can be considered as arising from periodic variations in ore fluid composition. The work of Scott (1968) on the nature of sphalerite ⇌ wurtzite equilibria also shows that small changes in conditions near this phase boundary can cause alternate precipitation of sphalerite and wurtzite (see p. 00). Similar textures have been reported in chalcopyrite, but only rarely.

An unusual form of occurrence of sphalerite has been reported from Lake Kivu, East Africa, by Degens et al. (1972). Here, microcrystalline sphalerite is associated with resin globules suspended in the lake. The hollow resinous spheres are formed on degassing of the water at depth, and serve to selectively extract zinc from the lake waters. H_2S, which like the zinc is derived from hydrothermal springs seeping into the lake bottom, reacts with the zinc forming sphalerite. The resin may also promote growth of sphalerite crystals through an epitaxial or catalytic function. On the basis of an average 2 ppm zinc in the lake water, one million tons of sphalerite has formed in Kivu.

Few mineral textures have resulted in a more voluminous literature than pyrite "framboids". This term describes microcrystalline spheroidal pyrite aggregates with a cellular structure, which occur in many base metal sulfide deposits. The occurrence of framboids and the associated literature has been reviewed by Love and Amstutz (1966). The suggestion that this texture is an indication of biogenic processes active in pyrite formation has resulted in controversy (Vallentyne, 1963; Love, 1965). Rickard (1970), on the basis of physical and crystal chemical arguments, concluded that framboids arise from pseudomorphism of a pre-existing spherical body. These may be gaseous vacuoles or organic globules, so that framboids are not necessarily indicators of a sedimentary environment, although they probably form at low temperature. Recently, laboratory synthesis of framboidal pyrite (Berner, 1969a; Farrand, 1970; Sunagawa et al., 1971) has conclusively demonstrated that biological control is not a requirement. In a detailed study of pyrite formation in the laboratory, Sweeney and Kaplan (1973) observed pyrite spheres and framboids as final products of a series of compositional and textural changes following the initial precipitation of mackinawite (or amorphous $FeS_{0.9}$). This mackinawite, which may change to hexagonal pyrrhotite, is sulfidized first to greigite and then pyrite. The greigite assumes spherical character and pyrite framboids only form on these spherical nuclei.

Mineral compositions as an indication of formation conditions

In addition to the study of the conditions (such as T, P, f_{S_2}, f_{O_2}, Eh, pH) under which certain mineral assemblages coexist in equilibrium, the study of compositional variations of certain single phases has also been undertaken as a function of such variables. Two major approaches have been attempted. One is the empirical or semi-empirical use of a minor or trace element in a mineral to distinguish its mode of origin (e.g., sedimentary vs. magmatic).[1] The other is the use of a compositional variation (e.g., Fe/S ratio of pyrrhotite, Fe content of sphalerite) to estimate temperature, pressure, or some

---

[1] Editor's note: see Chapter 1 by Mercer.

other variable at the time of formation. The latter obviously involves careful experimental calibration of the geothermometer, geobarometer, or other indicator.

One of the earliest examples of the first approach was the use of selenium contents or S/Se ratios in sulfides as indicators of chemical and physical formation conditions. Because sulfur and selenium have a similar chemistry (both members of subgroup VIb of the periodic table) and their ionic radii are similar in their divalent anions (S^{2-} = 1.84 Å, Se^{2-} = 1.98 Å), substitution of Se for S occurs in minerals. However, an important difference in the geochemistry of selenium to that of sulfur is the virtual absence of selenium in the oceans. This is because sulfur is readily oxidized to sulfate and transported in solution, but selenium is not readily oxidized (Coleman and Delevaux, 1957). Because selenium is less mobile, sulfur and selenium begin to separate at the onset of weathering. Consequently, the S/Se ratio of sea water is ~232,000/1 by weight compared to an average for igneous rocks of ~600/1. Thus, S/Se ratios appear to be a means of distinguishing sulfides derived from sedimentary processes and those derived from igneous (including hydrothermal) processes. As early as 1935, Goldschmidt and Strock showed that certain pyrite samples of obvious sedimentary origin had S/Se ratios of 200,000/1, others of clear igneous–hydrothermal origin had ratios of ~15,000/1. However, further work showed the situation to be much more complex since the local abundance of selenium is another important factor. Thus, sediments derived from source areas high in selenium can have S/Se ratios comparable to those from igneous origins, and conversely, samples from an igneous province poor in selenium can have a ratio considered in the sedimentary range. Thus, the S/Se ratio cannot be cited as unequivocal evidence of sedimentary or igneous origin of a sulfide ore, but can nevertheless serve as a useful indicator when used with caution (e.g., studies of S/Se ratios in the ores of the Tasman geosyncline by Loftus-Hills and Solomon, 1967).

Minor and trace quantities of other elements have been proposed as indicators of formation conditions. For example, thallium in sulfides (mainly FeS_2) is reportedly lower by an order of magnitude in sedimentary as against hypogenic phases (Voskresenskaya, 1969), although enriched in the sediments of "geosynclines" as against platform regions. The bismuth and antimony content of galena and the Sb/Bi ratio reportedly reflects aspects of the temperature and pressure of formation and to a lesser extent the ore solution and country rock chemistry (Malakhov, 1968). In particular, a low Sb/Bi ratio (<0.06) indicates galena formed at high temperature and a high ratio (>6.0–13.0) indicates a low temperature of formation. Presumably, this relates in part to the solid solution between galena and matildite at high temperatures, which would result in higher Bi contents with increasing temperature. Cobalt and nickel contents of sulfides, particularly pyrite, have also been used to discriminate between magmatic–hydrothermal and sedimentary environments (e.g., Loftus-Hills and Solomon, 1967). Pyrite of sedimentary origin is apparently richer in Co and Ni, and Fleischer (1955) concludes that there is good evidence that sedimentary pyrite generally has Co < Ni, whereas hydrothermal pyrite has Co > Ni more common. These distributions may be related to the metastable occurrence of (Fe,Co,Ni)S$_2$ compositions at lower temperatures. Also, in situations where pyrite and pyrrhotite co-exist, Co tends to be enriched over Ni in pyrite, and Ni over Co in pyrrhotite. This is due to the greater crystal field stabilization energy of Co^{2+} which is low-spin in pyrite, as against high-spin Ni^{2+} which favors the pyrrhotite lattice. At higher temperatures, this partitioning is enhanced and may be a potential indicator of metamorphic grade.

It should be noted that the use of Th, Sb/Bi, Co, and Ni as indicators of formation conditions is based largely on empirical correlations in which the contents or ratios for accepted sedimentary or igneous provinces are determined with a view to application in less clear-cut cases. However, most and probably all such techniques are not independent of variations in local abundance of the element or elements concerned, the problem discussed in the case of S/Se ratios.

The idea that compositional variation of a single sulfide could be used to determine certain aspects of the formation conditions probably became prominent with work on the FeS–ZnS system by Kullerud (1953). Kullerud demonstrated an apparent systematic relationship between the iron content of sphalerite and its temperature of formation — the basis of the "sphalerite geothermometer". It was necessary that the sphalerite co-existed with pyrrhotite and certain impurities be minor, and a pressure correction was necessary. Since these requirements were often fulfilled, widespread application of this geothermometer to natural ores followed. However, subsequent investigations (Barton and Kullerud, 1957, 1958; Barton and Toulmin, 1966) showed the phase relations to be more complex than anticipated, and the ZnS–FeS solvus to be greatly modified by variation in sulfur vapor pressure. The FeS content of sphalerite is, therefore, not useful in geothermometry without data on f_{S_2} at the time of deposition. This work led to interest in the three phase assemblage, sphalerite + pyrite + pyrrhotite, in which f_{S_2} is fixed at a given temperature by the co-existing iron sulfides (Scott and Barnes, 1971). However, below 550°C, the addition of pyrite to the system restricts the sphalerite composition to ~20 mole % FeS, but the pressure effect on the FeS content is large, suggesting its use as a "geobarometer" (Scott and Barnes, 1971; Scott, 1973; see Fig. 6). But even the applicability of these results to natural systems is being questioned. Browne and Lovering (1973) have studied drill-cores from a geothermal field in which sphalerite co-exists with pyrite and pyrrhotite at known temperature and pressure. They suggest that the geobarometer is inapplicable below 300°C.

Other geothermometers and geobarometers have been plagued by similar difficulties. The composition of pyrrhotite in equilibrium with pyrite was shown to vary as a function of temperature between 325 and 743°C (Arnold, 1962, and others) and to be relatively independent of pressure. However, the pyrite/pyrrhotite phase relations below 300°C are very complex and are still being evaluated (Kissin and Scott, 1972). The composition of pyrrhotite formed at low temperatures bears no relation to this high-temperature solvus, and even pyrite/pyrrhotite assemblages formed at high temperatures frequently appear to re-equilibriate or be modified later.

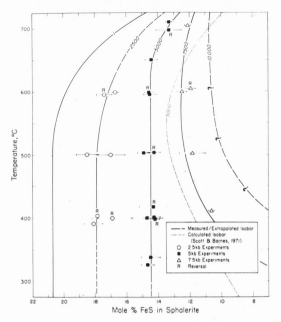

Fig. 6. $T-X$ projection of the sphalerite + pyrite + hexagonal pyrrhotite solvus isobars. Pressures are in bars (from Scott, 1973). (Reproduced from *Econ. Geol.*, 1973, vol. 68, p. 469, with the publisher's permission.)

Pb, Zn, Cu, AND Fe SULFIDES IN AQUEOUS SYSTEMS

The equilibrium relations amongst the minerals discussed in the previous section are crucial to any understanding of their genesis. In particular, for the sedimentary-type ores with which this volume is concerned, chemical equilibria at low temperatures and pressures in aqueous media are most relevant. The stabilities of the various mineral species may be studied as functions of a large number of variables including temperature, pH, Eh, and the partial pressures of various gases in the solution (O_2, CO_2, H_2S, etc.). Different methods for presenting the data may then be employed which result in a considerable array of diagrams and tables. In this section, Eh—pH and Eh—pH—partial pressure (or pH—partial pressure, Eh—partial pressure) diagrams will be used predominantly, although other special representations will be introduced where appropriate. The principles of chemical equilibrium as applied to aqueous systems are discussed in many textbooks (e.g., Garrels and Christ, 1965; Krauskopf, 1967; Stanton, 1972) and will not be considered. As always, the equilibrium relationships determined for *pure* phases in the laboratory must be applied to natural systems with caution.

Eh–pH and partial pressure diagrams

The methods for constructing Eh–pH diagrams, and diagrams involving the partial pressures of various gases, are given by Pourbaix (1949) and by Garrels and Christ (1965), from whose work many of the diagrams appearing in the literature originate. Eh–pH diagrams can be drawn up so as to illustrate different aspects of the system, but it is important to note the assumptions used in drawing up any particular diagram. For example, in Fig. 8B, which represents equilibria in the $Pb–CO_2–O_2–S–H_2O$ system, the assumptions are: (1) $a_{H_2O} = 1.0$; (2) $T = 25°C$; (3) $P = 1$ atm; (4) total sulfur is 10^{-1} molal; (5) $P_{CO_2} = 10^{-4}$; (6) the boundaries between the solids are drawn assuming sum of the ionic activities is 10^{-6}.

In studying Eh–pH diagrams, it is interesting to compare the stability fields of particular phases with the limits of Eh and pH so far found in natural environments. In Fig. 7, the limits taken from the compilation of Baas Becking et al. (1960) are shown, together with regions on the diagram corresponding to environments of particular interest in ore formation. Although pH may reach extreme values, common near-surface environments have pH ranges of 4.0–9.0. The pH of the open sea is ~8.0, but in restricted basins this may drop to 6.0, and in marsh and swamp areas, values of 3.0–6.0 are common. On the other hand, pH values of up to 10.0 are found in brine lakes and seas. The limits of Eh are broadly established by the values where water breaks down to form hydrogen and oxygen (Krauskopf, 1967, p. 245).

Equilibria amongst lead compounds. A series of figures illustrating equilibria amongst lead compounds, including galena, have been published in Garrels and Christ (1965, pp. 233–238) based on work by W. McIntyre. Fig. 8A is for the $Pb–O_2–S–H_2O$ system. Even though the dissolved sulfur is only 10^{-5} molal, galena has quite a large field of stability, whereas native lead is virtually excluded. In this diagram, the boundaries between fields have been drawn for an *ionic* activity of 10^{-2}. Fig. 8B includes the effect of

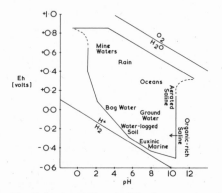

Fig. 7. Limits of Eh and pH in the natural environment.

CO_2 (P_{CO_2} = 10^{-4}) with the dissolved sulfur increased to 10^{-1} molal. This system resembles the conditions of surface oxidation with P_{CO_2} approximately that of the atmosphere. Both of these figures clearly demonstrate the wide pH range of stability of galena in reducing environments.

Equilibria amongst zinc compounds. In considering the low-temperature equilibria involving zinc sulfides, the relationship between sphalerite and wurtzite is of particular interest. Scott's (1968) work demonstrated that wurtzite is a stable phase, sulfur-deficient relative to sphalerite. From higher temperature experimental data, Scott has projected the sphalerite $\rightleftharpoons$ wurtzite equilibrium as a function of f_{S_2} and T to 25°C, and reduced uncertainties in the extrapolation using arguments based on observed assemblages. Fig. 9 shows that "best" sphalerite–wurtzite equilibrium as a function of log f_{S_2} and temperature, together with the pyrite–pyrrhotite solvus extrapolated from the data of Toulmin and Barton (1964). Since wurtzite is found with pyrite or marcasite in natural low-temperature assemblages, the sphalerite–wurtzite boundary crosses the pyrite–pyrrhotite solvus. Also, if as this figure suggests, the stability field of wurtzite above 150°C lies at very low f_{S_2} values not normally observed in nature, wurtzite must characteristically be a low-temperature mineral in ore deposits. Clearly Fig. 9 does not include the effects of impurities on stability relationships but iron, cadmium, and manganese, which are the major impurities in ZnS, are all more soluble in wurtzite than sphalerite and stabilize wurtzite relative to sphalerite under any given conditions (Barton and Toulmin, 1966). The sphalerite–wurtzite boundary should be displaced to higher f_{S_2} values by these impurities, increasing the temperature range of wurtzite stability in the presence of pyrite.

Relationships among sphalerite, wurtzite, and zinc oxide phases in an aqueous system as a function of log f_{O_2}, pH, and total dissolved zinc and sulfur (at 25°C and 1 atm) have been calculated by Scott (1968). These relationships are shown in Fig. 10. Increasing the total dissolved sulfur will expand the fields of wurtzite and sphalerite, and decreasing it will have the opposite effect. The presence of large amounts of wurtzite and the absence of oxidized phases indicates low f_{O_2} during deposition. Fig. 10 shows that both wurtzite and sphalerite have large fields of stability at pH within three units of neutrality. This observation has been confirmed by Hallberg (1972a), who synthesized sphalerite from zinc ions and sulfide derived from sulfate reducing bacteria and found no dependence on pH over the range 6–8 (see Trudinger's Chapter 6). Sphalerite was also found to form without intermediate phases such as the "monosulfide" /mackinawite/greigite encountered in the Fe–S system.

Equilibria amongst copper and copper–iron compounds. Aspects of equilibria in aqueous copper-bearing systems have been discussed by Pourbaix (1949), Garrels and Christ (1965), Burkin (1966), and Rickard (1970). Fig. 11A shows the Eh–pH relations in the $Cu–H_2O–O_2–S$ system at 25°C and 1 atm after Rickard (1970). Because thermody-

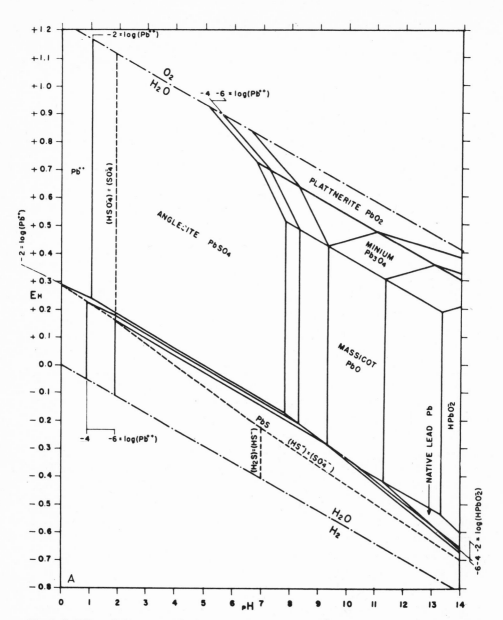

Fig. 8. Stability relations among lead compounds in water at 25°C and 1 atm total pressure. A. Total dissolved sulfur $= \cdot 10^{-5}$. B (p. 337). Total dissolved sulfur $= 10^{-1}$, $P_{CO_2} = 10^{-4}$. Boundaries of solids at total ionic activity $= 10^{-6}$. (Reproduced from Garrels and Christ, *Solutions, Minerals and Equilibria*, Harper and Row, 1965, with the authors' permission and courtesy of W. McIntyre.)

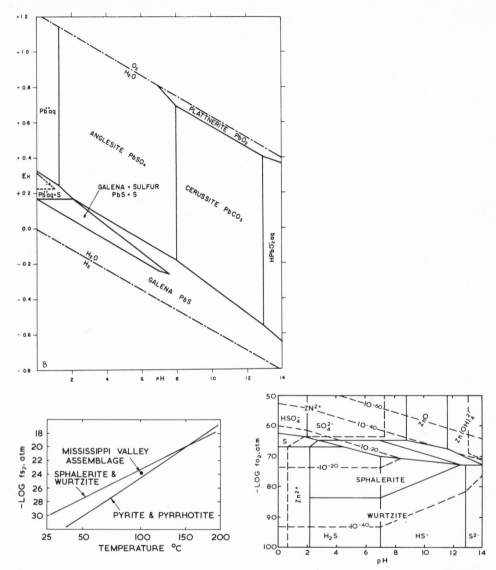

Fig. 9. The "best" sphalerite–wurtzite boundary and the pyrite–pyrrhotite solvus extrapolated to 25°C together with calculated value for f_{S_2} in Mississippi Valley-type assemblage (from Scott, 1968, reproduced with the authors' permission). Recent work, in fact, shows that the pyrite–pyrrhotite solvus lies at slightly higher f_{S_2} than shown in this figure, and is also influenced by the presence of monoclinic pyrrhotite which may displace the curve to higher f_{S_2} at low temperature (Prof. S.D. Scott, personal communication, 1975).

Fig. 10. Distribution of aqueous ions and solids (heavy lines) in the Zn–S–O system at total sulfur = 10^{-3} and 25°C. Solid boundaries between zinc minerals and aqueous species are drawn for total zinc = 10^{-6}. Dashed boundaries are for total zinc = 10^{-3}. The solid light lines represent boundaries between regions of predominance of sulfur-containing aqueous species (see Fig. 17). Dashed light lines are sulfur fugacites contoured in atmospheres. (From Scott, 1968, reproduced with the author's permission.)

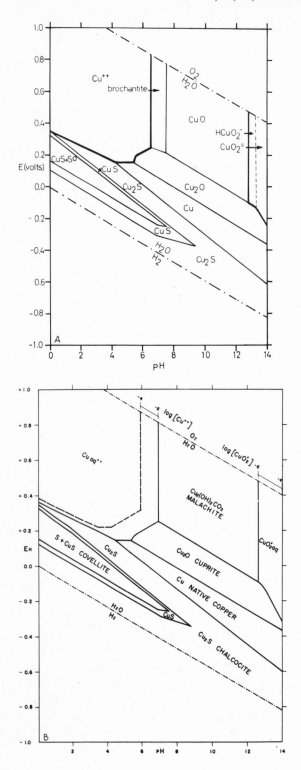

namic data are available only for stoichiometric covellite and chalcocite, other binary copper sulfides (djurleite, anilite, digenite, etc.) do not appear on the diagrams, and little is known of their stability fields in aqueous systems. It is interesting to note that copper still retains a large stability field even at fairly large sulfur activities. The diagram shows that oxidation of chalcocite to sulfur must proceed through covellite. Fig. 11B shows the effect of adding CO_2 to the system, and is taken from Garrels and Christ (1965, p. 240). The diagram is much the same except that malachite has displaced cuprite. In both figures, the existence of chalcocite over a wide range of pH in reducing conditions and the appearance of covellite at more acid pH is clearly shown. However, both the lack of data on the other copper sulfides and the frequent association of copper sulfides with Cu–Fe and other mixed metal sulfides limits applications of these data.

Cu–Fe–S–H_2O mineral relations at 25°C and 1 atm in the presence of 10^{-1} molal and 10^{-4} molal dissolved sulfur have been calculated by Natarajan and Garrels (Garrels and Christ, 1965, p. 231–232), and the figure for $\Sigma S = 10^{-1} M$ is reproduced here (Fig. 12). The best thermodynamic data available at the time were used, but, of course, the more recently described phases in the low-temperature Cu–S and Cu–Fe–S systems are not included.

The specific question of formation of sulfides, such as chalcopyrite and bornite at low temperatures, has been considered by a number of authors. In particular, Roberts (1963) has prepared chalcopyrite from CuS and FeS precipitates (heated at 150°C in distilled water) and converted the chalcopyrite to bornite by further heating (at 100°C) with copper sulfate solution. The author proposes that chalcopyrite and bornite may be formed from simple sulfides precipitated from solutions by sulfate reducing bacteria, and under conditions expected during diagenesis of sediments (see Chapter 6 by P.A. Trudinger).

Equilibria amongst iron compounds. Not surprisingly, iron minerals are the most extensively studied mineral group in terms of their Eh–pH, Eh–pH–partial pressure (of species such as sulfur, oxygen, CO_2) stability relations. Pourbaix (1949), Garrels and Christ (1965), Krauskopf (1967), Curtis and Spears (1968), and Berner (1971) are amongst the authors who have contributed in this area. Numerous approaches to the representation of equilibria have been tried, only some of which can be illustrated here.

In Fig. 13, the stability relations of hematite, magnetite, pyrrhotite, and pyrite are shown as a function of Eh, pH, and log P_{S_2} at 25°C and 1 atm total pressure in the presence of water. This well known figure, taken from Garrels and Christ (1965, p. 212)

Fig. 11. A. Eh–pH diagram for the system Cu–H_2O–O_2–S ($p[S] = 1, p[Cu] = 6$) at 25°C and 1 atm (from Rickard, 1970). (Reproduced from *Stockh. Contrib. Geol.*, 1970, vol. 23, p. 14, with the publisher's permission.) B. Eh–pH diagram for the system Cu–H_2O–O_2–S–CO_2 at 25°C and 1 atm. $P_{CO_2} = 10^{-3.5}$ and total dissolved sulfur = 10^{-1} (Reproduced from Garrels and Christ, *Solutions, Minerals and Equilibria*, Harper and Row, 1965, with the authors' permission and courtesy of J. Anderson.)

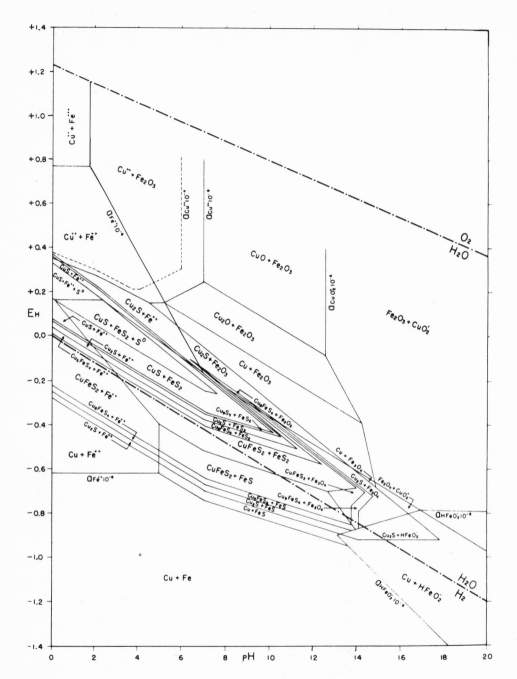

Fig. 12. Part of the Cu–Fe–S–O–H system at 25°C and 1 atm total pressure. Total dissolved sulfur = 10^{-1} M. (Reproduced from Garrels and Christ, *Solutions, Minerals and Equilibria,* Harper and Row, 1965, with the authors' permission and courtesy of R. Natarajan and R. Garrels.)

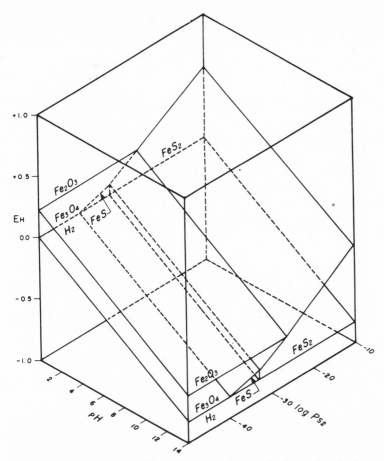

Fig. 13. Stability relations of hematite, magnetite, pyrrhotite, and pyrite in terms of Eh, pH, and log P_{S_2} at 1 atm total pressure in presence of water. (Reproduced from Garrels and Christ, *Solutions, Minerals and Equilibria*, Harper and Row, 1965, with the publisher's permission.)

shows the small field of stability of pyrrhotite compared to pyrite and the iron oxides, and the association of pyrite, pyrrhotite, and magnetite even at small values of P_{S_2}. The same authors (p. 218) also present the stability fields of pyrrhotite and pyrite relative to magnetite and hematite in solutions with total dissolved sulfur 10^{-1} molal. They show that pyrrhotite is not stable in water containing this much dissolved sulfur (the FeS/FeS_2 boundary is below the H_2O/H_2 boundary) and pyrite is the stable species.

Another system of geological interest for which Eh–pH equilibria at fixed concentrations have been determined is that involving iron oxides, sulfides, and carbonates (Fig. 14, from Garrels and Christ, 1965, p. 224). The figure shows relations for total dissolved carbonate (10 molal) and sulfur (10^{-6} molal) under equilibrium conditions. Thus, in order for siderite to have an important field of stability, dissolved carbonate must be very high

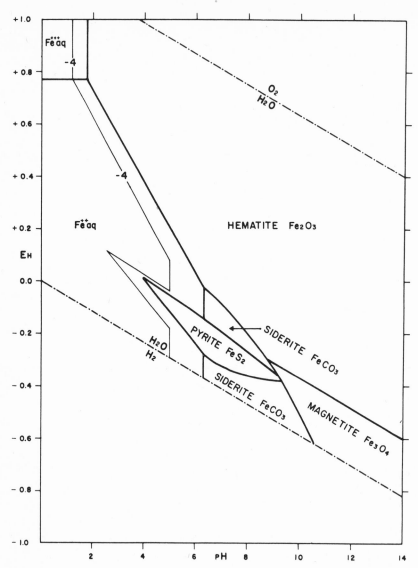

Fig. 14. Stability relations of iron oxides, sulfides, and carbonate in water at 25°C and 1 atm total pressure. Total dissolved sulfur = 10^{-6} M, total dissolved carbonate 10^0 M. (Reproduced from Garrels and Christ, *Solutions, Minerals and Equilibria,* Harper and Row, 1965, with the publisher's permission.)

and reduced sulfur very low. Pyrrhotite is eliminated under these conditions, but a considerable pyrite field remains. The Fe–C–O_2–S–H_2O system at 75–800°C and 1–2,500 atm has also been studied by Seguin (1971).

Berner (1971) points out that a disadvantage of the use of Eh–pH diagrams in de-

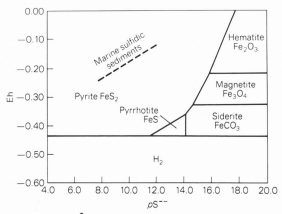

Fig. 15. Eh–pS^{2-} diagram for pyrite, pyrrhotite, hematite, magnetite, and siderite at 25°C and 1 atm total pressure. pH = 7.37, log P_{CO_2} = 2.40 (from Berner, 1971). (Reproduced from Berner, *Principles of Chemical Sedimentology*, McGraw-Hill, 1971, with the publisher's permission.)

scribing equilibria in sediments is the relative constancy of pH, particularly in marine environments (pH ~ 7.0–8.0). Because the concentration and distribution of dissolved sulfur species are more likely to vary due to their dependence on bacterial activity, Berner chooses to represent equilibria in terms of Eh, p_{CO_2}, and pS^{2-} (where pS^{2-} is the negative logarithm of the activity of sulfide ion). Thus in Fig. 15, an Eh–pS^{2-} diagram for pyrite, pyrrhotite, siderite, magnetite, and hematite is given. Again, the large stability field of pyrite relative to pyrrhotite (as in Fig. 13) is notable. In this figure, a line representing measurements in natural sulfidic marine sediments is plotted, and the area marked H_2 corresponds to regions beyond the H^+/H_2 boundary in Eh–pH diagrams. A series of diagrams of this type has been used by Curtis and Spears (1968) in discussing the formation of sedimentary iron minerals.

Berner (1967) has also discussed the stability of mackinawite and greigite in relation to pyrrhotite and pyrite. The free energy data, calculated from measured solubility equilibrium constants, show that mackinawite and greigite are unstable with respect to pyrite and stoichiometric pyrrhotite. Fig. 16 (after Berner, 1967) applies to recent sediments in which pyrite and pyrrhotite have not yet formed from mackinawite and greigite. The close proximity of the mackinawite–greigite boundary to the range of naturally measured Eh–pS^{2-} conditions supports determinations that the black iron sulfide of many recent sediments is a mixture of mackinawite and greigite. During diagenesis, many such deposits may be oxidized first to greigite and then pyrite. Particular aspects of the relationship between iron monosulfide and pyrite have been discussed by a number of authors. For example, Roberts et al. (1969) stated that pyrite forms most rapidly when S_2^{2-} ions are present in solution, whereas reaction of FeS and elemental S takes place more slowly. Berner (1970a) has discussed pyrite formation from FeS + S and identifies three main steps relevant to pyrite formation in sediments. These are: (1) bacterial sulfate reduction;

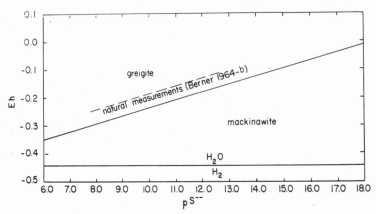

Fig. 16. Eh–pS^{2-} diagram for mackinawite and greigite at $25°C$ and 1 atm total pressure. pH = 7.5. (Reproduced from Berner, 1967, *Am. J. Sci.*, 265, p. 784, with the permission of the publisher and Prof. R.A. Berner.)

(2) reaction of H_2S with iron minerals to form monosulfide (mackinawite?); (3) reaction FeS + S → pyrite. The importance of the concentration of sulfur in regulating the mono-sulfide–disulfide equilibria has been emphasized by Skripchenko (1969) and the effect of T, P, and P_{O_2} examined by Roberts and Buchanan (1971). Presence of oxygen apparent-ly favors pyrite formation as against pyrrhotite formation from initial monosulfide (mackinawite?).

TRANSPORT AND DEPOSITIONAL PHENOMENA INVOLVING Pb, Zn, Cu AND Fe SULFIDES

The average abundances of Pb, Zn, Cu, and Fe in different geochemical environments have already been outlined at the beginning of this chapter (Table II). The mechanisms by which these elements form local concentrations which are sufficiently rich in metals to constitute an ore deposit are the central problem of ore genesis. Many aspects of this problem can only be considered with regard to the geological history of individual deposits — particularly problems concerned with the sources of ore metals.[1] Krauskopf (1971) has discussed the question of the sources of ore metals and pointed to the special circumstances requisite for their concentration. A specific example relevant to this work is the study by Helgeson (1967) of silicate metamorphism in sediments and the genesis of hydrothermal solutions. He has shown that interstitial seawater in buried arkosic sedi-ments (at ~200°C) may derive >1 ppm lead from potassium feldspar as the solution phase equilibrates with its environment. It can be shown that such a solution would provide sufficient lead to form an ore accumulation.

---

[1] See Chapter 2, Vol. 1, for examples.

However, there are several areas of this complex question where geochemical and mineralogical studies have proved enlightening. In the dissolution, transport, and redeposition in greater concentration of metals such as Pb, Zn, Cu, and Fe, geochemical studies of their solubilities and of the solubilities and nature of the complex ions which they form in various solutions, have proved informative. Also, the importance of reaction phenomena at solid–liquid interfaces and of the diffusion rates in sediments are being realized.

Solubilities and complex formation

The solubility of most metal sulfides in pure water is extremely small. The solubilities of PbS and ZnS in pure water at temperatures up to 350°C are 0.1 mg/l and 1.0 mg/l, respectively (Vukotic, 1961), both figures being largely temperature independent. Similarly, the solubility of covellite even in H_2S saturated water at up to 200°C is <1.5 mg/l (Romberger, 1968) and a maximum value obtained for chalcopyrite under similar conditions (at 50°C) is 1.9 mg/l. Haas and Barnes (1965) have studied pyrite solubilities with similar results.

In cases where metal ions in solution in oceans, seas, or even percolating solutions have reacted with sulfide ions (such as might be produced by biogenic reduction of sulfate) to form sulfide deposits "in situ", the insolubility of metal sulfides does not present a problem. However, in many cases, there is evidence to suggest that metal and sulfide were transported to the site of deposition in the same solution. Here, the role played by complex ions becomes significant and chloride and polysulfide complexes in particular have been discussed as possible media for the transport of the metals.

Helgeson (1964) in particular has advocated the importance of chloride complexes which can develop in the presence of high concentrations of Cl^- from chlorides, such as NaCl, HCl, KCl, generally in acid solutions. Alternatively, Barnes and Czamanske (1967) prefer to consider the transport of metals as sulfide complexes in alkaline solutions more likely. Most of the data and literature up to 1967 are covered in these two references, so that only a brief outline of the major points and the results of later work will be mentioned here.

Helgeson (1964) undertook a detailed study of equilibria in the system PbS–NaCl–HCl–H_2O at elevated temperatures. He demonstrated that galena concentrations of ~20–100 ppm can be carried in chloride-rich solutions at ~125°C and pH ≈ 5, the lead being principally as $PbCl^+$ and $PbCl_4^{2-}$ complexes. At lower temperatures, the $PbCl_4^{2-}$ complex dominates in concentrated solutions and $PbCl^+$ in dilute solutions, at higher temperatures $PbCl^+$ is dominant in both concentrated and dilute solutions. Barnes and Czamanske (1967) have been critical of the importance of chloride complexes in lead transport, pointing out that high chloride, low sulfur, and excessively acid conditions appear to be required. Nriagu and Anderson (1970) have calculated the solubilities of PbS, CuS, and a number of other sulfides in concentrated brine solutions containing such ligands as Cl⁻, S^{2-}, HS⁻, and H_2S. They state that chloride complexing is responsible for the large solubility of PbS in neutral to weakly acid solution, and that the brine solutions are capable of carrying enough reduced sulfur to precipitate a sizeable fraction of the ore metal. Studies of the geochemistry of the Cheleken thermal brines (Lebedev et al., 1971) show that most of the lead and zinc occurs as complex chloride anions, $PbCl_3^-$ and $PbCl_4^-$ being suggested as the dominant solution species for lead, with zinc as $ZnCl_3^-$. Clearly, these observations lend support to the importance of chloride complexes in transporting lead and zinc. Nriagu and Anderson (1971) have also examined the stability of Pb(II) complexes at elevated temperatures and suggest that $PbCl_2$ is the dominant species in moderately dilute solutions at lower temperatures, with $PbCl^+$ becoming dominant at >120°C. In concentrated chloride solutions <200°C, they state that $PbCl_4^{2-}$ is the most important species in agreement with Helgeson (1964).

The equilibrium relations between the predominant aqueous sulfur-containing species are shown in Fig. 17. These are the species of importance within the geological limits of pH and f_{O_2} and which may therefore contribute to complex formation. Barnes and Czamanske (1967) have summarized data on sulfide complexes in the transport of lead and other metals. At higher pH (alkali solutions), the solubility of lead can be increased

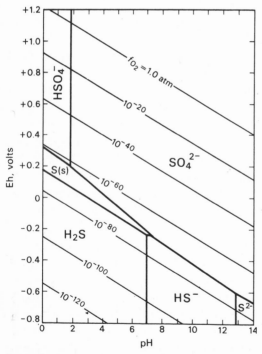

Fig. 17. Stabilities of the predominant aqueous sulfur-containing species as functions of Eh and pH at total dissolved sulfur = 0.1 M (25°C). (Reproduced from Barnes and Czamanske, 1967, in: H.L. Barnes (Editor), *Geochemistry of Hydrothermal Ore Deposits*, p. 342, Holt, Rinehart and Winston, with the authors' permission.)

by a factor of about 10, probably due to the formation of $Pb(HS)_3^-$. Subsequent studies include the work of Nriagu (1971) on the stability of Pb(II) thiocomplexes in the system $PbS-NaCl-H_2S-H_2O$. In solutions *saturated* with H_2S, chloride complexing is dominant at very acid pH (<2.5). At pH of 2.5–6.0, the solubility can be attributed to a neutral complex $PbS \cdot 2H_2S$, and the increase in solubility beyond pH = 6.0 is due to the formation of $Pb(HS)_3^-$ complex ions. These thiocomplexes are potential ore carriers and can carry >10 ppm Pb in neutral to slightly acid solutions, even if total reduced sulfur is low (0.2 molal). Listova (1966) has indicated the importance of complexes formed by lead with the "intermediate oxidation products" of sulfur under conditions prevalent in surface waters, and Malyshev and Khodakovskii (1964) discuss the importance of bisulfide, bicarbonate, and chloride complexes of lead in formation of the Zambarak Deposit (USSR).

With regard to ZnS, the importance of sulfide complexing and relative insignificance of chloride complexing of zinc has been argued by Barnes and his co-workers (see Barnes and Czamanske, 1967). In the Eh–pH region of HS^- formation (Fig. 17), a complex $Zn(HS)_3^-$ forms, resulting in solubilities of 2,700 mg/l (at 25°C, 7 atm P_{H_2S}, pH $\approx$ 8.2), although in even weakly acidic solutions the solubility is considerably decreased. Helgeson (1964) has maintained that chloride complexing makes an important contribution to sphalerite solubility in chloride solutions when sulfur is low, the sulfide complexes being important only in high-sulfur environments, whereas Barnes (1967) has proposed that chloride complex transport requires a solution that is virtually sulfide-free. Again, the work of Lebedev et al. (1971) on the Cheleken brines emphasizes the importance of such species as $ZnCl_3^-$, whereas fluid-inclusion studies by Sawkins (1964) suggest chloride complexes are unimportant in sphalerite transport (see Chapter 4, Vol. 2). Investigations of the solubilities of Zn and Cu in aqueous chloride–sulfide solutions have been undertaken by Melent'yev et al. (1969) between 100° and 180°C at pH = 1–8. High solubilities, from the geological point of view, were recorded but no inferences were drawn regarding the nature of the complexes present.

As with ZnS, Barnes and Czamanske (1967) contend that in order for chloride complexing to account for copper transport at low temperatures, an unreasonably low sulfide-ion activity is required. In studies of the solubility of CuS in sulfide solutions, Romberger and Barnes (1970) determined that covellite solubility is a function of HS^- concentration, which is in turn a function of H_2S pressure and temperature. Solutions of pH = 3.5–13.5 at temperatures 20–200°C were studied. At 25°C, $Cu(HS)_3^-$ is the important complex at pH < 7.3 and $CuS(HS)_3^{3-}$ at pH > 7.3. At 200°C, the complexes are $Cu(HS)_4^{2-}$ (pH < 6.6) and $CuS(HS)_3^{3-}$ (pH > 6.6). The authors maintain that if a solubility of >10 ppm Cu is necessary for significant ore transport, bisulfide complexes are potential ore carriers at 200°C when $\Sigma S > 0.25$ molal (in neutral to weakly alkaline pH).

Chloride complexing of Cu^+ is more effective than Cu^{2+}, although the order of stabilities of chloride complexes of the metals being considered are $Cu^{2+} \lesssim Zn^{2+} < Pb^{2+}$ at 25°C

(Helgeson, 1964). Kolonin and Aksenova (1970) have investigated complexing of copper and other metals in NaCl solutions at 20–90°C. At 90°C, a considerable amount of copper is in solution as the $CuCl_4^{2-}$ ion – however, these were solutions *free* of sulfur, so the sulfide complexes probably do play the major role in copper transport when sulfide ions are present.

Appreciable solubilities of pyrite appear not to be obtained through chloride or sulfide complexing, but a complex of dissociated pyrite with NH_3 seems to be a possibility (Barnes and Czamanske, 1967). Solubility of pyrite in this complex is extremely temperature dependent, increasing two orders of magnitude between 25 and 250°C.

Amongst recent studies of the effects of other species on transport by complexes might be mentioned the work of Rashid and Leonard (1973) on the increased solubility of metal sulfides (including Cu and Zn) in the presence of sedimentary humic acid (see Chapter 5, Vol. 2). In weakly alkaline conditions, organo–metallic complexes are formed which should enable migration and enrichment of metals. Accumulation of metals could be enhanced by these mechanisms in areas rich in organic matter. Also, Lebedev and Nikitina (1971) and Lebedev et al. (1971) have postulated that lead migrates in the form of chloro–carbonate and carbonate complexes such as $(PbCO_3)_2Cl^{3-}$ and $Pb(CO_3)_2^{2-}$ at depth in the Cheleken thermal brines.

Depositional reactions, depositional systems, and surface phenomena

Apart from the chemical precipitation of insoluble sulfides in a basin of deposition by reaction of metal with sulfide produced from bacteriological sulfate reduction, causes of deposition, particularly from sulfide complexes, have been reviewed by Barnes and Czamanske (1967). Changes in the physical environment of a solution carrying metals as sulfide, chloride or other complexes could include cooling or heating, dilution or reaction with other solutions, reaction with solids, or adiabatic expansion. Experimental results show that decreasing temperature decreases the solubility of pyrite in NH_4^+–H_2S solutions (Haas and Barnes, 1965) and covellite in HS^- solutions (Romberger and Barnes, 1965) enabling deposition. However, the solubility of sphalerite in HS^- solutions is nearly independent of temperature (Barnes, 1960) and the solubility of some mineral complexes increases with decreasing temperature. Pressure may similarly either increase or decrease solubility. For example, increasing H_2S pressure will increase the solubility of $Zn(HS)_3^-$ complexes (Barnes and Czamanske, 1967). The physical and chemical factors causing deposition of sulfides from sulfide complexes have been ranked in terms of *chemical efficiency* by Barnes and Czamanske (1967). Oxidation is most effective followed by decrease in pH, whereas dilution or decrease in pressure or temperature may cause no precipitation and is unlikely to cause complete precipitation.

The reaction of solutions with solid surfaces, whether at or near the sediment–water interface or by a solution percolating through a consolidated sediment, is an important aspect of ore deposition in sediments. Temple and LeRoux (1964) have conducted experiments in which absorption of Pb, Zn, Cu, and Fe ions by gel substances followed by desorption and precipitation as sulfides were shown to be possible mechanisms for metal sulfide concentration in sediments. Similar experiments by Weiss and Amstutz (1966) established the possible importance of ion exchange reactions on clay minerals in concentrating Pb, Zn, and other metals in sedimentary environments and Pantin (1965) has suggested that adsorption effects may affect the attainment of equilibrium. The example of sphalerite concentration in Lake Kivu (Degens et al., 1972) has already been described. Further work in this area, using recent theories of crystal chemistry, surface chemistry, and chemisorption may prove worthwile.

A related problem of considerable importance concerns the diffusion of components either during diagenesis or post-depositional alteration of sediments. Berner (1969b) has discussed the migration of iron and sulfur in reducing sediments during early diagenesis and, on the basis of various models, derived a number of tentative conclusions. Amongst these are that variability of organic matter concentration in an otherwise homogeneous sediment can cause migration of iron and sulfur, and iron sulfide bands may form by interdiffusion of dissolved iron and sulfide. Pyrite layers may also form by upward diffusion of Fe^{2+} from underlying sediments to an oxygenated sediment—water interface. Lambert and Bubela (1970) succeeded in producing monomineralic sulfide bands in sediments during laboratory experiments, and concluded that ionic migration processes probably occur in conjuction with deposition of metalliferous mud layers in forming ores of the McArthur River (North Australia) type. Study of the copper-rich zone of the basal Nonesuch Shale, Michigan, by Brown (1971) was accompanied by modelling of infiltration and diffusion processes. The mineralization of the shale, soon after deposition, seems to be supported by the modelling studies. Reactions below the sediment—water interface are important in all of these examples. (Cf. Chapter 2 by Duursma and Hoede, Vol. 2)

The subject of the involvement of micro-organisms in the formation of sulfide ores in sediments is considered in detail elsewhere in this book (see Chapter 6). However, since this aspect of sulfide ore genesis has received considerable attention, some major points will be mentioned here. Generally, the importance of micro-organisms lies in their action of reducing inorganic sulfate to sulfide, which may then react with metals such as Pb, Zn, Cu, and Fe to precipitate metal sulfides. The feasibility of the biochemical aspects has been examined by Temple (1964) and stated to be adequate for the syngenesis of sulfide ores. The suggestion that high concentrations of metals such as lead and copper may be toxic toward, or inhibit the activity of micro-organisms, has been rejected by Temple (1964). Sadler and Trudinger (1967) have investigated the effects of heavy metals on micro-organisms and noted the resistance of some bacteria to toxic metals, and the reduction of toxicity of other metals in natural environments by binding to natural chelating compounds.

A number of attempts have been made to construct models for the deposition of sulfides in sedimentary systems. The work of Berner (1969b) has been mentioned and recently Hallberg (1972b), and Rickard (1973) have both attempted more complete

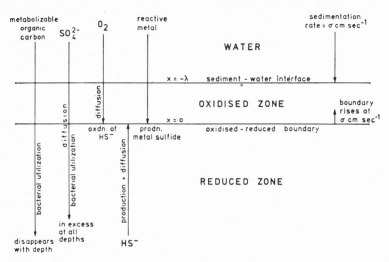

Fig. 18. Representation of a model for sedimentary metal sulfide formation in which the oxidized—reduced boundary is within the sediment (from Rickard, 1973). (Reproduced from *Econ. Geol.*, 1973, vol. 68, p. 607, with the publisher's permission.)

models. Hallberg used a semi-quantitative approach involving energy symbols combined in a circuit system. He concluded that sedimentary sulfide formation could take place through direct precipitation of metals by sulfide from bacterially reduced sulfate, or by later sulfidation of a metal oxide sedimentary layer. Thus, the presence of a sulfide-rich layer in a fossil sediment does not necessarily imply that the contemporaneous bottom water was chemically reduced. Furthermore, certain metals may show tendencies to follow one or other of these two routes and metal ratios (e.g., Cu/Zn) may serve as fossil indicators of the mechanism of sulfide formation and of the depositional environment. In an attempt to further quantify the factors involved in syngenetic formation of sediment-ary sulfides, Rickard (1973) performed calculations using estimates of maximum sulfide and metal flux, organic carbon concentration, sedimentation rate, porosity, etc. A schematic representation of the model is reproduced in Fig. 18. The calculations suggest at least 0.1% dry weight organic carbon is required to produce sulfide deposits with >1% metal, and such concentrations are known to be widespread in sediments. However, a metal flux greater than that normal in seawater is essential, and probably associated volcanism rather than just erosion of continental metal-rich source rocks is required.

EXAMPLES OF SEDIMENTARY ENVIRONMENTS CONTAINING Pb, Zn, Cu, AND Fe SULFIDES

Finally in this chapter, actual concentrations of Pb, Zn, Cu, and Fe sulfides in sedi-mentary rocks or recent sediments will be discussed. Only a limited number of examples can be considered, and by the use of the term "sedimentary environment" a syngenetic origin of these deposits is not necessarily implied. Since a discussion of the genesis of a particular mineral deposit requires evidence marshalled from many different sources, no attempt is made to argue particular cases here. This is left to chapters devoted to specific deposits or mineralized areas. However, ideas based on the sulfide mineralogy and geochemistry outlined in the preceding sections will be put forward. Sedimentary environments in which Pb, Zn, Cu or Fe sulfides are being formed at the present time will be considered first.

Modern environments

In a number of places in the world, ore-grade accumulations of metal sulfides are known to be forming in sediments. Examples include the stagnant or euxinic environments of certain landlocked seas, fjords or deep basins, the areas where warm brines are expelled into the sea or into sedimentary rocks, and even in some fairly open coastal waters.

Synsedimentary sulfide formation. The present-day formation of appreciable amounts of metal sulfides as part of normal sedimentary processes is virtually confined to the deposition of iron sulfides. Many authors have discussed the genesis of pyrite in modern sediments of seas and coastal waters, e.g., Volkov (1961), Roberts et al. (1969), Berner (1970a,b, 1971), Volkov et al. (1971), etc.. As Figs. 13 and 15 show, pyrite is the stable iron mineral under the low Eh and low pS^{2-} conditions characteristic of fine-grained marine sediments, rich in the organic matter which enables bacterial sulfate reduction. Berner (1970a, 1971) has discussed pyrite (and/or marcasite) formation in terms of the sources of iron and sulfur, the factors limiting formation and the mechanism of formation.

The main source of iron seems to be detrital iron minerals. Dissolution of such minerals can take place after deposition, through bacterial or inorganic processes – reduction of ferric oxides to soluble ferrous iron being important. Pyrite can form from reaction between dissolved H_2S and the iron minerals. Bacterial sulfate reduction is the major source of H_2S, with the breakdown of organic sulfur compounds from dead organisms providing a lesser source. Thus, the main limiting factors in pyrite formation are the concentration and reactivity of detrital iron minerals, the availability of dissolved sulfate, and the concentration of organic compounds which can be utilized by sulfate reducing bacteria (called "metabolized organic matter" by Berner).

Berner (1971) suggests that the concentration of metabolized organic matter is the most important factor, since iron compounds and dissolved sulfate are common constituents of sediments. This is supported by studies of pyrite formation in sediments off the coast of Connecticut (Berner, 1970a), where iron and sulfate are present to excess and the amount of iron converted to pyrite shows a direct relationship to the concentration of metabolized organic matter. However, in landlocked euxinic basins of which the Black Sea is a classic example, it may be the availability of *reactive* iron which controls sulfide formation (Ostroumov et al., 1961). Results of a recent detailed study of the Black Sea will also be available in the near future (Degens and Ross, 1974).

Much has already been said in this chapter about the importance of mackinawite and greigite as the first products of reaction between iron and sulfide in aqueous solution. Where it has been possible to identify the fine black sulfides of recent sediments, greigite and/or mackinawite have been found (Berner, 1964; Jedwab, 1967). The thermodynamic instability of greigite and mackinawite relative to pyrite and pyrrhotite (Berner, 1967) suggest that they should be transformed to the latter during diagenesis, and this is supported by the disappearance with depth of the black coloration due to fine-grained mackinawite and greigite. The reaction of FeS and Fe_3S_4 with elemental sulfur is believed to be the mechanism (Volkov, 1961; Sweeney and Kaplan, 1973). This is supported by observations of Volkov et al. (1971) of iron sulfide microconcretions in Black Sea sediments with compositions transitional between FeS/Fe_3S_4 and pyrite. When insufficient sulfur is available to convert mackinawite and greigite to pyrite, then they are retained. The Black Sea provides an interesting example of this situation (Volkov, 1961; Berner, 1970b) in which black FeS-rich clays are interbedded with gray pyrite-rich clays containing no FeS. This has been attributed to a reduction in elemental S and H_2S in the bottom waters at the time of deposition of the black clays, a suggestion supported by the lower reduced sulfur content of the black clays (Berner, 1970b). The reduction in sulfur content may have been due to desalination during a Pleistocene glacial period.

Clearly then, the formation of iron sulfides as a part of normal sedimentary processes is a widely observed phenomenon. However, the formation of appreciable quantities of other metal sulfides by such processes may require special circumstances — some of which will now be considered.

Geothermal brines. Several examples of heavy-metal rich geothermal brines have been recognized and studied recently. These modern hydrothermal fluids include the Red Sea brines (Degens and Ross, 1969, 1974 and Chapter 4, Vol. 4 of the present publication), the Salton Sea brines (White et al., 1963; Skinner et al., 1967), and the Cheleken brines (Lebedev, 1967; Lebedev and Nikitina, 1968).

The Red Sea brines occur in three pools in adjacent depressions in the center of the median valley of the Red Sea. Apart from the high salinity of these sea bottom pools, they are notable for their high metal contents. In Table IV, the contents of major and important minor ions in the brines are tabulated and compared with ocean water. The metal contents of some sediment samples recovered in these regions reach 21% ZnO, 4% CuO, 0.8% PbO, 85.0% Fe_2O_3, and 5.7% Mn_3O_4. The mineralogy of the deposits has been discussed in detail by Bischoff (1969) who describes, for example, a sulfide facies occurring as a homogeneously black bed at the bottom of the Atlantis II Deep. Sphalerite with lesser amounts of chalcopyrite and pyrite are the major minerals. Bischoff states that

TABLE IV

Ion concentrations in the Red Sea brine compared to ocean waters[1] (values in g/kg)

	Red Sea Atlantis II Deep	Ocean water
Na^+	92.60	10.76
K^+	1.87	0.39
Ca^{2+}	5.15	0.14
Mg^{2+}	0.76	1.29
Sr^{2+}	0.04	0.008
Cl^-	156.03	19.35
Br^-	0.13	0.066
SO_4^{2-}	0.84	2.71
HCO_3^-	0.14	0.72
Si	0.03	0.004
Fe	0.08	0.00002
Mn	0.08	0.00001
Zn	0.005	0.000005
Cu	0.0003	0.00001
Co	0.0002	
Pb	0.0006	
Salinity	257.76	35.71
Temperature	56.5°C	
Density	1.178	1.03

[1] After Emery et al. in Degens and Ross (1969), p. 557. Reproduced with the publisher's permission.

much "X-ray amorphous iron monosulfide" (fine-grained mackinawite?) must also be present. Further studies of sulfides from other areas (Stephens and Wittkopp, 1969) emphasize the importance of the sulfides as carriers of the base metals and state that marcasite was the most common sulfide with chalcopyrite and iron-rich sphalerite occurring in many of the samples (with up to 27% Fe). Some suggested processes involved in mineral precipitation in the Atlantis II Deep have been diagrammatically presented by Bischoff (1969) and are reproduced in Fig. 19. The Red Sea brines may be of meteoric rather than volcanic origin, and may have acquired their salinity by percolating through evaporite beds. The subsurface temperatures of the brines are suggested to be ~100–150°C (White, 1968), and it is of particular interest with reference to the role of chloride rather than sulfide complexing of metals, that the sulfide ion concentration is low relative to available metal ions.

The Red Sea brines are of particular interest because of their intimate involvement with sediments at the actual time of their deposition. Both the Salton Sea and Cheleken thermal brines were discovered whilst drilling for geothermal power, and can be considered as in the more general class of hydrothermal solution, i.e., as producing mineralization by subsequent alteration of sedimentary (or other) rock sequences. Data on the compositions of these brines may be obtained from Lebedev and Nikitina (1968), and Muffler and

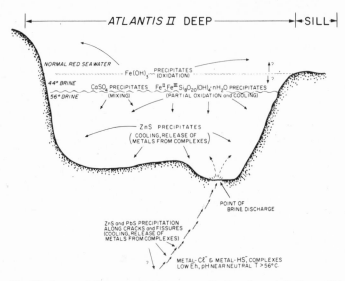

Fig. 19. Some of the suggested processes of mineral precipitation within the Atlantis II Deep (from Bischoff, 1969). (Reproduced from *Hot Brines and Recent Heavy Metal Deposits in the Red Sea* (Degens and Ross, Editors), 1969, p. 399, Springer, with the publisher's permission.)

White (1969), respectively. In contrast to the Red Sea brines, the Salton Sea brines exhibit much higher temperatures (300–350°C) and have a much greater heavy metal content. A siliceous scale deposited on discharge pipes at the rate of 2–3 tons/month contained ~20% Cu and <6% Ag (Skinner et al., 1967). The mineralogy is also fairly complex with sulfides including digenite, bornite, chalcocite, stromeyerite, arsenopyrite, tetrahedrite, chalcopyrite, and pyrite. The brines are in equilibrium with a sulfide assemblage in the reservoir rocks, and data on the iron content of sphalerite (16.6 mol. % FeS) in equilibrium with the brine (which has estimated $a_{S_2} = 10^{-10.2}$, $T = 325°C$) are in good agreement with the projected high-temperature data of Barton and Toulmin (1966) for the Fe–Zn–S system. It has been suggested (Skinner et al., 1967; White, 1968) that the brines are chiefly of local surface origin, and the salts and heavy metals are derived from the sediments of the brine reservoir. Again, it is remarkable that the heavy metals content greatly exceeds the presence of sulfide sulfur such that less than 5% of the total sulfide required to precipitate all the chalcophile metals as sulfides is available. The metals themselves are presumably present as chloride complexes.

The situation at Cheleken in the USSR closely parallels that at the Salton Sea. Again, a natural reservoir of geothermal saline waters, probably chiefly of surface origin, is depositing large quantities of metals, particularly lead and zinc (Lebedev, 1967; Lebedev and Nikitina, 1968). Subsurface water-temperatures reported are ~80°C, and detailed investigations of the role of chloride complexing in the Cheleken brines have already been described (p. 347).

Copper and iron sulfide concentrations in sedimentary environments

There are a number of examples of ore-grade sulfide concentrations, rich in copper minerals, which have been formed in sedimentary environments. In some cases, such as Mount Isa in Queensland, Australia, there appears to be a strong link with submarine volcanism, but in others there is no such clear indication of a distinct source for ore materials. Two much studied examples of the latter class are the Kupferschiefer of north central Europe and the deposits of the Northern Rhodesian Copperbelt (cf. Chapter 6 and 7, Vol. 6). We shall consider the Kupferschiefer as an example.

The Kupferschiefer is a thin bed of marine bituminous marl of Upper Permian Age and occurring in northern Germany and Poland, Holland, and N.E. England. It has been described in detail by a number of authors (e.g., Eisenhuth and Kautsch, 1954; Wedepohl, 1964, 1971, and this book) as containing detrital silicates, carbonates, and bituminous carbon (averaging 6%) and sulfides of Pb, Zn, and Cu. In fact, the average Pb + Zn concentration is ten times greater than that of copper, and high concentrations of the metals are very localized in the formation, apparently related to areas of sea bottom stagnation.

Pyrite is the most abundant sulfide mineral over the whole area, although the range of iron concentrations is the same as non-bituminous marls and shales. The high Pb and Zn concentrations occur as galena and ZnS, reported as sphalerite. High copper concentrations occur as bornite, chalcopyrite, chalcocite, covellite, and idaite in decreasing order of abundance. Texturally, the sulfides occur as very small grains ($20-100$ μ). Rarer sulfides include argentite, niccolite, skutterudite, maucherite, and molybdenite, and partly as a consequence of this, a number of other elements show an unusually high concentration in the marls (As, Sb, Bi, Ag, Se, Cd, Tl, Au, and Pt-metals; also, V, Mo, Ni, Cr, U, and Co). Although copper concentrations of >0.3% Cu occur in only about 1% and zinc concentrations of >0.3% Zn in about 5% of the total area covered by this marl, the Kupferschiefer is nevertheless a major ore deposit. The total copper is comparable to the Congo— Rhodesian copperbelt, but conditions for mining are much less favorable in the Kupferschiefer (Wedepohl, 1971).

Much of the literature regarding the genesis of the Kupferschiefer has been discussed by Dunham (1964). Not surprisingly, most workers have regarded it as a syngenetic deposit, but there have been contrary suggestions. Notable amongst these have been the ideas put forward by Davidson (1962, 1965) who argues for the introduction of metals by brines derived from associated evaporites, which leach base metals from magmatic sulfides at depth and redeposit them in the host Kupferschiefer sediments. A major objection to this idea is the very large lateral extent of the Kupferschiefer deposits.

The more widely accepted syngenetic theories have various distinctive features, notably regarding the origin of the metals. The existence of euxinic bottom conditions at the time of deposition, and of organic matter suitable for enabling bacterial sulfate reduction is widely acknowledged. However, by analogy with present-day sulfide formation in such environments, enrichment in sulfides other than iron sulfides requires an unusual source of Cu, Pb, and Zn. In the past, suggestions that the unusual amounts of Cu, Pb, and Zn were derived from red-bed formations and/or sulfide mineral veins of the surrounding and underlying land masses have been countered by suggestions that heavy metal enrichment was due to submarine exhalations, like those of the present-day Red Sea area. Dunham (1964), for example, favors the latter view, whereas Wedepohl (1971) favors an origin purely from surrounding red beds — pointing out that red-bed country rocks trangressed by a sea with reducing bottom conditions and a low rate of detrital accumulation is rare in earth history.

   In other recent studies, Haránczyk (1970) has examined Kupferschiefer lead-bearing shales in Poland and reconstructed the patterns of sedimentation and redox conditions over the area. Hallberg (1972b) in discussing models for sedimentary sulfide formation, proposes that the high Cu/Zn ratio in the oldest beds of the Kupferschiefer and decreasing values in the younger strata, result from better oxygenated conditions in the later bottom waters. Serkies et al. (1967) and Oberc and Serkies (1968) have studied Polish Kupferschiefer copper mineralization and suggest the derivation of Cu from leaching of underlying sediments.

Lead–zinc sulfide ores

   Any discussion of lead or zinc sulfide ore deposits has to include the so-called "Mississippi Valley" or "Alpine"-type mineralization occurring in limestone regions (see separate chapters in Vol. 5 and 6). Although it seems unlikely that such mineralization was contemporaneous with limestone deposition, the inclusion of these deposits in a book on sedimentary ores is appropriate. Important examples occur in Missouri, Kansas, Oklahoma, Tennessee (USA); Siberian platform and Kazakhstan (USSR); Pine Point, N.W.T. (Canada); Eastern Alps; Silesia and other areas of Europe including the English Pennines and central Ireland. Deposits range in age chiefly from Lower Palaeozoic to Late Mesozoic. The majority occur within sedimentary carbonate rocks, with some in associated sands and shales. The carbonates usually form part of a reef. Many of the deposits are also near major regional faults and are frequently restricted to particular units of the local succession. Ores also tend to be localized in sedimentary and tectonic structures and the fine details of their textures are highly variable.

   The mineralogy is characteristically simple, although complex paragenetic sequences have been taken by some authors to indicate numerous repetitive periods of mineralization (Hagni and Grawe, 1964). Galena and sphalerite (also wurtzite) in various proportions dominate the ore minerals, but marcasite and pyrite are major accessories and chalcopyrite a minor accessory mineral. Large amounts of barites and fluorspar are usually present. The sphalerite is usually low in iron and manganese, but it does contain high cadmium contents and greenockite (CdS) is sometimes present. Galena is usually very low in silver. Some aspects of the ore textures have been used by European authors (Amstutz, 1959; Zimmerman, 1970) to support a syngenetic–diagenetic origin. Other aspects of the ore textures have already been mentioned, particularly the "colloform" and "schalen-blende" texture types. Large, beautifully formed crystals, lining the walls of solution cavities, are also found. Fluid-inclusion studies suggest formation temperatures of ~100–150°C and frequently less than 100°C (see Chapter 4).

The literature describing deposits of this type and discussing their genesis is considerable. Important examples of recent literature include the monograph edited by Brown (1967) and studies by Beales and Jackson (1966), Dunham (1966), Heyl (1969), Brown (1970), Dozy (1970), Doe and Delevaux (1972), and Vols. 5 and 6.

The various opinions on the genesis of these ores is perhaps best summarized by Brown (1970), who shows that North American opinion favors a "dominantly connate marine but epigenetic ore fluid with probably minor additions from deeper sources". Perhaps partly because of the different results of lead isotope studies and the less attention devoted to fluid inclusions, European opinion is ."divided almost equally between proponents of syngenesis–diagenesis, and of magmatic epigenetic origin". Thus, the balance of opinion favors an epigenetic origin. Also, the studies of thermal brines discussed in the previous section have encouraged the view that epigenetic hypersaline connate waters may have transported the metals as chloride complexes which could pass into the permeable limestones and react with reservoirs of trapped H_2S gas (Beales and Jackson, 1966) or biogenically reduced sulfur (Gerdemann and Meyers, 1972).

CONCLUDING REMARKS

In this chapter, an attempt has been made to review recent knowledge regarding the mineralogy and geochemistry of Pb, Zn, Cu, and Fe sulfides of sedimentary affiliation. The Cu and Fe sulfides in particular are clearly more complex than was previously believed, and the stability relationships among these phases and their role in the sedimentary geochemistry of Cu and Fe requires further study. Much more remains to be learned about the processes of diffusion and surface phenomena during diagenesis and post-depositional alteration. The use of the available chemical and physical data in developing quantitative models for the genesis of "sedimentary" ores would seem the best approach to clarifying their mode of origin.

ACKNOWLEDGEMENTS

Part of this article was completed whilst the author was a Research Associate at the Massachusetts Institute of Technology, and the support of grants from NASA and NSF are gratefully acknowledged. Discussions with Prof. R.G. Burns (MIT) and the technical assistance of Ms. Roxanne Regan proved valuable at this stage.

I should also like to thank Professors J.R. Craig (VPI) and D.D. Hawkes (Aston Univ.) for their critical comments on the chapter. Mrs. Montgomery and Mrs. Espley (Aston Univ.) helped in the preparation of the final manuscript.

REFERENCES

Amstutz, G.C., 1959. Syngenetic zoning in ore deposits. *Geol. Assoc. Can. Proc.*, 11: 95–113.

Arnold, R.G., 1962. Equilibrium relations between pyrrhotite and pyrite from 325 to 743°C. *Econ. Geol.*, 57: 72–90.

Baas Becking, L.G.M., Kaplan, I.R. and Moore, D., 1960. Limits of the natural environment in terms of pH and oxidation–reduction potentials. *J. Geol.*, 68: 243–284.

Barnes, H.L., 1960. Ore solutions. *Carnegie Inst., Wash., Yearb.*, 59: 137–141.

Barnes, H.L., 1967. Sphalerite solubility in ore solutions of the Illinois–Wisconsin District. *Econ. Geol. Monogr.*, 3: 326–332.

Barnes, H.L. and Czamanske, G.K., 1967. Solubilities and transport of ore minerals. In: H.L. Barnes (Editor), *Geochemistry of Hydrothermal Ore Deposits*. Holt Rinehart and Winston, New York, N.Y., pp. 334–381.

Barton, P.B., 1973. Solid solutions in the system Cu–Fe–S. Part I: the Cu–S and CuFe–S joins. *Econ. Geol.*, 68: 455–465.

Barton Jr., P.B. and Kullerud, G., 1957. Preliminary report on the system Fe–Zn–S and implications regarding the use of the sphalerite geothermometer. *Bull. Geol. Soc. Am.*, 68: 1699 (Abstract).

Barton Jr., P.B. and Kullerud, G., 1958. The Fe–Zn–S system. *Carnegie Inst. Wash., Yearb.* 57: 227–229.

Barton Jr., P.B. and Toulmin, P., III, 1966. Phase relations involving sphalerite in the Fe–Zn–S system. *Econ. Geol.*, 61: 815–849.

Beales, F.W. and Jackson, S.A., 1966. Precipitation of lead–zinc ores in carbonate reservoirs as illustrated by Pine Point ore field, Canada. *Trans. Inst. Miner. Metal.*, 75: B278–B285.

Behre Jr., C.H., 1939. Wisconsin–Illinois–Iowa lead–zinc district. In: E.S. Bastin (Editor), *Contributions to a Knowledge of the Lead and Zinc Deposits of the Mississippi-Valley Region. Geol. Soc. Am., Spec. Pap.*, 24: 114–118.

Berner, R.A., 1964. Iron sulfides formed from aqueous solution at low temperatures and atmospheric pressure. *J. Geol.*, 72: 293–306.

Berner, R.A., 1967. Thermodynamic stability of sedimentary iron sulfides. *Am. J. Sci.*, 265: 773–785.

Berner, R.A., 1969a. The synthesis of framboidal pyrite. *Econ. Geol.*, 64: 383–384.

Berner, R.A., 1969b. Migration of iron and sulfur within anaerobic sediments during early diagenesis. *Am. J. Sci.*, 267: 19–42.

Berner, R.A., 1970a. Sedimentary pyrite formation. *Am. J. Sci.*, 268: 1–23.

Berner, R.A., 1970b. Pleistocene sea levels possibly indicated by buried black sediments in the Black Sea. *Nature*, 227: 700.

Berner, R.A., 1971. *Principles of Chemical Sedimentology*. McGraw-Hill, New York, N.Y., 256 pp.

Berry, L.G. and Thompson, R.M., 1962. X-ray powder data for ore minerals: the Peacock Atlas. *Geol. Soc. Am., Mem.*, 85: 281 pp.

Bischoff, J.L., 1969. Red Sea geothermal brine deposits: their mineralogy, chemistry, and genesis. In: E.T. Degens and D.A. Ross (Editors), *Hot Brines and Recent Heavy Metal Deposits in the Red Sea*. Springer, New York, N.Y., pp. 368–401.

Brett, R. and Yund, R.A., 1964. Sulfur-rich bornites. *Am. Mineral.*, 49: 1084–1098.

Brown, A.C., 1971. Zoning in the White Pine copper deposit, Ontonagon County, Michigan. *Econ. Geol.*, 66: 543–573.

Brown, J.S., 1967. *Genesis of Stratiform Lead–Zinc–Barite–Fluorite Deposits. Econ. Geol. Monogr.*, 4: 443 pp.

Brown, J.S., 1970. Mississippi-Valley type lead–zinc ores. *Miner. Deposita*, 5: 103–119.

Browne, P.R.L. and Lovering, J.F., 1973. Composition of sphalerites from the Broadlands Geothermal Field and their significance to geothermometry and geobarometry. *Econ. Geol.*, 68: 381–387.

Burkin, A.R., 1966. *The Chemistry of Hydrometallurgical Processes.* Spon., London, 157 pp.

Burns, R.G., 1970. *Mineralogical Applications of Crystal Field Theory.* Cambridge Univ. Press, London, 224 pp.

Cabri, L.J., 1973. New data on phase relations in the Cu–Fe–S system. *Econ. Geol.*, 68: 443–454.

Cabri, L.J. and Hall, S.R., 1972. Mooihoekite and haycockite, two new copper–iron sulfides and their relationship to chalcopyrite and talnakite. *Am. Mineral.*, 57: 689–708.

Chukrov, F.V., Genkin, A.D. and Soboleva, S.V., 1965. Smythite from the iron ore deposits of the Kerch Penninsula. *Geochim. Int.*, 2: 372–381.

Clark, A.H., 1969. Preliminary observations on chromian mackinawite and associated native iron, Mina do Abessedo, Vinhais, Portugal. *Neues Jahrb. Mineral., Monatsh.*, 6: 282–288.

Clark, A.H. and Sillitoe, R.H., 1971. Cuprian galena solid solutions, Zapallar mining district, Atacama, Chile. *Am. Mineral.*, 56: 2142–2145.

Coey, J.M.D., Spender, M.R. and Morrish, A.H., 1970. The magnetic structure of the spinel, Fe_3S_4. *Solid State Comm.*, 8: 1605–1608.

Coleman, R.C. and Delevaux, M., 1957. Occurrence of selenium in sulfides from some sedimentary rocks of the Western United States. *Econ. Geol.*, 52: 499–527.

Cotton, F.A. and Wilkinson, G., 1972. *Advanced Inorganic Chemistry.* Interscience, New York, N.Y., 959 pp.

Craig, J.R., 1967. Phase relations and mineral assemblages in the Ag–Bi–Pb–S system. *Miner. Deposita*, 1: 278–306.

Curtis, C.D. and Spears, D.A., 1968. The formation of sedimentary iron minerals. *Econ. Geol.*, 63: 257–270.

Davidson, C.F., 1962. The origin of some stratabound sulfide ore deposits. *Econ. Geol.*, 57: 265–275.

Davidson, C.F., 1965. A possible mode of origin of strata-bound copper ores. *Econ. Geol.*, 60: 942–954.

Degens, E.T. and Ross, D.A., 1969. *Hot Brines and Recent Heavy Metal Deposits in the Red Sea.* Springer, New York, N.Y., 600 pp.

Degens, E.T. and Ross, D.A. (Editors), 1974. *The Black Sea: Its Geology, Chemistry and Biology. Mem., Am. Assoc. Pet. Geol.*

Degens, E.T., Okada, H., Honjo, S. and Hathaway, J.C., 1972. Micro-crystalline sphalerite in resin globules suspended in Lake Kivu, East Africa. *Miner. Deposita*, 7: 1–12.

Dell, C.I., 1972. An occurrence of greigite in Lake Superior sediments. *Am. Mineral.*, 57: 1303–1304.

Doe, B.R. and Delevaux, M.H., 1972. Source of lead in southeast Missouri galena ores. *Econ. Geol.*, 67: 409–425.

Dozy, J.J., 1970. A geological model for the genesis of the lead–zinc ores of the Mississippi Valley, U.S.A. *Trans. Am. Inst. Min. Metall. Pet. Eng.*, 79: B163–B175.

Dunham, K.C., 1964. Neptunist concepts in ore genesis. *Econ. Geol.*, 59: 1–21.

Dunham, K.C., 1966. Role of juvenile solutions, connate waters and evaporitic brines in the genesis of lead–zinc–fluorine–barium deposits. *Trans. Am. Inst. Min. Metall. Pet. Eng.*, 75: B226–B229.

Eisenhuth, K.H. and Kautsch, E., 1954. *Handbuch für den Kupferschieferbergbau.* Fachbuchverlag, Leipzig, 632 pp.

Erd, R.C., Evans Jr., H.T. and Richter, D.H., 1957. Smythite, a new iron sulfide, and associate pyrrhotite from Indiana. *Am. Mineral.*, 42: 309–333.

Evans Jr., H.T., Milton, C., Chao, E.C.T., Adler, I., Mead, C., Ingram, B. and Berner, R.A., 1964. Valleriite and the new iron sulfide, mackinawite. *U.S. Geol. Surv. Prof. Pap., 475-D*: D64–D69.

Farrand, M., 1970. Framboidal sulfides precipitated synthetically. *Miner. Deposita*, 5: 237–247.

Fleischer, M., 1955. Minor elements in some sulfide minerals. *Econ. Geol.*, Fiftieth Anniv. Vol: 970–1024.

Garrels, R.M. and Christ, C.L., 1965. *Solutions, Minerals, and Equilibria*. Harper and Row, New York, N.Y., 450 pp.

Gerdemann, P.E. and Meyers, H.E., 1972. Relationship of carbonate facies patterns to ore distribution and to ore genesis in the Southeast Missouri lead district. *Econ. Geol.*, 67: 426–433.

Goldschmidt, V.M., 1958. *Geochemistry*, Oxford Univ. Press, London, 730 pp.

Haas, J.L. and Barnes, H.L., 1965. Ore solution transport of pyrite. *Trans. Am. Geophys. Union*, 46: 183 (Abstract).

Hagni, R.D. and Grawe, O.R., 1964. Mineral paragenesis in the Tri-State District, Missouri, Kansas, Oklahoma. *Econ. Geol.*, 59: 449–457.

Hallberg, R.O., 1972a. Iron and zinc sulfides formed in a continuous culture of sulfate reducing bacteria. *Neues Jahrb. Mineral., Monatsh.* 11: 481–500.

Hallberg, R.O., 1972b. Sedimentary sulfide mineral formation – an energy circuit system approach. *Miner. Deposita*, 7: 189–201.

Haránczyk, C., 1970. Zechstein lead-bearing shales in the Fore–Sudetian monocline in Poland. *Econ. Geol.*, 65: 481–495.

Helgeson, H.C., 1964. *Complexing and Hydrothermal Ore Deposition*. Pergamon, New York, N.Y., 128 pp.

Helgeson, H.C., 1967. Silicate metamorphism in sediments and the genesis of hydrothermal ore solutions. *Econ. Geol. Monogr.*, 3: 333–342.

Heyl, A.V., 1969. Some aspects of genesis of zinc–lead–barite–fluorite deposits in the Mississippi Valley, U.S.A. *Trans. Am. Inst. Min. Metall. Eng.*, 78: B148–B160.

Jedwab, J., 1967. Mineralization en greigite de debris vegetaux d'une vase recente (Grote Geul). *Soc. Belge. Geol. Bull.*, 76: 1–19.

Karup-Møller, S., 1971. On some exsolved minerals in galena. *Can. Mineral.*, 10: 871–876.

Kissin, S.A. and Scott, S.D., 1972. Phase relations of intermediate pyrrhotites. *Geol. Soc. Am., Ann. Meet. Abstr.*, p. 562.

Kolonin, G.R. and Aksenova, T.P., 1970. Investigation of complexing of some heavy metals in NaCl solutions at 20–90°C. *Geochim. Int.*, 7: 973–978.

Krauskopf, K.B., 1967. *Introduction to Geochemistry*. McGraw-Hill, New York, N.Y., 721 pp.

Krauskopf, K.B., 1971. The source of ore metals. *Geochim. Cosmochim. Acta*, 35: 643–660.

Kullerud, G., 1953. The FeS–ZnS system, a geological thermometer. *Nor. Geol. Tidsskr.*, 32: 61–147.

Kullerud, G., 1967. Sulfide studies. In: P.H. Abelson (Editor), *Researches in Geochemistry*, 2. Wiley, New York, N.Y., p. 286–321.

Lambert, I.B. and Bubela, B., 1970. Banded sulfide ores: the experimental production of monomineralic sulfide bands in sediments. *Miner. Deposita*, 5: 97–102.

Lebedev, L.M., 1967. Contemporary deposition of native lead from hot Cheleken thermal brines. *Dokl. Akad. Nauk. SSSR (Earth Sci. Trans.)*, 174: 173–176.

Lebedev, L.M. and Nikitina, I.B., 1968. Chemical properties and ore content of hydrothermal solutions at Cheleken. *Dokl. Akad. Nauk. SSSR. (Earth Sci. Trans.)*, 183: 180–182.

Lebedev, L.M. and Nikitina, I.B., 1971. The chemical complexes in which lead migrates in the Cheleken thermal brines at depth. *Dokl. Akad. Nauk. SSSR*, 197: 229–230.

Lebedev, L.M., Baranova, N.N., Nikitina, I.B. and Vernadskiy, V.I., 1971. On the forms of occurrence of lead and zinc in the Cheleken thermal brines. *Geochim. Int.*, 8: 511–516.

Listova, L.P., 1966. Experimental data on the solubility of lead sulfide under oxidizing conditions. *Geochim. Int.*, 3: 25–33.

Loftus-Hills, G. and Solomon, M., 1967. Cobalt, nickel, and selenium in sulfides as indicators of ore genesis. *Miner. Deposita*, 2: 228–242.

Love, L.G., 1965. Micro-organic material with diagenetic pyrite from the Lower Proterozoic Mount Isa shale and a carboniferous shale. *Proc. Yorks. Geol. Soc.*, 35: 187–202.

Love, L.G. and Amstutz, G.C., 1966. Review of microscopic pyrite. *Fortschr. Mineral.*, 43: 273–309.

Malakhov, A.A., 1968. Bismuth and antimony in galenas as indicators of some conditions of ore formation. *Geochim. Int.*, 5: 1055–1068.

Malyshev, B.I. and Khodakovskii, I.L., 1964. Some geochemical characteristics of transport and deposition of lead by hydrothermal solutions as illustrated by the Zambarak deposit. *Geochim. Int.*, 3: 421–428.

Melent'yev, B.N., Ivanenko, V.V. and Pamfilova, L.A., 1969. Solubility of some ore-forming sulfides under hydrothermal conditions. *Geochim. Int.*, 2: 416–460.

Moh, G.H., 1964. Blaubleibender covellite. *Carnegie Inst. Wash., Yearb.*, 63: 208–209.

Morimoto, N. and Kullerud, G., 1961. Polymorphism in bornite. *Am. Mineral.*, 46: 1270–1282.

Morimoto, N. and Koto, K., 1970. Phase relations of the Cu–S system at low temperatures: stability of anilite. *Am. Mineral.*, 55: 106–117.

Muffler, L.J.P. and White, D.E., 1969. Active metamorphism of Upper Cenozoic sediments in the Salton Sea geothermal field and the Salton Trough, southeastern California. *Geol. Soc. Am. Bull.*, 80: 157–182.

Nickel, E.H., 1965. A review of the properties of zinc sulfide. *Mines Branch Inf. Circ.*, IC 170, Ottawa.

Nriagu, J.O., 1971. Experimental investigation of a portion of the system PbS–NaCl–HCl–H_2O at elevated temperatures. *Am. J. Sci.*, 271: 157–169.

Nriagu, J.O. and Anderson, G.M., 1970. Calculated solubilities of some base-metal sulphides in brine solutions. *Trans. Am. Inst. Min. Metall. Eng.*, 79: B208–B212.

Nriagu, J.O. and Anderson, G.M., 1971. Stability of the lead (II) chloride complexes at elevated temperatures. *Chem. Geol.*, 7: 171–183.

Oberc, J. and Serkies, J., 1968. Evolution of the Fore–Sudetian copper deposit. *Econ. Geol.*, 63: 372–379.

Ostroumov, E.A., Volkov, I.I. and Fomina, L.C., 1961. Distribution pattern of sulfur compounds in the bottom sediments of the Black Sea. *Tr. Akad. Nauk. SSSR, Inst. Okeanol.*, 7: 70–90.

Pantin, H.M., 1965. The effect of adsorption on the attainment of physical and chemical equilibrium in sediments. *N.Z. J. Geol. Geophys.*, 8: 453–464.

Pourbaix, M.J.N., 1949. *Thermodynamics of Dilute Aqueous Solutions*. Arnold, London, 136 pp.

Radusinovic, D.S., 1966. Greigite from the Lojane chromium deposit, Macedonia. *Am. Mineral.*, 51: 209–215.

Ramdohr, P., 1969. *The Ore Minerals and their Intergrowths*. Pergamon, Oxford, 1174 pp.

Rashid, M.A. and Leonard, J.D., 1973. Modifications in the solubility and precipitation behavior of various metals as a result of their interaction with sedimentary humic acid. *Chem. Geol.*, 11: 89–97.

Rickard, D.T., 1969. The chemistry of iron sulfide formation at low temperatures. *Stockh. Contrib. Geol.*, 20: 67–95.

Rickard, D.T., 1970. The origin of framboids. *Lithos*, 3: 269–293.

Rickard, D.T., 1970a. The chemistry of copper in natural aqueous solutions. *Stockh. Contrib. Geol.*, 23: 1–64.

Rickard, D.T., 1970b. Djurleite synthesis in low-temperature aqueous solution. *Acta Chem. Scand.*, 24: 2236–2238.

Rickard, D.T., 1972. Covellite formation in low-temperature aqueous solutions. *Miner. Deposita*, 7: 180–188.

Rickard, D.T., 1973. Limiting conditions for synsedimentary sulfide ore formation. *Econ. Geol.*, 68: 605–617.

Roberts, W.M.B., 1963. The low-temperature synthesis in aqueous solution of chalcopyrite and bornite. *Econ. Geol.*, 58: 52–61.

Roberts, W.M.B.. and Buchanan, A.S., 1971. The effects of temperature, pressure and oxygen on copper and iron sulfides synthesized in aqueous solution. *Miner. Deposita*, 6: 23–33.

Roberts, W.M.B., Walker, A.L. and Buchanan, A.S., 1969. The chemistry of pyrite formation in aqueous solution and its relation to the depositional environment. *Miner. Deposita*, 4: 18–29.

Roedder, E., 1968. The noncolloidal origin of "colloform" textures in sphalerite ores. *Econ. Geol.*, 63: 451–471.

Romberger, S.B., 1968. *Solubility of Copper in Aqueous Sulfide Solutions Coexisting with Covellite from 25 to 200°C with Geologic Applications*. Thesis, Pennsylvania State Univ.

Romberger, S.B. and Barnes, H.L., 1965. Solubility of covellite in synthetic ore solutions. *Geol. Soc. Am., Spec. Pap.*, 82: 165 pp.

Romberger, S.B. and Barnes, H.L., 1970. Ore solution chemistry, III. Solubility of CuS in sulfide solutions. *Econ. Geol.*, 65: 901–919.

Roseboom Jr., E.H., 1966. An investigation of the system Cu–S and some natural copper sulfides between 25 and 700°C. *Econ. Geol.*, 61: 641–672.

Sadler, W.R. and Trudinger, P.A., 1967. The inhibition of micro-organisms by heavy metals. *Miner. Deposita*, 2: 158–168.

Sawkins, F.J., 1964. Lead–zinc ore deposition in the light of fluid inclusion studies, Providencia Mine, Zacatecas, Mexico. *Econ. Geol.*, 59: 883–919.

Schwarz, E.J. and Vaughan, D.J., 1972. Magnetic phase relations of pyrrhotite. *J. Geomagn. Geoelectr.*, 24: 441–458.

Scott, S.D.. 1968. *Stoichiometry and Phase Changes in Zinc Sulfide*. Thesis, Pennsylvania State Univ.

Scott, S.D., 1973. Experimental calibration of the sphalerite geobarometer. *Econ. Geol.*, 68: 466–474.

Scott, S.D. and Barnes, H.L., 1971. Sphalerite geothermometry and geobarometry. *Econ. Geol.*, 66: 653–669.

Seaman, D.M. and Hamilton, H., 1950. Occurrence of polymorphous wurtzite in western Pennsylvania and eastern Ohio. *Am. Mineral.*, 35: 43–50.

Seguin, M.K., 1971. Phase relations in the $Fe–C–O_2–S–H_2O$ system and its geological application. *Chem. Geol.*, 7: 5–18.

Serkies, J., Oberc, J. and Idzikowski, A., 1967. The geochemical bearings of the genesis of Zechstein copper deposits in southwest Poland as exemplified by the studies on the Zechstein of the Leszczyna syncline. *Chem. Geol.*, 2: 217–232.

Shannon, R.D. and Prewitt, C.T., 1969. Effective ionic radii in oxides and fluorides. *Acta. Cryst.*, B25: 925–946.

Shimazaki, H. and Clark, L.A., 1970. Synthetic $FeS_2–CuS_2$ solid solution and fukuchilite-like minerals. *Can. Mineral.*, 10: 648–664.

Skinner, B.J., Erd, R.C. and Grimaldi, F.S., 1964. Greigite, the thiospinel of iron: a new mineral. *Am. Mineral.*, 49: 543–555.

Skinner, B.J., White, D.E., Rose, H.J. and Mays, R.E., 1967. Sulfides associated with the Salton Sea geothermal brine. *Econ. Geol.*, 62: 316–330.

Skripchenko, N.S., 1969. Iron disulfide–monosulfide relationships in low-temperature mineral formation. *Geochim. Int.*, 1: 114–118.

Springer, G., Schachner-Korn, D. and Long, J.V.P., 1964. Metastable solid solution relations in the system $FeS_2–CoS_2–NiS_2$. *Econ. Geol.*, 59: 475–491.

Stanton, R.L., 1972. *Ore Petrology*. McGraw-Hill, New York, N.Y., 713 pp.

Stephens, J.D. and Wittkopp, R.W., 1969. Microscopic and electron-beam microprobe study of sulfide minerals in Red Sea mud samples. In: E.T. Degens and D.A. Ross (Editors), *Hot Brines and Recent Heavy Metal Deposits in the Red Sea*. Springer, Berlin, p. 441–447.

Sunagawa, I., Endo, Y. and Nakai, N., 1971. Hydrothermal synthesis of framboidal pyrite. *Soc. Min. Geol. Jap., Spec. Issue*, 2: 10–14.

Sweeney, R.E. and Kaplan, I.R., 1973. Pyrite framboid formation: laboratory synthesis and marine sediments. *Econ. Geol.*, 68: 618–634.

Takeno, S., 1965. Thermal studies on mackinawite. *J. Sci., Hiroshima Univ.*, Ser. C4: 455–478.

Takeno, S., Zoka, H. and Niihara, T., 1970. Metastable cubic iron sulfide – with special reference to mackinawite. *Am. Mineral.*, 55: 1639–1649.

Taylor, L.A., 1970. Smythite $Fe_{3+x}S_4$ and associated minerals from the Silverfields Mine, Cobalt, Ontario. *Am. Mineral.*, 55: 1650–1658.

Taylor, L.A. and Williams, K.L., 1972. Smythite $(Fe,Ni)_9S_{11}$ – a redefinition. *Am. Mineral.*, 57: 1571–1577.

Temple, K.L., 1964. Syngenesis of sulfide ores: an evaluation of biochemical aspects. *Econ. Geol.*, 59: 1473–1491.

Temple, K.L. and LeRoux, N.W., 1964. Syngenesis of sulfide ores: desorption of absorbed metal ions and their precipitation as sulfides. *Econ. Geol.*, 59: 647–655.

Toulmin, P.T. III and Barton, P.B., 1964. A thermodynamic study of pyrite and pyrrhotite. *Geochim. Cosmochim. Acta*, 28: 641–671.

Turekian, K.K., 1969. Distribution of elements in sea water. In: K.H. Wedepohl (Editor), *Handbook of Geochemistry*. Springer, New York, N.Y., pp. 297–320.

Turekian, K.K. and Wedepohl, K.H., 1961. Distribution of the elements in some major units of the earth's crust. *Geol. Soc. Am. Bull.*, 72: 175–192.

Uda, M., 1967. The structure of synthetic $Fe_3\overset{.}{S}_4$ and the nature of the transition to FeS. *Z. Anorg. Allg. Chem.*, 350:105–109.

Uytenbogaardt, W. and Burke, E.A.J., 1971. *Tables for Microscopic Identification of Ore Minerals*. Elsevier, Amsterdam, 430 pp.

Vallentyne, J.R., 1963. Isolation of pyrite spherules from recent sediments. *Limnol. Oceanogr.*, 8: 16–30.

Vaughan, D.J., 1969. Nickelian mackinawite from Vlakfontein, Transvaal. *Am. Mineral.*, 54: 1190–1193.

Vaughan, D.J. and Ridout, M.S., 1971. Mossbauer studies of some sulfide minerals. *J. Inorg. Nucl. Chem.*, 33:741–747.

Vaughan, D.J., Burns, R.G. and Burns, V.M., 1971. Geochemistry and bonding of thiospinel minerals. *Geochim. Cosmochim. Acta*, 35: 365–381.

Vaughan, D.J., Tossell, J.A. and Johnson, K.H., 1974. The bonding of ferrous iron to sulfur and oxygen: a comparative study using SCF-X_α scattered wave molecular orbital calculations. *Geochim. Cosmochim.* 38: 993–1005.

Vinogradov, A.P., 1962. Average contents of chemical elements in the principal types of igneous rocks of the earth's crust. *Geochem.*, 7: 641–664.

Volkov, I.I., 1961. Iron sulfiedes, their interdependence and transformation in the Black Seabottom sediments. *Tr. Akad. Nauk. SSSR Inst. Okeanol.*, 50: 68–92.

Volkov, I.I., Rozanov, A.G. and Yagodinskaya, T.A., 1971. Pyrite microconcretions in Black Sea sediments. *Dokl. Akad. Nauk. SSSR*, 197: 202–205.

Voskresenskaya, N.T., 1969. Thallium in sedimentary sulfides. *Geochim. Int.*, 2: 225–236.

Vukotic, S., 1961. Contribution à l'étude de la solubilité de la galene, de la blende et de chalcopyrite dans l'eau en presence d'hydrogene sulfure entre 50 et 200°C. *Bull. Bur. Rech. Geol. Min.*, 3: 11–27.

Ward, J.C., 1971. The structure and properties of some iron sulfides. *Rev. Pure Appl. Chem.*, 20: 175–206.

Wedepohl, K.H., 1964. Untersuchungen am Kupferschiefer in Nordwestdeutschland; ein Beitrag zur Deutung der Genese bituminöser Sedimente. *Geochim. Cosmochim. Acta*, 28: 305–364.

Wedepohl, K.H. (Editor), 1969. *Handbook of Geochemistry*. Springer, Berlin, 442 pp.

Wedepohl, K.H., 1971. "Kuperferschiefer" as a prototype of syngenetic sedimentary ore deposits. *Soc. Min. Geol., Jap. Spec. Issue*, 3: 268–273.

Weiss, A. and Amstutz, G.C., 1966. Ion exchange reactions on clay minerals and cation selective membrane properties as possible mechanisms of economic metal concentration. *Miner. Deposita*, 1: 60–66.

White, D.E., 1968. Environments of the generation of some base metal ore deposits. *Econ. Geol.*, 63: 301–335.

White, D.E., Anderson, E.T. and Grubbs, D.K., 1963. Geothermal brine well: mine deep drill hole may tap ore-bearing magmatic water and rocks undergoing metamorphism. *Science*, 139: 919–922.

Ypma, P.J.M., 1968. Pyrite group: an unusual member $Cu_{0.60}Ni_{0.14}Co_{0.03}Fe_{0.23}S_2$. *Science*, 159: 194.

Zimmerman, R.A., 1970. Sedimentary features in the Meggen barite–pyrite–sphalerite deposit and a comparison with the Arkansas barite deposits. *Neues Jahrb. Mineral., Abh.*, 113: 179–214.